METHODS IN MOLECULAR BIOLOGY™

Series Editor
John M. Walker
School of Life Sciences
University of Hertfordshire
Hatfield, Hertfordshire, AL10 9AB, UK

For other titles published in this series, go to
www.springer.com/series/7651

METHODS IN MOLECULAR BIOLOGY™

Phospho-Proteomics

Methods and Protocols

Edited by

Marjo de Graauw, PhD

*Division of Toxicology, Leiden/Amsterdam Center for Drug Research (LACDR),
Leiden University, Leiden, Netherlands*

Editor
Marjo de Graauw, PhD
Division of Toxicology
Leiden/Amsterdam Center for Drug Research (LACDR)
Leiden University, Leiden, Netherlands

ISBN: 978-1-60327-833-1   e-ISBN: 978-1-60327-834-8
ISSN: 1064-3745   e-ISSN: 1940-6029
DOI: 10.1007/978-1-60327-834-8

Library of Congress Control Number: 2008944207

Printed on acid-free paper

springer.com

Preface

Phosphorylation of proteins on specific amino acid residues is a key regulatory mechanism in cells. Protein phosphorylation controls many basic cellular processes, such as cell growth, differentiation, migration, metabolism, and cell death, and is in itself regulated by the activity of kinases and phosphatases. Identification of differentially phosphorylated proteins by means of phospho-proteomics therefore increases our insight into the signal transduction pathways that are activated in cells in response to different stimuli, such as growth factor stimulation or exposure to toxicants.

Phospho-proteomics presents both well-established protocols and some of the newest strategies for the identification and evaluation of protein phosphorylation on Tyr, Ser, and Thr residues. Detailed protocols and methodologies are included that focus on twodimensional gel electrophoresis and protein phosphorylation, enrichment of phosphoproteins and peptides, quantitative analysis of phosphorylation by labeling and MS analysis, and antibody and kinase arrays. In addition, this volume contains three chapters on bioinformatics and phosphosite prediction and two reviews providing an overview of recent advances in the identification of phosphoproteins by mass spectrometry and an outline of different chemical tagging strategies.

Phospho-proteomics is aimed at those who wish to gain insight into signal transduction pathways by studying and identifying protein phosphorylation. It is written for a broad audience ranging from researchers who are new to phosphoproteomic techniques and for those with more experience in the field. This book covers those proteomic methods and technologies that are widely used and have been well-established in the prominent proteomic labs. In addition, several protocols have been included that point to a new direction in the phospho-proteomics field.

Finally, I thank my colleagues at the Division of Toxicology, LACDR, Leiden, for their support and all the authors who kindly contributed their time, expertise, suggestions, and enthusiasm.

Leiden, The Netherlands   *Marjo de Graauw, PhD*

Contents

Contributors

GANESH K. AGRAWAL • *University of Missouri, Division of Biochemistry, Columbia, MO, USA*

ANGELA AMORESANO • *Department of Organic Chemistry and Biochemistry, Federico II University of Naples, Naples, Italy*

ROLAND S. ANNAN • *Proteomics and Biological Mass Spectrometry Laboratory, GlaxoSmithKline, King of Prussia, PA, USA*

NIKOLAJ BLOM • *Technical University of Denmark, Center for Biological Sequence Analysis, Lyngby, Denmark*

HONG-LIN CHAN • *Cancer Proteomics Laboratory, EGA Institute for Women's Health, UCL, London, UK*

RAGHOTHAMA CHAERKADY • *Mckusick-Nathans Institute of Genetic Medicine and the Department of Biological Chemistry; Institute of Bioinformatics. International Technology Park, Bangalore, India*

Y. EUGENE CHIN • *Department of Surgery, Rhode Island Hospital, Providence, RI, USA*

ALICIA S. CHUNG • *Department of Genetics and Complex Diseases, Harvard School of Public Health, Boston, MA, USA*

CLAUDIA CIRULLI • *Department of Organic Chemistry and Biochemistry, Federico II University of Naples, Naples, Italy*

HELEN J. COOPER • *School of Biosciences, University of Birmingham, Birmingham, UK*

KEVIN DIERCK • *Department of Clinical Chemistry, University Medical Center, Hamburg-Eppendorf, Germany*

Mᴬ CARMEN DURAN • *Cancer Proteomics Laboratory, EGA Institute for Women's Health, UCL, London, UK*

PHILIP R. GAFKEN • *Proteomics Facility, Fred Hutchinson Cancer Research Center, Seattle, WA, USA*

KRIS GEVAERT • *Department of Medical Protein Research and Department of Biochemistry VIB, Ghent University, Ghent, Belgium*

BRIAN D. HALLIGAN • *Biotechnology and Bioengineering Center, Medical College of Wisconsin, Milwaukee, WI, USA*

ANTON ILIUK • *Department of Biochemistry, Medicinal Chemistry and Molecular Pharmacology, Purdue University, West Lafayette, IN, USA*

OLE N. JENSEN • *Department of Biochemistry and Molecular Biology, University of Southern Denmark, Odense, Denmark*

RALF KRÜGER • *University Hospital Mainz, Institute of Clinical Chemistry and Laboratory Medicine, Mainz, Germany*

MARTIN R. LARSEN • *Department of Biochemistry and Molecular Biology, University of Southern Denmark, Odense, Denmark*

ROBIN E.C. LEE • *Ottawa Health Research Institute, The Ottawa Hospital, Ottawa, ON, Canada*

WOLF D. LEHMANN • *Molecular Structure Analysis, German Cancer Research Center (DKFZ), Heidelberg, Germany*

ALEXANDER LEITNER • *Department of Analytical Chemistry and Food Chemistry, University of Vienna, Vienna, Austria*

WOLFGANG LINDNER • *Department of Analytical Chemistry and Food Chemistry, University of Vienna, Vienna, Austria*

KAZUYA MACHIDA • *Department of Genetics and Developmental Biology, University of Connecticut, Farmington, CT, USA*

GENNARO MARINO • *Department of Organic Chemistry and Biochemistry, Federico II University of Naples, Naples, Italy*

BRUCE J. MAYER • *Department of Genetics and Developmental Biology, University of Connecticut, Farmington, CT, USA*

DEAN E. MCNULTY • *Proteomics and Biological Mass Spectrometry Laboratory, GlaxoSmithKline, King of Prussia, PA, USA*

LYNN A. MEGENEY • *Ottawa Health Research Institute, The Ottawa Hospital, Ottawa, ON, Canada*

MARTIN L. MILLER • *Technical University of Denmark, Center for Biological Sequence Analysis, Lyngby, Denmark*

GIANLUCA MONTI • *Department of Organic Chemistry and Biochemistry, Federico II University of Naples, Naples, Italy*

THOMAS A. NEUBERT • *Department of Pharmacology, New York University School of Medicine, New York, NY, USA*

PETER NOLLAU • *Department of Clinical Chemistry, University Medical Center, Hamburg-Eppendorf, Germany*

AKHILESH PANDEY • *Oncology and Pathology, Johns Hopkins University School of Medicine, Baltimore, MD, USA*

KAUSHAL PARIKH • *Department of Cell Biology, University Medical Center Groningen, Groningen, The Netherlands*

MAIKEL P. PEPPELENBOSCH • *Department of Cell Biology, University Medical Center Groningen, Groningen, The Netherlands*

GENARO PIMIENTA • *Department of Biological Chemistry, McKusick-Nathans Institute of Genetic Medicine, Baltimore, MD, USA*

LAWRENCE G. PUENTE • *Ottawa Health Research Institute, The Ottawa Hospital, Ottawa, ON, Canada*

ERIC QUEMENEUR • *CEA-VALRHO, DSV-DIEP-SBTN, Service de Biochimie post-génomique & Toxicologie Nucléaire, Bagnols-sur-Cèze, France*

TITA RITSEMA • *Plant–Microbe Interactions, Institute of Environmental Biology, Utrecht University, Utrecht, The Netherlands*

W. CARL SAXINGER • *Center for Cancer Research, National Cancer Institute, Fort Detrick, Frederick, MD, USA*

STEVE M. M. SWEET • *University of Birmingham, School of Biosciences, Birmingham, UK*

W. ANDY TAO • *Department of Biochemistry, Medicinal Chemistry and Molecular Pharmacology, Purdue University, West Lafayette, IN, USA*

JAY J. THELEN • *Division of Biochemistry, University of Missouri, Columbia, MO, USA*

TINE E. THINGHOLM • *Department of Biochemistry and Molecular Biology, University of Southern Denmark, Odense, Denmark*

JOHN F. TIMMS • *Cancer Proteomics Laboratory, EGA Institute for Women's Health, UCL, London, UK*

AVIVA M. TOLKOVSKY • *Department of Biochemistry, University of Cambridge, Cambridge, UK*

JOËL VANDEKERCKHOVE • *Department of Medical Protein Research and Department of Biochemistry VIB, Ghent University, Ghent, Belgium*

ANDREAS WYTTENBACH • *University of Southampton, Southampton Neuroscience Group, Southampton, UK*

GUOAN ZHANG • *Department of Pharmacology, New York University School of Medicine, New York, NY, USA*

NICO ZINN • *German Cancer Research Center (DKFZ), Molecular Structure Analysis, Heidelberg, Germany*

Part I

Protein Phosphorylation and 2D Gel Electrophoresis

Chapter 1

A High-Resolution Two Dimensional Gel- and Pro-Q DPS-Based Proteomics Workflow for Phosphoprotein Identification and Quantitative Profiling

Ganesh K. Agrawal and Jay J. Thelen

Summary

The two-dimensional (2-D) gel-based proteomics platform remains the workhorse for proteomics and is fueled by a number of key improvements, including fluorescence-based stains for detection and quantification of proteins and phosphoproteins with high sensitivity and linear dynamic ranges. One such stain is Pro-Q diamond phosphoprotein stain (Pro-Q DPS), which binds to the phosphate moiety of phosphoproteins irrespective of the phosphoamino acid. We recently introduced a modified Pro-Q DPS protocol to detect phosphoprotein spots on 2-D gels with very low background addressing some prime concerns, including high cost and reproducibility of Pro-Q DPS. The major modifications were a threefold dilution of Pro-Q DPS and the use of threefold less volume of the diluted staining solution. In this chapter, use of the modified Pro-Q DPS protocol along with the 2-D gel-based proteomics for phosphoprotein detection and quantification is described in detail. This 2-D gel- and Pro-Q DPS-based proteomics workflow has seven major steps: preparation of total protein, separation of proteins by 2-D gel electrophoresis, detection of phosphoprotein and total protein, image analysis and quantitative expression profiling, excision of 2-D spots, mass spectrometry analysis, and data processing and organization. Involvement of the modified Pro-Q DPS protocol in this proteomics workflow alone reduces the overall cost by at least ninefold for conducting phospho-proteomics analysis on a global scale, thereby making this entire process economically attractive to the scientific community.

Key words: Two-dimensional polyacrylamide gel electrophoresis, Reversible protein phosphorylation, Proteomics, Phospho-proteomics, Phosphoproteome, Phosphoproteins, Quantification, Method, Protocol.

Marjo de Graauw (ed.), *Phospho-Proteomics, Methods and Protocols, vol. 527*
© 2009 Humana Press, a part of Springer Science+Business Media, New York, NY
Book DOI: 10.1007/978-1-60327-834-8_1

1. Introduction

Two-dimensional gel electrophoresis (2-DGE) coupled with mass spectrometry (MS) is generally known as a 2-D gel-based proteomics platform. 2-DGE separates proteins by their isoelectric point (pI) through an immobilized pH-gradient (IPG) gel matrix in the first dimension and then by molecular weight in the second dimension (1). 2-D gel-based proteomics is now considered a mature and well-established technique and remains the most widely and routinely used technique in the proteomics field. This is mainly due to continuous improvements in separation and detection technologies, such as narrow-range IPG strips and fluorescence-based protein labeling and stains (1, 2).

Fluorescence-based stains were recently developed to detect proteins, including post-translationally modified proteins with high sensitivity and linearity. One such stain is the Pro-Q diamond phosphoprotein stain (Pro-Q DPS). Pro-Q DPS binds directly and specifically to the phosphate moiety of phosphoproteins at levels as low as 1 ng and is fully compatible with other stains and MS (3, 4). Therefore, a combination of 2-DGE and Pro-Q DPS allows quantitative analysis of phosphoproteins on a global scale, a major advancement in the phospho-proteomics field. Phospho-proteomics is an emerging field (5, 6), the goal of which is to map phosphorylation networks and understand protein phosphorylation events controlling numerous cellular processes. Quantitative and sensitive detection of phosphoproteins is integral to these objectives.

Recently we introduced a modified Pro-Q DPS protocol that addressed the prime concerns of high cost and reproducibility, which were limiting the utilization of this novel technology (7). The major changes to the manufacturer's protocol include a threefold dilution of Pro-Q DPS by diluting the stock solution, of which a third less volume, compared with that recommended, is used. This practical change in staining solution reduces the overall cost by ninefold for performing a large-scale phospho-proteomics analysis, making the 2-D gel- and Pro-Q DPS-based proteomics workflow more attractive for routine use. Moreover, the required volume of other solutions, such as fixation and destaining, was determined to be half of the recommended volume. The optimized conditions of this modified protocol are summarized for a large gel (size, 26 cm × 20 cm × 1 mm) in **Table 1** and have been utilized in both plant and animal phospho-proteomics analyses (8–10).

In this chapter, we describe the 2-D gel- and Pro-Q DPS-based proteomics workflow for identification and quantitative profiling of phosphoproteins, which is mainly based on two publications – the modified Pro-Q DPS protocol (7) and the

Table 1
A modified Pro-Q DPS protocol for phosphoprotein detection in large-format 2-D polyacrylamide gels

Step	Solution	Amount (mL)	Time (min)	Dark incubation	Reuse of solution
1	Fixation	250	2×30	Not required	Yes
2	Washing	250	2×15	Not required	No
3	Staining	150	120	Required	Yes
4	Destaining	250	4×30	Required	Yes
5	Washing	250	2×5	Required	No

application of this simple, economical protocol in large-scale identification and quantification of phosphoproteins expressed in developing rapeseed *(9)* – along with the ongoing work in our laboratory. This proteomics process involves seven major steps: (1) preparation of total protein, (2) separation of proteins by 2-D gel electrophoresis, (3) detection of phosphoproteins and total proteins, (4) image analysis and quantitative expression profiling, (5) excision of 2-D spots, (6) MS analysis, and (7) data processing and organization. The published reports and our unpublished data demonstrate that this system is highly suitable and reproducible for performing phosphoproteomic analyses on a large scale.

2. Materials

2.1. General

1. Fresh TEMED (N,N,N',N'-tetramethyl-ethylenediamine) is important for polymerization of gel (*see* **Note 1**).
2. Unless otherwise stated, all solutions were prepared in deionized water (18.2 MΩ conductivity).
3. Filter sterilization of solutions using either 25-mm syringe filter (0.2 µm, nylon; Fisher Scientific, Houston, TX) or steritop (0.22 µm; Millipore Corporation, Billerica, MA) depending on the solution volume.

2.2. Materials and Storage

1. Collect biological materials, immediately freeze with liquid nitrogen, and store at –80°C.

2.3. Total Protein Extraction: Phenol–Ammonium Acetate/Methanol Method

1. Mortar and pestle.

2. Extraction buffer: 100 mM Tris–HCl, pH 8.8, 10 mM (w/v) ethylenediamine tetra-acetic acid (EDTA), 900 mM (w/v) sucrose, 0.4% (v/v) 2-mercaptoethanol. Store at 4°C (*see* **Note 2**).

3. Phenol buffered with Tris–HCl, pH 8.8. Store at 4°C (*see* **Note 3**).

4. Extraction buffer plus phosphatase and protease inhibitors: Add 5 mM (w/v) sodium vanadate (or sodium meta vanadate), 5 mM (w/v) sodium fluoride, 25 mM (w/v) glycerophosphate disodium salt pentahydrate, and protease inhibitor cocktail complete EDTA-free tablet (1 tablet/10-mL solution) to the extraction buffer (*see* **step 2** of **Subheading 2.3**) and mix until dissolved (*see* **Note 4**).

5. Phenol extraction buffer: Add equal volume of extraction buffer (plus phosphatase and protease inhibitors solution from **step 4**) and buffered phenol (*from* **step 3**) in a 15-mL Falcon tube and mix before use.

6. Ammonium acetate/methanol solution: 100 mM (w/v) ammonium acetate in 100% methanol. Store at 4°C.

7. 80% (v/v) acetone in deionized water. Store at 4°C.

8. 70% (v/v) ethanol in deionized water. Store at 4°C.

9. Isoelectric focusing (IEF) extraction solution: 8 M (w/v) urea, 2 M (w/v) thiourea, 2% (w/v) CHAPS, 2% (v/v) Triton X-100, and 50 mM (w/v) DTT. Store in single-use aliquots of 1 mL at –20°C.

10. Balance and load shaking tray (Nutator; Becton Dickinson, Franklin Lakes, NJ).

11. Dounce tissue grinder (15 mL, Wheaton, Millville, NJ).

2.4. Protein Quantification: A Modified Bradford Method

1. SDS (sodium dodecyl sulfate) running buffer: 25 mM Tris-base (w/v), 192 mM (w/v) glycine, and 0.1% (w/v) SDS. Prepare 10× SDS running buffer and keep at room temperature (RT).

2. 0.5× SDS running buffer. Store at RT.

3. 1 mg/mL bovine serum albumin (BSA) in 0.5× SDS running buffer. Store at –20°C in aliquots (*see* **Note 5**).

4. Bradford dye: Dilute the Bradford protein assay dye (Bio-Rad, Hercules, CA) fivefold in deionized water before use.

5. Microtiter plate spectrophotometer (a 96-well plate reader; Thermo Electron Corporation, Multiskan MCC, Type 355).

2.5. Isoelectric Focusing of Total Proteins

1. Protean IEF Cell, Protean IEF system 24 cm disposable trays, and isoelectric point gel (IPG) focusing tray are from Bio-Rad.

2. Linear IPG strips (pH 4–7, 24 cm; GE Healthcare, Piscataway, NJ) (*see* **Note 6**).

3. IPG buffer (pH 4–7, 24 cm; GE Healthcare, Piscataway, NJ).

2.6. SDS-PAGE

1. Ettan DALTtwelve electrophoresis unit (GE Healthcare, Piscataway, NJ).

2. Acrylamide stock (30.8% T): 30% (w/v) acrylamide and 0.8% *N*,*N*′-methylene bis-acrylamide. Filter sterilize and store at 4–8°C (*see* **Note 7**).

3. 1.5 M Tris–HCl, pH 8.8. Filter sterilize and store at 4°C.

4. 0.5 M Tris–HCl, pH 6.8. Filter sterilize and store at 4°C.

5. 10% (w/v) SDS in deionized water. Filter sterilize and store at RT.

6. 10% (w/v) ammonium persulfate in deionized water. Use freshly prepared solution.

7. Displacement solution: 375 mM Tris–HCl, pH 8.8, 50% (v/v) glycerol, and trace of bromophenol blue in deionized water. Store at 4°C.

8. PeppermintStick phosphoprotein molecular weight standards (Invitrogen, Carlsbad, CA). Store in aliquots at –20°C (*see* **Note 8**).

9. SDS equilibration buffer: 50 mM Tris–HCl, pH 8.8, 6 M (w/v) urea, 30% (v/v) glycerol, and 4% (w/v) SDS. Filter sterilize and store in single-use aliquots at –20°C in 15-mL Falcon tubes.

10. Reduction solution: 2% (w/v) DTT in SDS equilibration buffer. Prepare right before use.

11. Alkylation solution: 2.5% (w/v) iodoacetamide in SDS equilibration buffer. Prepare fresh and protect from light.

12. Electrode wicks (Protean IEF system; Bio-Rad Laboratories, Hercules, CA).

13. Agarose sealing solution: Add 125 mg agarose to 1× SDS running buffer plus a few grains of bromophenol blue.

14. Kimwipes.

15. Rocking platform.

2.7. Phosphoprotein Detection with Pro-Q DPS

1. Large gel-staining tray, size 26 cm × 20 cm × 10 cm (Daigger, Vernon Hills, IL).

2. Orbital shaker.

3. Pro-Q DPS (Invitrogen, Carlsbad, CA). Store at 4°C (*see* **Note 9**).

4. Fixation solution: 50% (v/v) methanol, 10% (v/v) acetic acid in deionized water.

5. Washing solution: Deionized water.

6. Staining solution: Threefold diluted Pro-Q DPS (v/v) in deionized water.

7. Destaining solution: 50 mM sodium acetate–acetic acid, pH 4.0, 20% (v/v) acetonitrile in deionized water. To prepare 1 L of destaining solution, combine 50 mL of 1.0 M sodium acetate, pH 4.0, 750 mL of deionized water, and 200 mL of acetonitrile (*see* **Note 10**).

2.8. Total Protein Detection with Colloidal CBB

1. Colloidal Coomassie Brilliant Blue (CBB) solution: 800 mL of ethanol, 3.2 g of brilliant blue G-250, 64 mL of phosphoric acid, 320 g of ammonium sulfate to prepare 4 L in deionized water (*see* **Note 11**).

2. Gel storage solution: To prepare 250 mL, use 25 mL of CBB solution, 2.5 mL of 100× sodium azide (2% (w/v) stock), and 222.5 mL of deionized water. Use 250 mL per 24 cm of large-format gel, shake for 5 min, and store at 4°C (up to 3–4 months).

2.9. Image Acquisition of 2-D Gels

1. Fluorescence image capture device: FLA 5000 laser scanner (Fuji Medical Systems, Stamford, CT) for scanning phosphoprotein gels.

2. Protein image capture device: ScanMaker 9800XL (Microtek, Carson, CA).

2.10. Image Analysis

1. Image quantification software: Image Gauge Analysis software (Fuji, Stamford, CT).

2. ImageMaster 2D Platinum software version 5 or version 6 (GE Healthcare, Piscataway, NJ)

2.11. 2-D Spots Excision and In-Gel Digestion

1. Spot picker (1.5 mm; The Gel Company, San Francisco, CA).

2. MultiScreen Solvinert – a 96-well filtration system (0.45 μm low-binding hydrophilic PTFE; Millipore, Bedford, MA).

3. Vacuum manifold system (Millipore, Bedford, MA).

4. Microplate shaker.

5. Polypropylene 96-well V-bottom sample collection plate.

6. 100 mM ammonium bicarbonate, pH 8.0 in MilliQ water. Store at RT.

7. In-gel wash solution: 50% acetonitrile and 50 mM ammonium bicarbonate, pH 8.0. 100 mL of in-gel wash solution: 50 mL of 100 mM ammonium bicarbonate, pH 8.0 and 50 mL of acetonitrile. Use freshly prepared solution.

8. In-gel trypsin solution: Use sequencing grade trypsin (Promega, Madison, WI). One vial contains 20 μg. Dissolve 20 μg in 5.1 mL of 50 mM ammonium bicarbonate, pH 8.0. Use freshly prepared solution. Store on ice until use.

9. In-gel extraction solution: 60% (v/v) acetonitrile and 1% (v/v) formic acid in MilliQ water.

10. CentriVap Console.

2.12. Mass Spectrometry and Data Analysis

1. 0.1% (v/v) formic acid in MilliQ water.

2. LTQ ProteomeX XL linear ion trap LC-MS/MS instrument and associated software for data processing such as SEQUEST algorithm as part of the BioWorks software suite (Thermo Fisher Scientific, Waltham, MA).

3. Methods

To identify and quantify phosphoproteins in large-format 2-D gels, it is important to prepare a high-quality protein sample that is free from interfering compounds such as carbohydrates and nucleic acids. Phenol–ammonium acetate/methanol method has been successfully used to extract proteins from developing seeds (rich in oil and carbohydrate) of soybean and rapeseed suitable for generating high-resolution 2-D gel reference maps *(9, 11, 12)*. The phenol-based extraction method has also been used by many other labs including Wolfram Weckwerth's group (Germany) for large-scale phosphoprotein analysis *(13)*. Based on these and other gel-based proteomics studies, it can be stated that the phenol–ammonium acetate/methanol method provides high-quality protein suitable for performing high-resolution 2-DGE while minimizing proteolysis and blocking endogenous phosphatase activity during protein isolation. Nevertheless, we cannot rule out the possibility of very low levels of endogenous phosphatase or protease activity. Taking this into account, it is always good to add phosphatase and protease inhibitors (*see* **step 4** of **Subheading 2.3**).

Isolated proteins are then separated by 2-DGE. Both biological and experimental replications are required for downstream 2-D gel image analysis and quantification. The 2-DGE technique has been combined with the modified method of Pro-Q DPS to detect and quantify phosphoproteins on a large scale. **Table 1** summarizes all steps including the required volumes of solutions for large-size gels, total incubation times and conditions, and whether solutions can be reused. Poststaining of the gel with total protein stains such as colloidal CBB (mentioned in this chapter), silver nitrate, or SYPRO Ruby provides insight into dynamics of proteins and phosphoproteins and also a landmark for excision of phosphoprotein spots.

Upon image analysis of 2-D gels and expression profiling, excised 2-D spots are in-gel digested with trypsin. Extracted peptides are then subjected to MS and database analysis for identification of phosphoproteins. We have attempted to provide step-by-step information in this chapter. However, it is difficult to provide details on all the steps involved in this analysis. For example, it is almost impossible to describe the steps involved in the image analysis of 2-D gels using ImageMaster 2D Platinum software. In such cases, readers are encouraged to utilize the user manual for the ImageMaster software. We should emphasize that the detection of all distinct spots and manual editing of those spots are important steps in the 2-D gel image analysis. Recently, ImageMaster software has been updated and version 6 of this software is now being used.

3.1. Total Protein Extraction: Phenol–Ammonium Acetate/Methanol Method

1. Grind 250 mg of plant materials with liquid nitrogen in a mortar and pestle to obtain a powder (*see* **Note 12**).

2. Add 10 mL of phenol extraction buffer and continue grinding for an additional 30 s.

3. Transfer to a 15-mL Falcon tube and agitate on a nutator for 30 min at 4°C.

4. Centrifuge for 30 min at 2,800 g (4°C).

5. Transfer the upper-phase solution to a fresh 50-mL Falcon tube (*see* **Note 13**).

6. Add 5 volumes of ammonium acetate/methanol solution (ice cold), vortex, and incubate at –20°C overnight to precipitate the phenol-extracted proteins.

7. Centrifuge for 30 min at 2,800 g (4°C) to collect the precipitate.

8. Wash the pellet twice with ice-cold ammonium acetate/methanol solution, then twice with ice-cold 80% acetone solution, and finally with cold 70% ethanol solution (*see* **Note 14**).

9. Dry the final pellet (after removing the wash solution) for 20 min at 37°C (*see* **Note 15**).

10. Resuspend the final pellet in IEF extraction solution (*see* **Note 16**).

11. Centrifuge for 15 min at 28,000 g at RT.

12. Carefully transfer clear supernatant to a new 1.5-mL microfuge tube. Store in single-use aliquots at –80°C.

3.2. Protein Quantification: A Modified Bradford Method

1. Dilute the protein sample with 0.5× SDS running buffer in a 1.5-mL microfuge tube, and mix (*see* **Note 17**).

2. Use 0.5× SDS running buffer as blank and add 6 µL to the first column of the microtiter plate.

3. Add 1, 2, 4, and 6 μL of 1 mg/mL BSA standard in triplicate to wells and bring the final volume to 6 μL with 0.5× SDS running buffer.

4. Add 1 μL of tenfold diluted protein sample in triplicate to wells plus 5 μL of 0.5× SDS running buffer to make a total volume of 6 μL (*see* **Note 18**).

5. Add 200 μL of diluted Bradford dye to wells with protein.

6. Mix and incubate at RT for 5 min.

7. Vortex the microtiter plate on the spectrophotometer and measure absorbance at 595 nm (*see* **Note 19**).

8. Calculate the protein concentration.

3.3. Isoelectric Focusing of Total Proteins

1. Add 1 mg protein to a 1.5-mL microfuge tube and bring the final volume to 450 μL with IEF extraction buffer.

2. Add 2.25 μL (final concentration of 0.5%) of the correct IPG buffer to the same tube.

3. Mix the sample by vortexing and centrifuging at 28,000 g for 5 min to remove insoluble materials.

4. Pipette the supernatant into the IPG focusing tray starting from one end to the other (*see* **Note 20**).

5. Peel apart a dehydrated IPG strip (pH 4–7; 24 cm) using forceps and place the dried acrylamide side face down into the sample well in the IPG focusing tray (*see* **Note 21**).

6. Cover the focusing tray with the lid and keep inside a plastic bag to minimize evaporation.

7. Allow the IPG strip to rehydrate for 90 min at RT.

8. Overlay the IPG strip with mineral oil (~2.5 mL) to prevent dehydration (*see* **Note 22**).

9. Place the IPG focusing tray into Protean IEF cell unit.

10. Perform active rehydration (10 h at 50 V), followed by three-step focusing protocol: 100 V for 100 V h, 500 V for 500 V h, and 8,000 V for 99 kV h.

3.4. Assembling the Ettan DALTtwelve Gel Caster Unit

1. Tilt the DALTtwelve unit back so that it rests on its support legs (*see* **Note 23**).

2. Place a thicker separator sheet against the back wall to easily remove the last cassette from the gel caster unit after polymerization.

3. Clean each glass plate carefully with deionized water and ethanol using Kimwipes.

4. Fill the caster by alternating cassettes with separator sheets. End with a separator sheet, and then use the thicker separator

sheets to bring the level of the stack even with the edge of the caster.

5. Lubricate the foam gasket with a small amount of GelSeal and place it in the groove on the faceplate.

6. Turn four black-knobbed screws into the four threaded holes across the bottom until they are well engaged. Usually two to three full turns are enough.

7. Place the faceplate carefully onto the caster with the bottom slots resting on their respective screws. Screw the four remaining black-knobbed screws into the holes at the sides of the faceplate and tighten all eight evenly. The assembled unit is now ready to cast gels (*see* **Note 24**).

3.5. Casting 12% SDS-PAGE into the Ettan DALTtwelve Gel Caster

1. To cast twelve gels, add 300 mL of acrylamide stock, 188 mL of 1.5 M Tris–HCl, pH 8.8, 7.5 mL of 10% SDS, and 252 mL of deionized water in 1-L sidearm flask. This is the separating gel solution.

2. Place the flask on stir plate, add a medium size stir bar, and stir the solution.

3. Connect sidearm to vacuum, cover top opening with solid rubber stopper, and apply vacuum for 30 min.

4. Turn off vacuum, remove rubber stopper, and disconnect hose.

5. Add 3.6 mL of 10% ammonium persulfate while stirring the solution.

6. Add 120 μL of TEMED and continue to stir for 30 s. Move quickly to cast the gels.

7. Slowly pour the gel solution into the caster through the hydrostatic balance chamber until it is about 2 cm below the desired gel height.

8. Pour the displacement solution into the chamber until it is 0.25 cm below the surface of glass plates, and then immediately place the feed tube into the grommet to stop the flow.

9. Very slowly overlay each gel with 1.5 mL of deionized water.

10. Cover the upper portion of the unit with plastic wrap to prevent dehydration.

11. Allow polymerization for 16 h.

12. Bring the unit near a sink, carefully disassemble, and scrape off access acrylamide (*see* **Note 25**).

13. Remove the gel cassette from the caster by pulling forward on the separator sheets, rinse the outer surface of each gel cassette with deionized water to remove any polyacrylamide particles, and place gel cassettes in a cassette rack.

14. Prepare stacking gel solution by combining 10.6 mL of acrylamide stock, 20 mL of 0.5 M Tris–HCl, pH 6.8, 0.8 mL of 10% SDS, and 48.8 mL of deionized water in a 200-mL beaker.

15. Mix thoroughly by stirring the solution on a stir plate.

16. Add 0.6 mL of 10% ammonium persulfate while stirring the solution.

17. Add 40 μL of TEMED and continue to stir for 30 s.

18. Place enough stacking gel solution on top of the separating gel using a plastic transfer pipette to give 2 cm height after polymerization.

19. Overlay each gel with 1 mL isobutanol (*see* **Note 26**).

20. Allow to polymerize for ~1 h.

21. Once polymerized, pour off isobutanol and wash several times with deionized water to remove any traces of isobutanol.

22. Add 1× SDS running buffer on top of the stacking gel to prevent dehydration (*see* **Note 27**).

3.6. Reduction/Alkylation of Proteins in the IPG Strips

1. Remove the IPG strips from the IPG focusing tray by holding one end of the strip with forceps and blotting off excess mineral oil using Kimwipes.

2. Use Protean IEF system 24-cm disposable tray to place the IPG strip in a well carrying 2.5 mL reduction solution, facing the gel side up.

3. Incubate on rocking platform at medium speed for 15 min at RT.

4. Transfer the IPG strips to a new disposable tray with 2.5 mL alkylation solution in each well.

5. Incubate again on rocking platform at medium speed for 15 min at RT.

6. Transfer the IPG strips again to a new disposable tray carrying 2.5 mL 1× SDS running buffer in each well for a brief wash. The IPG strips are now ready for SDS-PAGE.

3.7. SDS-PAGE

1. Remove the IPG strip from the tray by holding one end of the strip with forceps and placing the IPG strip carefully on the surface of stacking gel (*see* **Note 28**).

2. Place a small square of electrode wick (0.5× 0.5 cm²) containing 1–2 μL of PeppermintStick phosphoprotein standards next to the acidic end of the strip (*see* **Note 29**).

3. Quickly overlay the IPG strip and the electrode wick with 2–3 mL agarose overlay solution (*see* **Note 30**).

4. Pour about 7.5 L of 1× SDS running buffer to the lower chamber of the separation unit (*see* **Note 31**).

5. Once agarose is solidified, insert the gel cassette into the separation unit through the buffer seal slots flanked by rubber gaskets (*see* **Note 32**).

6. Pour about 2.0 L of 2× SDS running buffer until the solution reaches the marked upper level on the separation unit.

7. Run the electrophoresis unit at 2 W/gel until dye migrates off the gel (*see* **Note 33**).

3.8. Phosphoprotein Detection with Pro-Q DPS and Image Acquisition

1. Take out the gel cassettes one by one from the separation unit, open the cassette, and carefully transfer the gel to a large gel-staining tray.

2. Wash the gel twice with deionized water for 10 min each. Incubate at RT with constant shaking on an orbital shaker at a speed of 35 rpm in all steps of this modified Pro-Q DPS procedure, unless stated otherwise (*see* **Note 34**).

3. Decant deionized water and immerse the gel in 200 mL of fixation solution. Decant the fixation solution and repeat (*see* **Note 35**).

4. Immerse the gel in 250 mL of washing solution for 15 min. Decant wash solution and repeat.

5. Incubate the gel in 150 mL of staining solution in the dark for 2 h and decant solution.

6. Immerse the gel in 250 mL of destaining solution and incubate in the dark for 30 min. Decant the destaining solution and repeat this step three more times. The total required destaining time is 2 h.

7. Wash the gel with 250 mL of deionized water in the dark for 5 min. Decant deionized water and repeat.

8. Scan the gel using a laser imager with 532-nm excitation and 580-nm bandpass emission filter (Fujifilm FLA 5000). Collect and analyze data as 100-μm resolution, 16-bit TIFF files (Image Gauge Analysis software, Fuji, Stamford, CT) (*see* **Notes 36** and **37**).

9. Use Image Gauge Analysis software (Fuji, Stamford, CT) and ImageMaster 2D Platinum software version 5 or 6 (GE Healthcare, Piscataway, NJ) to display and analyze data.

3.9. Total Protein Detection with Colloidal CBB

1. Once gel imaging is finished, immerse the gel in 250 mL of colloidal CBB with agitation for 16 h to detect total proteins and then destain in deionized water (*see* **Note 38**).

2. Scan and analyze the gel using ScanMaker 9800XL (300 dpi resolution and 16-bit grayscale pixel depth) and ImageMaster software, respectively.

3. Decant destain solution and add 250 mL of gel storage solution to store the gel at 4°C for up to a few months.

3.10. Phosphoprotein and Protein Spots Excision and In-Gel Digestion

1. Overlay the images of phosphoproteins and proteins as false colors using Adobe photoshop by aligning the phosphoprotein and protein markers.

2. Manually excise the desired phosphoprotein spots using protein spots as landmarks with the help of 0.15-mm spot picker and transfer to a 96-well MultiScreen plate.

3. Use a vacuum manifold system to process in-gel digestion reactions.

4. Destain gel plugs with 200 μL in-gel wash solution for 15 min at RT with gentle agitation on a microplate shaker at a speed of 50 rpm. Evacuate the solution from the bottom of the filter plate using a vacuum manifold system. Repeat this step two more times or until all stain is removed.

5. Dehydrate the gel plugs with 100 μL of acetonitrile for 5 min and remove acetonitrile by vacuum evacuation.

6. Remove residual acetonitrile by blotting the plate gently with Kimwipes.

7. Place a 96-well V-bottom sample collection plate underneath the MultiScreen plate.

8. Add 50 μL of in-gel trypsin solution to wells to rehydrate the gel plugs, cover the plate with adhesive film, and place the cassette inside a plastic sealable bag.

9. Incubate at 37°C for 16 h.

10. Add 100 μL of in-gel extraction solution to wells and incubate for 10 min at RT with gentle agitation.

11. Centrifuge at $2,000 \times g$ for 2 min to collect trypsin-digested peptides into V-bottom polypropylene collection plate.

12. Repeat **steps 10** and **11** once more.

13. Dry the pooled extracted peptides using CentriVap Console and store at −80°C.

3.11. Mass Spectrometry and Data Analyses

1. Resuspend the dried pellet in 50 μL of 0.1% formic acid.

2. Load 15 μL for mass spectral analysis on an LTQ ProteomeX linear ion trap LC-MS/MS instrument following standard procedures (*see* **Note 39**).

3. Search MS/MS data against the suitable database using the SEQUEST algorithm as part of the BioWorks 3.2 software suite.

4. Assign search parameters for the database and then assign protein assignment criteria to identify phosphoproteins.

4. Notes

1. The quality of TEMED may decline with time after opening, and therefore gels take a longer time to polymerize. It is recommended to buy a small bottle of TEMED and to store at 4°C.

2. 2-Mercaptoethanol should be added just before preparation of phenol extraction buffer.

3. Prepared buffered phenol solution should be kept at –20°C in 40 mL aliquots in 50-mL polypropylene tubes.

4. It takes ~30 min to dissolve the chemicals.

5. Any protein standard can be used.

6. It is important to use medium or narrow-range IPG strips to avoid spot overlap to obtain distinct phosphoprotein spots.

7. Unpolymerized acrylamide is a neurotoxin; so care should be taken to avoid exposure.

8. PeppermintStick phosphoprotein molecular weight standards carry two phosphorylated (ovalbumin and bovine β-casein of 45.0 and 23.6 kDa, respectively) and four nonphosphorylated proteins (β-galactosidase, BSA, avidin, and lysozyme of 116.25, 66.2, 18.0, and 14.4 kDa, respectively). Such protein markers are essential to include in any Pro-Q DPS-based experiments to exclude false-positive identification by normalizing the detected phosphoprotein bands or spots against positive and negative phosphoprotein markers. It is the authors' experience that any protein standard with ovalbumin and/or casein can be used for this purpose.

9. In our experience, storage of Pro-Q DPS at 4°C prolongs its stability by at least twofold.

10. It is recommended to prepare a stock solution of 1.0 M sodium acetate, pH 4.0. Use acetic acid to adjust the pH to 4.0. Store at RT.

11. To prepare this solution, add CBB G-250 to ethanol, add a suitable size stirring bar, and stir the solution overnight on stir plate. Next morning add phosphoric acid while stirring and then add ammonium sulfate dissolved in deionized water. Bring the final volume to 4 L, stir for an additional few hours, and leave it at RT to enable settling down of undissolved particles, if any. Transfer the solution slowly into a brown glass bottle and store at RT.

12. Protein extraction should be performed in the fume hood.

13. Do not remove the white interface between the phenol and aqueous layer.

14. It is important to completely resuspend the pellet each time by vortexing, and if necessary sonication (this usually takes longer with the first wash). Place the resuspended sample at −20°C for 20 min in between each wash.

15. Do not over-dry the protein pellet. A completely dry pellet is very difficult to resuspend in IEF extraction solution. Store at −80°C until further use.

16. Pipetting and/or vortexing at RT help in resuspending the pellet. If necessary, incubate the sample for 30–60 min at RT with gentle agitation. Do not heat the sample under any circumstances, as this will lead to protein carbamylation.

17. We usually prepare 10- and 20-fold dilutions.

18. It is advisable to check whether 1 μL produces a signal level that comes within the range of BSA standards; if not, a 20-fold dilution is required.

19. Based on the loading of blanks and protein standards, a program can be created in the instrument along with the parameters that vortex the plate before measuring the absorbance.

20. Care must be taken to avoid introducing any bubbles.

21. Remove bubbles if trapped by lightly tapping the upper surface of the IPG strip. Laying the IPG strip down from one side to the other gradually should prevent bubble formation.

22. Mineral oil should be applied dropwise from the center of the IPG strip until the entire strip is covered.

23. We recommend reading the user manual for Ettan DALT-twelve electrophoresis system.

24. Make sure the sealing gasket is compressed evenly by the faceplate and forms a tight seal with the caster. Do not over-tighten the screws.

25. Start this step 2 h before the completion of IPG focusing experiment.

26. Isobutanol is added to obtain uniform and straight layer on the gel, which is necessary in order to place the IPG strip.

27. It is important to keep the ready gels moist to prevent any dehydration. Gels can also be stored for a week at 4°C, if wrapped with plastic wrap and moist paper towels to prevent dehydration.

28. Add 1× SDS running buffer on top of the stacking gels. It helps in sliding the strip between the plates and in positioning the strip on the gel surface. Again, avoid trapping air bubbles between the strip and the gel. The acidic end of the strip should be on the left. The gel face of the strip should not touch the opposite glass plate.

29. We recommend dipping the electrode wick containing PeppermintStick standards into the agarose sealing solution to prevent the diffusion of standards into liquid. Usually we place electrode wick near the acidic end of the IPG strips. It is important to remove excess liquid before placing the electrode wick.

30. Agarose overlay solution should be ca. 37°C. Again there should not be any air bubbles.

31. The Ettan DALTtwelve electrophoresis unit requires a total of about 9.5 L of SDS running buffer. (About 7.5 L of 1× SDS running buffer for the lower chamber and about 2.0 L of 2× SDS running buffer for the upper chamber.)

32. Unoccupied slots should be filled with blank cassettes. Use 1× SDS running buffer from a squirt bottle to wet the surface of gel or blank cassettes before inserting the cassettes, as it helps to slide easily into the unit.

33. It usually takes overnight (about 15–16 h) for complete electrophoresis.

34. To decant the solution, it is important to wear powder-free nitrile gloves. Hold the tray with one hand and use the other to hold the gel in the tray. Tilt the tray to decant most of the solution. Be careful not to press the gel hard, otherwise it may break or have a finger impression.

35. **Table 1** summarizes all steps including the required volumes of solutions for large-size gels, total incubation times and conditions, and whether solutions can be reused.

36. Gels should be imaged immediately after Pro-Q DPS staining.

37. Any flatbed fluorescent imager can be used for image capture. Pro-Q DPS has ~555/580 nm excitation/emission maxima. Therefore, stained gels can be best imaged using excitation lasers or LEDs with a range of 532–560 nm coupled with a ~580 nm longpass or a ~600 nm bandpass emission filter.

38. After gel imaging, the gel can be directly stained with a total-protein stain, such as colloidal CBB, silver stain, or SYPRO Ruby protein gel stain. This should be performed immediately after gel imaging. The total-protein stain provides a landmark for excising the phosphoproteins for their identification by MS and helps in determining the relative phosphorylation state of a given protein.

39. Details are described by Agrawal and Thelen (9). Though we have used LTQ linear ion trap, any LC-MS/MS MS system and associated software can be used for protein identification.

Acknowledgments

This research was supported by National Science Foundation Plant Genome Research grants DBI-0332418 and DBI-0445287 (to J.J.T.). The authors have no conflicts-of-interest with the manufacturer of Pro-Q DPS, other mentioned products, or equipment.

References

1. Gorg, A., Weiss, W., and Dunn, M. J. (2004) Current two-dimensional electrophoresis technology for proteomics. *Proteomics* 4, 3665–3685

2. Miller, I., Crawford, J., and Gianazza, E. (2006) Protein stains for proteomic applications: Which, when, why? *Proteomics* 6, 5385–5408

3. Steinberg, T. H., Agnew, B. J., Gee, K. R., Leung, W.-Y., Goodman, T., Schulenberg, B., Hendrickson, J., Beechem, J. M., Haugland, R. P., and Patton, W. F. (2003) Global quantitative phosphoprotein analysis using Multiplexed Proteomics technology. *Proteomics* 3, 1128–1144

4. Schulenberg, B., Goodman, T. N., Aggeler, R., Capaldi, R. A., and Patton, W. F. (2004) Characterization of dynamic and steady-state protein phosphorylation using a fluorescent phosphoprotein gel stain and mass spectrometry. *Electrophoresis* 25, 2526–2532

5. Mann, M. and Jensen, O. N. (2003) Proteomics analysis of post-translational modifications. *Nat. Biotechnol.* 21, 255–261

6. Jensen, O. N. (2006) Interpreting the protein language using proteomics. *Nat. Rev. Mol. Cell Biol.* 7, 391–403

7. Agrawal, G. K. and Thelen, J. J. (2005) Development of a simplified, economical polyacrylamide gel staining protocol for phosphoproteins. *Proteomics* 5, 4684–4688

8. Zahedi, R. P., Begonja, A. J., Gambaryan, S., and Sickmann, A. (2006) Phosphoproteomics of human platelets: A quest for novel activation pathways. *Biochem. Biophys. Acta Proteins Proteomics* 1764, 1963–1976

9. Agrawal, G. K. and Thelen, J. J. (2006) Large scale identification and quantitative profiling of phosphoproteins expressed during seed filling in oilseed rape. *Mol. Cell. Proteomics* 5, 2044–2059

10. Chitteti, B. R. and Peng, Z. H. (2007) Proteome and Phosphoproteome differential expression under salinity stress in rice (Oryza sativa) roots. *J. Proteome. Res.* 6, 1718–1727

11. Hajduch, M., Ganapathy, A., Stein, J. W., and Thelen, J. J. (2005) A systematic proteomic study of seed filling in soybean. Establishment of high-resolution two-dimensional reference maps, expression profiles, and an interactive proteome database. *Plant Physiol.* 137, 1397–1419

12. Hajduch, M., Casteel, J. E., Hurrelmeyer, K. E., Song, Z., Agrawal, G. K., and Thelen, J. J. (2006) Proteomic analysis of seed filling in *Brassica napus*. Developmental characterization of metabolic isozymes using high-resolution two-dimensional gel electrophoresis. *Plant Physiol.* 141, 32–46

13. Wolschin, F. and Weckwerth, W. (2005) Combining metal oxide affinity chromatography (MOAC) and selective mass spectrometry for robust identification of in vivo protein phosphorylation sites. *Plant Methods* 1, 9

Chapter 2

Differential Phosphoprotein Labelling (DIPPL) Using ^{32}P and ^{33}P

Aviva M. Tolkovsky and Andreas Wyttenbach

Summary

Differential labelling techniques like differential in-gel electrophoresis (DIGE) enable mixing a control with an experimental sample prior to protein separation, thereby reducing complexity and greatly improving the resolution and analysis of changes in protein expression. Although the shift caused by phosphorylation to a more acidic pI can, in principle, reveal phosphorylation events using DIGE, analysis and verification of the phosphorylation are fraught with problems. Here we describe a differential phospho-labelling technique that obtains the same advantages as DIGE, which we named DIPPL, for differential phosphoprotein labelling. The technique involves labelling two samples, one with ^{32}Pi (orthophosphate) and the other with ^{33}Pi (orthophosphate). The two samples are mixed and proteins are separated on a single gel. Dried gels are exposed twice: once so that total radiation from ^{32}P and ^{33}P is collected on a film or screen; then acetate sheets are interposed between the gel and the screen such that ^{33}P radiation is filtered out leaving ^{32}P radiation to filter through. We demonstrate the utility of this approach by studying the MEK/ERK-dependent changes in stathmin phosphorylation induced by NGF in primary sympathetic neurons.

Key words: Phosphorylation, Differential protein labelling, Metabolic labelling.

1. Introduction

The phospho-proteome is an ever-expanding entity whose limits are still unknown. It is estimated that there are about 100,000 potential phosphorylation sites in the human proteome of which a fewer than 2,000 are currently known (1). In addition to identification of phosphorylation sites, quantification of phosphorylation is needed to understand the regulation of signal transduction. Identification of phospho-proteins or phospho-peptides by mass spectrometry requires large amounts of protein and is not always feasible, especially

Marjo de Graauw (ed.), *Phospho-Proteomics, Methods and Protocols, vol. 527*
© 2009 Humana Press, a part of Springer Science + Business Media, New York, NY
Book DOI: 10.1007/978-1-60327-834-8_2

when using rare cell types like neurons. Moreover, even when using antibodies to detect specific phospho-protein modifications *(2)*, it may be useful to acquire an image of the entire phospho-proteome of a sample prior to homing in on specific proteins.

Labelling cells with phosphorus-32 (^{32}P) has long been used as a means for identifying phospho-proteins *(3)*. The rapidity with which ^{32}P is incorporated into ATP (in the γ position), the relatively high principle emission energy (1.709 MeV), and low cost make it a label of choice for studying phosphorylation. Although it is possible to run simultaneous gels to compare changes in phospho-protein profiles between differentially-treated samples, it would be an advantage to discriminate changes in phosphorylation between two samples on a single gel, just as DIGE is used to highlight changes in protein expression between two samples while eliminating the variability of protein separation patterns. We previously noted that it is possible to prelabel cellular proteins metabolically with [^{35}S]methionine (principle emission, 0.167 MeV) and then label the same cells with ^{32}P to detect which of these proteins are phosphorylated; as ^{35}S emission has lower energy it is possible to screen out the lower energy using a simple device such as an acetate sheet while retaining ^{32}P radiation, which penetrates through the screen *(4)*. The maximum emission energy for phosphorus-33 (^{33}P) is 0.249 MeV, which is about 6.8 times lower than that of ^{32}P. Hence, we reasoned that it might be possible to use a similar configuration to that of DIGE by mixing differentially treated samples, one labelled with ^{32}P and the other labelled with ^{33}P, and running them on a single (2D) gel. This approach would maximize the yield (especially when dealing with small samples, such as rare tissue, e.g., primary neurons) while eliminating ambiguity in spot detection due to differences between the patterns of two 2D gels. We have demonstrated the utility of this method by focusing on phospho-stathmin as our test protein and primary rat superior cervical ganglion (SCG) neurons as our relatively rare cell type *(5)*. In the protocol below we have generalized the method so that it is applicable to all cells. We envisage two scenarios: (a) in one sample the kinase is inactive, whereas the other sample contains the active kinase (for e.g., one could add TPA, serum, insulin, cyclic AMP, etc. to one of the samples to stimulate kinase activity); (b) the kinase is activated in both samples, but one sample contains a kinase inhibitor, thus screening for substrates for a particular protein kinase. The sample containing the inactive kinase (either through lack of stimulation or through the action of the inhibitor) should always be the one labelled with ^{32}P, because in the absence of the acetate screens, the images are a sum of ^{32}P and ^{33}P radiation, whereas the acetate sheets only allow ^{32}P radiation through. Hence it is the ^{32}P-labelled sample that should contain the inactive kinase.

2. Materials

2.1. Radioisotopes and Labelling Media

1. Phosphorus-32 radionuclide ([^{32}P]H$_3$PO$_4$ or ^{32}Pi) #NEX-053H025MC available from Perkin Elmer (las.perkinelmer.com), #64,014 10 mCi Carrier free, available from MP Biomedicals (www.mpbio.com). (*see* **Notes 1** and **2** for neutralization and overview of safety requirements).

2. Phosphorus-33 radionuclide ([^{33}P]H$_3$PO$_4$ or ^{33}Pi) Perkin Elmer (#NEZ08001) or MP Biomedials (#58,300) (*see* **Notes 1** and **2**). The final specific activity of ^{33}Pi should be about 3× higher than that of ^{32}P (*see* **Subheading 3.1**).

3. Phosphate-free DMEM or RPMI-1,640 containing 0.2–0.6 mCi/mL neutralized ^{32}Pi or ^{33}Pi.

4. Phosphate-free (dialyzed) serum (*see* **Note 3**). Keep at –20°C in aliquots.

5. Optional: MEK inhibitor U0126 20 mM stock in DMSO, store in small aliquots at –20°C.

6. PBS for cell wash: 9 g/L NaCl, 10 mM phosphate buffer, pH 7.4. Keep at 4°C.

2.2. Cell Lysis and Electrophoresis Solutions

1. 1D gel lysis buffer: 50 mM Tris–HCl, pH 7.4, 1% NP-40, 120 mM NaCl, 1 mM EDTA, 25 mM sodium fluoride, 40 mM sodium-glycerophosphate, 1 mM sodium orthovandate, 1 mM benzamidine, 5 mM sodium pyrophosphate. Store aliquots at –20°C. Add fresh protease inhibitor cocktail just before use. *Note:* this buffer separates nuclei from cytosol and leaves proteins, such as histones in the nuclear pellet.

2. IEF lysis buffer: 7 M urea, 2 M thiourea, 4% CHAPS, 1.2% ampholytes, pH 3–10, 20 mM DTT, 10 mM Tris–HCl, pH 8, 0.01% bromphenol blue. Store aliquots at –20°C. *Note:* never heat this buffer as hot urea will carbamylate proteins and shift their pI.

3. Loading buffer (4×): 50 mM Tris–HCl, pH 6.8, 2% SDS, 10% glycerol, 100 mM dithiothreitol (DTT) and 0.01% bromophenol blue. This buffer is added to 1D gel lysis buffer when running 1D SDS-PAGE, and is also used in its own right for preparing a whole-cell protein extract.

4. Transfer buffer: 50 mM Tris–HCl, pH 6.8, 2% SDS, 6 M urea, 30% glycerol, and 20 mM fresh DTT.

5. Optional: ^{14}C-Rainbow markers.

6. Coomassie blue gel stain solution: 2.50 g Coomassie Brilliant Blue R-250, 450 mL ethanol, 450 mL distilled water, 100 mL glacial acetic acid.

7. Gel destaining solution: 450 mL ethanol, 450 mL deionized/distilled water, 100 mL glacial acetic acid.

2.3. Equipment

1. Clear acetate sheets.

2. Polyetheleneimine-impregnated P81 phosphocellulose paper.

3. Immobilised pH-gradient (IPG) strips for IEF (for example, Immobiline™ 3–10 nonlinear or 4–7 linear drystrips, GE-Healthcare).

4. Phosphorimaging screens (*see* **Note 4**).

5. Phosphoimager system.

6. Isoelectric focusing equipment compatible with IPG strips.

7. Polyacrylamide slab gel running equipment for second dimension.

8. Perspex beta shields for plates, microfuge tubes, pipetman.

9. Gieger counters to monitor radioactivity.

10. Cell scrapers.

3. Methods

3.1. Equalising ^{32}P and ^{33}P Labelling Intensities and Checking for Efficiency of Acetate Sheets

It is useful to run a pilot experiment such as to ensure that the balance of intensities of ^{32}P and ^{33}P-labelled phospho-proteins will be about equal. Since the gel will be exposed twice, it is a good idea to collect the ^{33}P radiation as rapidly as possible to ensure that sufficient intensity of ^{32}P radiation remains for the second exposure. This prequel also checks that the acetate sheets block out the ^{33}P radiation (*see* **Note 5**).

1. Beginning with ~5 nCi/5 μL of each label, prepare a series of dilutions (e.g., 1:1, 1:2, 1:3, 1:4, 1:5).

2. Spot 5 μL of each solution onto P81 paper in two lanes. (*see* **Fig. 1**).

Fig. 1. Equalising intensities of labelling media. *B*, decreasing amounts but equal volumes of each radio-isotope (starting intensity, ~5 nCi; *bottom line*) were spotted onto Whatman poly-ethyleneimine-impregnated p81 paper. After drying, the paper was imaged for 24 h without (*left*) or with (*right*) two acetate sheets placed between the paper and the screen; the difference in intensities at this time was about 1:3 (^{32}P:^{33}P). Hence three times more ^{33}P isotope compared with ^{32}P isotope needs to be used for in vivo labeling. Note that there is only slight attenuation of ^{32}P intensity with acetate sheets in place. *dil*, dilution. © 2006 by The American Society for Biochemistry and Molecular Biology, Inc. Copied with permission from **ref.** *5*.

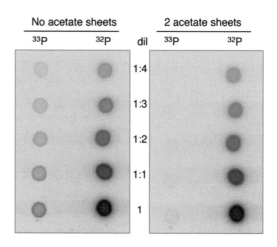

3. Dry paper thoroughly and expose to a phosphorimaging screen or to an X-ray film of choice.

4. Measure spot intensities (*see* **Note 6**).

5. Place acetate sheets (we use two sheets) between the dried paper and a clean screen (if using a phosphimaging screen remember to erase the first exposure) and expose again to ensure that your sheets are blocking ^{33}P radiation.

6. Adjust the concentrations of the two labelling media. Empirically, a 3:1 ratio of radioactivity of ^{33}P:^{32}P yielded spots of roughly equal intensity in our hands.

3.2. Metabolic Labelling

This protocol is for adherent cells, but it can be easily adapted for suspension cells (*see* **ref.** *6*).

1. Plate cells in 35-mm dishes so that at least 200,000 cells are present. If you are planning to stimulate the cells to activate the kinase, then it is crucial that the kinase in question is quiescent prior to stimulation. Hence, if the dialysed serum contains a kinase stimulant, it is important to maintain the cells in minimal serum.

2. Wash cells two times with a warmed and gassed medium having the same composition as that of the Pi-free medium that will be used for radiolabeling.

3. Add the radiolabeling media and incubate for a sufficient time to saturate the production of radiolabelled [γ-^{32}P]ATP (*see* **Note 7**).

4. If a kinase inhibitor is used, add it (in our example, the MEK1 inhibitor U0126 at 10 μM) to the plate containing ^{32}P and add an equivalent amount of solvent (in this case DMSO to 0.05%) to the control plate labelled with ^{33}P. As only ^{32}P will show up after insertion of the acetate sheets, it is important to add the inhibitor to the plate labelled with ^{32}P to demonstrate which phospho-proteins that were labelled with ^{33}P were inhibited. Likewise, it is the unstimulated control sample that should be labelled with ^{32}P and not the sample where the kinase is stimulated.

5. After a 30-min incubation, add the kinase stimulant (in our example, 100 ng/mL NGF was added from a 200-fold concentrate to maintain steady-state labelling) to both experimental and control dishes for the desired time (for neurons we used 3 h).

3.3. Sample Preparation

1. Remove the plates from the incubator onto a bed of ice placed in a tray behind a beta shield.

2. Leave plenty of room to manoeuvre.

3. Wash cells carefully but rapidly two times in cold medium containing the normal amount of unlabelled phosphate.

Avoid using medium with lots of serum at this point as this requires extensive cell washing afterwards, invariably contaminating gels with serum albumin.

4. Scrape off in 0.5 mL of same medium containing 0.01% BSA (to avoid adherence to microfuge tube walls) into cooled microfuge tubes (with screw top lids to avoid aerosol release during centrifugation). Tubes should be placed in a beta shield tube holder.

5. Spin for 3 min at $500 \times g$ in a cooled centrifuge to pellet cells.

6. Wash the pellet with ice-cold PBS (to which phosphatase inhibitors may be added) or serum-free medium and spin again.

7. Lyse the cells in half the volume that would be required to load one lane of a SDS-PAGE 1D gel or an IEF IPG gel strip as the samples labelled with ^{32}P and ^{33}P will be mixed prior to protein separation. Three lysis buffers for three types of analysis are listed in **Subheading 2.3, 1–3** (whose different purpose is explained in **Note 8**). Alternatively, lysis buffer can be added directly to the dish but it is then difficult to control the final volume to avoid over-dilution of small numbers of cells.

8. Vortex the pellet vigorously a few times for 30–60 s until complete solubilisation is achieved.

9. Spin at $25,000 \times g$ for 20–30 min in a cooled microfuge to remove any particulate material (it is especially important to remove particulate material when running IEF gels using IPG strips).

10. Optionally, take 1% of the volume into 5 mL of 10% ice-cold trichloroacetic acid (TCA) and measure the amount of ^{32}P and ^{33}P that was incorporated (*see* **Note 9**). This allows a final adjustment of the respective total label content between experimental and control samples before mixing both samples.

11. Mix the appropriate volumes of ^{32}P and ^{33}P-labeled samples. The volumes are determined by the capacity of the IPG strip (supplied by the manufacturer for each strip size). Samples should be used immediately, or the next day, to avoid loss of ^{32}P labelling intensity.

12. If the samples are to be used for 2D gel analysis, absorb the sample into an inverted Immobiline™ DryStrip during an overnight incubation at room temperature under mineral oil. (*see* **Note 8** for resolving samples using 1D SDS-PAGE).

13. Focus proteins as recommended for each strip type by the manufacturer (*see* **ref.** *7* for one protocol).

14. Equilibrate strips for 15 min with gentle shaking in a transfer buffer and place on top of a second dimension SDS-PAGE slab gel of the appropriate percentage acrylamide and bisacrylamide

to resolve proteins in the range sought. Secure slab in place using a melted mixture of the agarose solution. It is a good idea to leave enough space at the basic end to create one well to load protein size markers, such as radiolabeled Rainbow markers.

15. Separate proteins.

16. Stain (and fix) gels in Coomassie Blue solution, destain, and dry thoroughly. If using high percentage acrylamide gels, gels can become brittle when drying. To add flexibility, it may be useful to add glycerol (to 10% to the destain solution).

17. Expose to a phosphorimaging screen for the minimum time required to obtain a good detailed image of phosphorylated proteins (or to X-ray film).

18. Scan screens at maximum resolution.

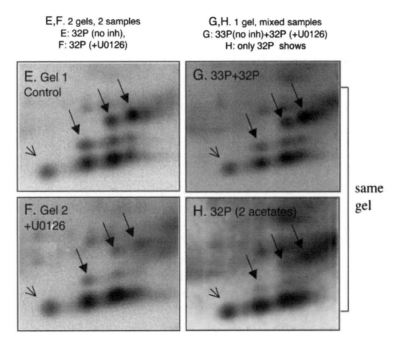

Fig. 2. Detection of MEK-dependent (U0126-inhibitable) phosphorylation sites on rat Stathmin (*see* **ref.** *5* for full details). *E* and *F*, comparison of the stathmin pattern obtained when two samples are independently labeled with ^{32}P and proteins are separated on two independent 2D gels. *E*, SCG neurons were pre-labeled with ^{32}P for 3 h, DMSO (U0126 solvent) was added for 0.5 h after which 100 ng/mL NGF was added for an additional 3 h. *F*, neurons were pre-labeled in the presence of as in *E* for 3 h, and then 10 μM U0126 was added for 0.5 h after which 100 ng/mL NGF was added for 3 h. *G* and *H*, comparison of the stathmin pattern obtained when two samples are independently labeled as in *E* and *F* except that ^{33}P was used to label cells in the absence of U0126, and ^{32}P was used to label cells in the presence of U0126. Samples were mixed and run on a single 2D gel. *G*, no acetate sheets (both radioisotopes imaged). *H*, re-exposure with two intervening acetate sheets. **Note** the similarity between *E* and *G* on the one hand and *F* and *H* on the other hand. *Black full arrows* indicate spots whose intensity has decreased in the sample treated with U0126. *Arrowhead* indicates a spot whose intensity has increased in the sample treated with U0126. This increase is due to the enrichment of the mono-phosphorylated form of the protein due to the loss of the phosphorylated form indicated by the *left-most filled arrow*. © 2006 by The American Society for Biochemistry and Molecular Biology, Inc. Copied with permission from **ref.** *5*.

19. Place two acetate sheets between an erased screen (or fresh X-ray film) and the dried gel, expose, and scan again (*see* **Fig. 2**).

20. It may be useful to obtain an image of the coomassie-stained gel as well. This is obtained by scanning with a flat-bed scanner.

21. Import raw images into an image analysis programme such as ImageJ to quantify intensity of bands/spots.

4. Notes

1. GE Heathcare (formerly Amersham Biosciences) has discontinued manufacture of short life radiochemicals including ^{32}P and ^{33}P. Similar products are available from Perkin Elmer (formerly NEN) or MP Biosciences (formerly ICN). H_3PO_4 is acidic even when supplied in water. The pH should be adjusted to around 7.5 with drop-wise addition of base (100 mM NaOH or Tris base) before use.

2. Precautions using ^{32}P and ^{33}P. Special safety procedures will be in place in your institution for handling these radiochemicals. Perspex (plexiglass) shields are available commercially for Gilson pipettes and eppendorf tubes. We recommend making Perspex holders for 35-mm tissue culture plates, where the plate is sunk in a snugly-fitting trough gouged out of a block of Perspex and is held in place with a plastic screw. A perspex lid that fits into the trough is used to shield the top. Double gloves are recommended. Some of this equipment can be bought from www.nalgenelabware.com.

3. Dialysed serum may be purchased commercially or is easily prepared by dialyzing 50 mL of FCS against twice 2 L of water (with one change) over 12 h. PBS will cause a precipitant to appear. Store at −20°C in aliquots.

4. X-ray film can be used although the dynamic range and linearity are inferior compared with phosphorimaging screens.

5. Minimal-time exposure is recommended to obtain the signal as the half-life of ^{32}P is only 14.3 days (whereas that of ^{33}P is 25.4 days). Once the dilutions that give equal signalling intensities are chosen, it is important to re-adjust the amount of ^{32}P and ^{33}P added to cells on subsequent days according to their half-lives, using the formula $A_t/A_0 = e^{(-0.693t/T)}$ where A_t is the amount of radioactivity on day t, A_0 is the present radioactivity, T is the half life value).

6. Image analysis software accompanies all phosphorimaging systems. Alternatively one can download image analysis freeware from the NIH (http://rsb.info.nih.gov/nih-image/).

7. It is important to reach labelling equilibrium, otherwise the specific activity of ATP may change upon stimulation leading to an increased intensity of spots that is not due to a true increase in kinase activity. This is especially pertinent if the comparison is between a stimulated and an unstimulated sample but is less relevant if one is stimulating both samples and using an inhibitor to delineate kinase targets).

8. If SDS-PAGE in one dimension on the entire proteome is to be carried out, lyse the pellet in 4× SDS-PAGE sample buffer (**Subheading 2.3 (#3)**).

9. After precipitation of proteins in the tube, decant onto a wetted Whatman GF-C glass fibre filter under vacuum, wash tube and filter four times with 5 mL of ice-cold 5% TCA, dry filter, insert into a scintillation vial, add scintillant (such as HiSafe, www.PerkinElmer.com) and count.

Acknowledgements

This work was supported by grants from the BBSRC and the Wellcome Trust.

References

1. Kalume DE, Molina H, Pandey A. (2003) Tackling the phosphoproteome: tools and strategies *Current opinion in chemical biology* 7(1), 64–9.

2. Kaufmann H, Bailey JE, Fussenegger M. (2001) Use of antibodies for detection of phosphorylated proteins separated by two-dimensional gel electrophoresis *Proteomics* 1(2), 194–9.

3. Mitrius JC, Morgan DG, Routtenberg A. (1981) In vivo phosphorylation following [32P]orthophosphate injection into neostriatum or hippocampus: selective and rapid labeling of electrophoretically separated brain proteins *Brain research* 212(1), 67–81.

4. Amess B, Tolkovsky AM. (1995) Programmed cell death in sympathetic neurons: a study by two-dimensional polyacrylamide gel electrophoresis using computer image analysis *Electrophoresis* 16(7), 1255–67.

5. Wyttenbach A, Tolkovsky AM. (2006) Differential phosphoprotein labeling (DIPPL), a method for comparing live cell phosphoproteomes using simultaneous analysis of (33)P- and (32)P-labeled proteins *Mol Cell Proteomics* 5(3), 553–9.

6. Sefton BM. (2001) Labeling cultured cells with 32Pi and preparing cell lysates for immunoprecipitation *Current Protocols in Cell Biology*, section 14.4. Wiley, NJ.

7. Tolkovsky AM. (2000) Analysis of gene expression by two-dimensional gel electrophoresis *Hunt SP, Livesey R (eds) Functional Genomics, a practical approach*, pp. 181–94.

Part II

Enrichment of Phosphoproteins and Peptides

Chapter 3

Identification of Oxidative Stress-Induced Tyrosine Phosphorylated Proteins by Immunoprecipitation and Mass Spectrometry

Mª Carmen Duran, Hong-Lin Chan, and John F. Timms

Summary

Oxidative stress is the result of an increased presence of reactive oxygen species (ROS) in cells and is able to promote, among others, protein and lipid oxidation, DNA damage, mutagenesis, oncogenic activation, or inhibition of tumour suppression, resulting in pathological processes such as myocardial dysfunction or carcinogenesis. External treatment of cells with oxidants such as H_2O_2 or high intracellular levels of ROS has been shown to trigger protein tyrosine phosphorylation. This occurs, at least in part, through the oxidation of reactive cysteine groups in protein tyrosine phosphatases resulting in an inhibition of their activities. Herein, we focus on the characterization of stress-induced protein tyrosine phosphorylation events in a cellular model of human mammary luminal epithelial cells (HB4a cells) stimulated with H_2O_2, in an attempt to better understand the mechanisms by which oxidative stress could promote such phenomena. Thus, immunoprecipitation with anti-phosphotyrosine antibodies and mass spectrometry have allowed us to identify a number of phosphorylated proteins that respond to oxidative stress and thereby further probe the effects of these changes on cellular function.

Key words: Oxidative stress, H_2O_2, Tyrosine phosphorylation, Immunoprecipitation, Mass spectrometry, LC-MS/MS.

1. Introduction

In the last decades, ROS have been shown to play a pivotal role in aging, degenerative diseases, immuno-defence, and cancer (1–3). Apart from the well-known defensive role assigned for these oxygen species in phagocytic cells (4), they also seem to participate in many diverse biological systems. For example,

Marjo de Graauw (ed.), *Phospho-Proteomics, Methods and Protocols, vol. 527*
© 2009 Humana Press, a part of Springer Science + Business Media, New York, NY
Book DOI: 10.1007/978-1-60327-834-8_3

ROS, including hydrogen peroxide (H_2O_2), have been shown to induce apoptotic proteins (5) and promote DNA damage and mutagenesis supporting oncogenic transformation (6). It is also recognised that many types of cancer cells exhibit increased H_2O_2 production that is linked to proliferative signalling and tumourigenesis (7–10), although low levels of H_2O_2 appear permissive for normal proliferative signalling (11). ROS are thought to contribute to cancer through interference with signalling pathways involving nuclear transcription factor kappa B (NFκB), activated protein-1 (AP-1), mitogen-activated protein kinases (MAPKs), and Akt and Jun kinase (12–15). In this regard, ROS have been found to promote a significant increase in protein phosphorylation, an essential event in these signalling cascades, by promoting protein kinase activation, but also by inactivating protein phosphatases by direct oxidation of the active site cysteine in these molecules (16, 17).

Our main goal in this work has been to determine the extent of the stress-induced phosphorylation in a cellular model of human mammary luminal epithelial cells (HB4a cells) (18) after stimulation with H_2O_2, in order to evaluate the processes taking place in an oxidative stress environment. We show that immunoprecipitation using anti-phosphotyrosine-specific antibodies is a good strategy for the enrichment of tyrosine-phosphorylated proteins. Furthermore, by application of mass spectrometry (MS) we were not only able to identify the proteins phosphorylated in response to H_2O_2, but also to detect some of the phosphorylation sites in the protein sequences.

2. Materials

2.1. Cell Culture and Hydrogen Peroxide Treatment

1. Cells: HB4a human mammary luminal epithelial cells (HMLECs) (18).

2. RPMI-1,640 growth medium: RPMI-1640 medium (with 25 mM HEPES and L-glutamine), supplemented with 10% fetal bovine serum (FBS), 2 mM L-glutamine, 100 μg/mL penicillin-streptomycin (all from Gibco/Invitrogen), 5 μg/mL hydrocortisone, and 5 μg/mL insulin (both from Sigma-Aldrich, St Louis, MO).

3. Solution of 0.25% trypsin and 1 mM EDTA (Gibco/Invitrogen).

4. Hydrogen peroxide treatment: 30% whydrogen peroxide solution (Sigma-Aldrich) diluted to 0.5 mM final concentration in the growth media.

2.2. Cell Lysis

1. Modified RIPA lysis buffer: 1% NP-40, 150 mM NaCl, 50 mM Tris–HCl, pH 7.4, 2 mM EDTA, 0.1% sodium dodecyl sulphate (SDS). Make up in ddH$_2$O. Store at 4°C.

2. Phosphatase inhibitors: Sodium orthovanadate (prepare 200 mM stock in ddH$_2$O, use at 2 mM final concentration), okadaic acid (prepare 1 mM stock in ethanol, use at 1 μM, final concentration), fenvalerate (prepare 5 mM stock in ethanol, use at 5 μM), bpV(phen) (bisperoxovanadium 1,10-phenanthroline) (prepare 5 mM stock in ddH$_2$O, use at 5 μM). Prepare and store aliquots at –20°C.

3. Protease inhibitors: Pepstatin A (prepare 1 mg/mL stock in ddH$_2$O, use at 1 μg/mL final concentration), leupeptin (prepare 4.8 mg/mL stock in ddH$_2$O, use at 4.8 μg/mL), AEBSF (prepare as 10 mg/mL stock in ethanol with heating at 60°C, use at 100 μg/mL). Store all at –20°C. Aprotinin stock solution stored at 4°C (dilute 100 times for a working solution of 17 μg/mL).

4. Phosphate-buffered saline: (PBS).

5. Coomassie Protein Assay Reagent (Pierce) and 96-well flat-bottomed assay plates for Bradford Assay.

2.3. Phosphotyrosine Enrichment

1. Protein A-Sepharose CL-4B medium (Amersham Biosciences, Sweden).

2. Anti-phosphotyrosine (4G10) monoclonal antibody, agarose conjugated (Upstate Cell Signalling Solutions).

3. Binding buffer: 50 mM Tris–HCl, pH 7.0.

4. 200 mM phenyl phosphate dissolved in RIPA buffer.

5. RIPA lysis buffer.

2.4. SDS-Polyacrylamide Gel Electrophoresis (SDS-PAGE)

1. Ammonium persulphate (APS): 10% solution in ddH$_2$O and store at 4°C for not more than 3 weeks.

2. Separating buffer: 30% acrylamide/bisacrylamide solution (Pierce), 1.5 M Tris–HCl, pH 8.8, 10% SDS, 10% APS, N,N,N,N'-tetramethyl-ethylenediamine (TEMED).

3. Stacking buffer: 30% acrylamide/bisacrylamide solution (Pierce), 1 M Tris–HCl, pH 6.8, 10% SDS, 10% APS, TEMED.

4. SDS sample buffer (5×): 250 mM Tris–HCl, pH 6.8, 10% SDS, 30% glycerol, 25% β-mercaptoethanol, 0.001% bromophenol blue. Aliquot and store at –20°C.

5. SDS-PAGE electrophoresis running buffer (10×): Tris–glycine SDS buffer 10× (Severn Biotech Ltd, UK). Store at room temperature (RT).

6. Pre-stained molecular weight markers: Full range Rainbow Molecular Weight Markers.

7. Hoeffer SE-400 gel rig system (GE Healthcare, UK).

2.5. Colloidal Coomassie Blue Staining and Image Scanning

1. Fixing solution: 50% (v/v) ethanol, 2% (v/v) phosphoric acid (from 85% stock; 2.35 mL/100 mL). Prepare in ddH$_2$O.

2. Staining solution: 34% (v/v) methanol, 17% (w/v) ammonium sulphate, 3% (v/v) phosphoric acid (from 85% stock; 3.53 mL/100 mL) (*see* **Note 1**).

3. Coomassie Blue G-250 (Merck, Germany).

4. BioRad GS800 scanning densitometer and QuantityOne software (BioRad Laboratories Inc. USA).

2.6. In-Gel Digestion

1. 50% and 100% acetonitrile (AcN) (HPLC grade, Fisher Scientific, UK). Store at RT.

2. 5 mM ammonium bicarbonate (Ambic), pH 8.0, in ddH$_2$O. Store at 4°C or prepare a 100 mM stock, aliquot and store at –20°C.

3. 10 mM dithiothreitol (DTT) in 5 mM Ambic. Prepare fresh.

4. 50 mM iodoacetamide (IAM) in 5 mM Ambic. Prepare fresh.

5. 5% trifluoroacetic acid (TFA) (Fisher Scientific) in 50% AcN. Prepare fresh.

6. 12.5 ng/µL sequencing grade modified trypsin (Promega) in 5 mM Ambic, pH 8.0. Prepare 500 ng stocks in buffer provided (50 mM acetic acid). Store at –20°C.

2.7. Liquid Chromatography-Electrospray Ionization-Mass Spectrometry (LC-ESI-MS/MS) Analysis

1. Ultimate nanoflow-HPLC system (LC Packings, Netherlands).

2. PepMap C18 75-µm inner diameter column (Dionex, UK).

3. Quadrupole Time-of-Flight I (QTOFI) mass spectrometer (Waters, UK).

4. Solvent A: 0.1% formic acid (FA) in HPLC grade H$_2$O.

5. Solvent B: 100% (v/v) AcN, 0.1% FA, in HPLC grade H$_2$O.

2.8. Immunoblotting and Antibodies

1. Tris-buffered saline with Tween 20 (TBS-T): 50 mM Tris–HCl, pH 8.0, 150 mM NaCl, 0.1% Tween 20.

2. Transfer buffer: 195 mM glycine, 25 mM Tris–HCL, 20% (v/v) methanol. Prepare 10× transfer buffer in ddH$_2$O without methanol and store at 4°C. Dilute to a 1× solution prior to transfer, adding 20% (v/v) methanol.

3. Polyvinylidene fluoride membrane (PVDF) (Immobilon P, Millipore, Bedford, MA) and 3 MM chromatography paper.

4. Blocking buffer: 5% w/v low-fat milk in TBS-T.

5. Enhanced chemiluminescence (ECL) reagents (Perkin-Elmer Life Sciences) and Fuji RX X-ray film.

6. Primary antibody: Mouse monoclonal anti-phosphotyrosine antibody (pTyr, PY99; Santa Cruz Biotechnology). Prepare a

dilution of 1:5,000 in TBS-T supplemented with 1.5% bovine serum albumin (BSA, Sigma) (*see* **Note 2**).

7. Secondary antibody: Anti-mouse IgG HRP-linked antibody (GE Healthcare). Prepare a dilution of 1:5,000 in TBS-T supplemented with 1.5% BSA.

3. Methods

3.1. Tissue Culture

3.1.1. Cell Growth

1. Culture HMLEC line HB4a in 15-cm tissue culture dishes in RPMI-1640 growth media at 37°C in a 10% CO_2-humidified incubator.

2. Split cells approximately 1:6 when confluent. Do not over split.

3. Prepare 20 × 15-cm dishes of cells.

3.1.2. Cell Treatment and Lysis

1. Add H_2O_2 solution to growth media of HB4a cells at ~80% confluence to a final concentration of 0.5 mM with gentle swirling or leave untreated (10 × 15 cm dishes per condition; ~10^8 cells per condition). Leave treated cells for 20 min.

2. After treatment, remove medium and wash cells twice with ice-cold PBS and add 500 μL of RIPA lysis buffer/plate. Place dishes immediately on ice. Proceed in the same way with untreated cells.

3. Scrape cells and collect them in labelled tube.

4. Homogenise by passage through a 25-gauge needle six times. Vortex and leave lysates on ice for 30 min with occasional mixing.

5. Remove insoluble material by centrifugation (13,000 × g/10 min/4°C) and transfer the supernatant to fresh tubes.

6. Determine the protein concentration using the coomassie protein assay reagent. Make a 50 mg/mL stock of BSA in RIPA buffer and prepare serial dilutions of 0, 0.25, 0.5, 0.75, 1.0, 2.5, 5.0, and 10.0 mg/mL to make a standard curve. Use a 96-well flat-bottomed assay plate and make triplicate measurements for the BSA standards and four replicates for the experimental samples. For this, add 2 μL of sample per well and 250 μL of assay reagent and mix without introducing bubbles. Use a plate reader at 595 nm wavelength and calculate protein concentrations using the standard curve.

3.2. Phosphotyrosine Enrichment

3.2.1. Packing of Protein A-Sepharose CL-4B Medium

1. Resuspend protein A-Sepharose in ddH_2O (0.1 g dried powder gives 500 μL final volume) by gentle swirling. Do not vortex.

2. Wash the resin with 200 mL ddH_2O per gram of powder and allow it to settle by gravity or spin down (1,000 × g, 2 min) and decant off the supernatant. Repeat this step three times.

3.2.2. Enrichment of Tyrosine-Phosphorylated Proteins by Immunoprecipitation

The strategy followed for affinity purification of phosphotyrosine-containing proteins is described in **Fig. 1**.

3. Prepare 50% slurry of the resin with binding buffer. Swollen resin can be stored at 4–8°C in 20% ethanol.

1. Incubate the whole-cell lysate (18 mg total protein) with 500 μL of a 50% slurry of protein A-Sepharose pre-washed three times with RIPA lysis buffer. Leave the mixture tumbling for 6 h at 4°C on a rotor to "pre-clear" the lysate (*see* **Note 3**).

2. Centrifuge (1,000 × *g*, 2 min) and transfer the supernatant to a fresh tube.

3. Incubate the supernatant with 300 μL of a 50% slurry of agarose-conjugated anti-phosphotyrosine (4G10) monoclonal antibody (1 mg/mL) and leave tumbling for 4–6 h at 4°C.

4. Collect the agarose beads by centrifugation (1,000 × *g*, 5 min, 4°C). Retain the supernatant as unbound fraction and wash the agarose-conjugated beads three times with RIPA buffer to remove non-specifically bound proteins.

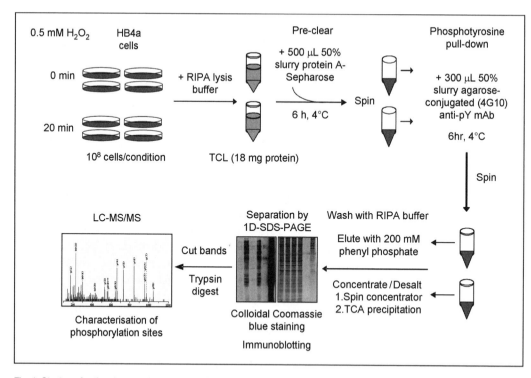

Fig. 1. Strategy for the characterisation of oxidative stress-induced protein tyrosine phosphorylation by immunoprecipitation and MS. HB4a cells are stimulated with 0.5 mM H$_2$O$_2$ for 20 min and lysed in RIPA buffer. The immunoprecipitation steps; pre-clearing with protein A-sepharose, incubation with conjugated anti-phosphotyrosine antibody and elution with 100 mM phenyl phosphate are presented. Finally, proteins are separated by SDS-PAGE, for analysis by MS and validation by immnuoblotting.

5. Elute bound antigens using one bed volume of 200 mM phenyl phosphate in RIPA buffer. Mix the agarose beads gently with the phenyl phosphate solution using a pipette and incubate for 10 min at 37°C with gentle mixing after 5 min. Collect the supernatant by centrifugation ($1,000 \times g$, 4 min, 4°C) and repeat this step once more, pooling the supernatants. Retain the agarose beads (*see* **Notes 4** and **5**).

6. Determine the protein concentration of all fractions (i.e. unbound, washed, and phosphotyrosine-enriched fraction) from both H_2O_2-treated and untreated cells, proceeding as stated in **Subheading 3.1.2, step 6**, using the appropriate buffer as a blank and to prepare standards (*see* **Note 6**).

3.2.3. Concentration and Desalting of Samples (see Note 7)

1. Add one volume of 100% TCA to four volumes of the sample. Prepare TCA by adding 500 g of TCA to 277 mL ddH$_2$O and store at RT.

2. Incubate 10 min at 4°C and spin at $14,000 \times g$ for 5 min. Remove the supernatant, leaving the protein pellet intact.

3. Wash the pellet with 200 mL of ice-cold acetone and spin at $14,000 \times g$, 5 min. Repeat once.

4. Air-dry the pellet and resuspend in 100 μL of SDS sample buffer *(19)*. Boil samples for 5 min.

3.3. SDS-PAGE

1. SDS-PAGE is performed according to Laemmli et al *(19)*. For our experiments we used a Hoeffer SE-400 gel system (16 cm), but any similar system can be employed.

2. Assemble the glass plates, spacers, and clamps according to manufacture's guidelines.

3. Prepare a 12% acrylamide/bisacrylamide resolving gel solution and pour a 1.5-mm thick gel, leaving space for the stacking gel. Overlay carefully with water. Leave gel to polymerise for at least 30 min.

4. Prepare the stacking gel (5% acrylamide/bisacrylamide), pour gel, and insert comb, avoiding air bubbles. Leave to polymerise.

5. Prepare 1× running buffer solution from the 10× stock in ddH2O.

6. Once the stacking gel has set, carefully remove the comb and use a 3-mL syringe fitted with a 19-gauge needle to wash the wells with running buffer.

7. Load samples and include one lane of pre-stained molecular weight markers.

8. Add running buffer to the upper and lower chambers and complete the assembly of the gel unit. Connect to a power

supply. Run gel either overnight at 10 mA or during the day (about 5 h) at 35 mA.

9. Stain gels (*see* **Subheading 3.4**) for band cutting and protein identification by MS or use for immunoblotting (*see* **Subheading 3.6**).

3.4. Colloidal Coomassie Blue Staining

1. Stain the gels with colloidal Coomassie Blue G-250. Briefly, fix the gels in 200 mL fixing solution from 3 h to overnight with gentle shaking. Wash three times with ddH$_2$O (30-min each) and incubate in 200 mL staining solution for 1 h. Add 0.5 g/L Coomassie Blue G-250 to the staining solution and gently shake until protein bands have stained (hours to days). Rinse gels 2–3 times in ddH$_2$O (*see* **Note 8**).

2. Scan the gels using an appropriate scanner such as a BioRad GS800 scanning densitometer. Export the images as 8-bit tiff files into Adobe Photoshop (Adobe Systems Inc.) for image editing and presentation. An example of the results is shown in **Fig. 2**.

3.5. Protein Identification by Mass Spectrometry (MS)

3.5.1. In-Gel Digestion

1. Ideally samples for MS analysis should be prepared in a clean room or other clean area to avoid keratin and other contaminations.

2. Use a scalpel to excise bands from each lane of the gel. Include all sections of the lane. Cut each band again into several pieces and place them into siliconised micro-centrifuge tubes (*see* **Note 9**) and cover with ddH$_2$O. It is possible to freeze the samples at this point or proceed directly to in-gel digestion.

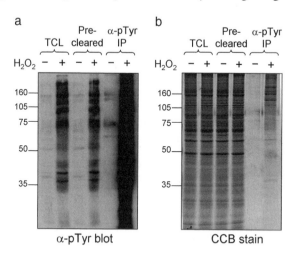

Fig. 2. Enrichment of tyrosine phosphorylated proteins. The figure shows the colloidal Coomassie blue stained gel (**a**) and anti-phosphotyrosine immunoblotting (**b**) results for the enrichment of tyrosine phosphorylated proteins from HB4a cells extracts after treatment with 0.5 mM H$_2$O$_2$ (+) and without treatment (–). The strong signal in the last lane of the immunoblot corroborates the efficiency of the phosphotyrosine enrichment step. TCL (total cell lysate); pre-cleared (fraction pre-incubated with protein-A sepharose) and α-pTyr IP (fraction immunoprecipitated with agarose-conjugated anti-phosphotyrosine monoclonal antibody).

3. The protocol of Shevchenco et al. *(20)* is used for in-gel digestion with minor modifications. All buffers should be prepared fresh prior to digestion.

4. Shake gel pieces in ddH$_2$O for 15 min. Replace water with 50% AcN and shake for a further 15 min. Repeat this step two or three times until gel pieces are completely destained.

5. Remove the 50% AcN and wash once with 100% AcN. The gel pieces should get whitish and shrink. Remove the AcN and dry in a speed vac for 15–20 min.

6. Reduce the samples by adding sufficient volume of 10 mM DTT (in 5 mM Ambic, pH 8.0) to cover the gel pieces and incubating for 45 min at 50°C, with gentle shaking.

7. Remove the DTT solution and alkylate by adding enough 50 mM IAM (in 5 mM Ambic, pH 8.0) to cover the gel pieces and incubating for 1 h at RT in the dark.

8. Remove the IAM solution and wash the gel pieces twice with 50% AcN, 15-min each.

9. Dry the gel pieces in a speed vac for approximately 10 min.

10. Digest the samples with trypsin. From a 500 ng/μL stock of trypsin in buffer, dilute 100 times (to 5 ng/μL) to provide sufficient volume for all samples (10 μL of trypsin at 50 ng per sample). Pipette up and down 3–4 times to ensure that the enzyme enters the gel pieces. Leave the samples 5 min on a shaker at 37°C. Add sufficient volume of 5 mM Ambic, pH 8.0, to cover the gel pieces. Place samples in an incubator or rocking heater block at 37°C and leave to digest overnight.

11. Briefly spin the samples, collect the supernatant and transfer to new siliconised tubes. Add sufficient 50% AcN/5% TFA to cover the gel pieces and shake for 20–30 min to aid peptide extraction. Remove the supernatant and pool together with the first. Repeat this step twice.

12. Speed vac to dryness. Samples can be stored at –20°C or be directly analysed by MS.

13. Resuspend peptides in 15 μL of 0.1% FA by gently shaking.

3.5.2. Protein Identification by LC-MS/MS

1. Chromatographic separation is accomplished by injecting 5 μL of digested peptide samples onto a reversed phase capillary column (PepMap, 75 μm × 150 mm) using a nanoflow HPLC system (Dionex, Ultimate) connected on-line to an ESI Q-TOF mass spectrometer (Waters).

2. For nanoflow HPLC, the flow rate is set at 200 nL/min and the separation/elution of peptides is achieved using a 5–45% gradient of solvent B (0.1% FA, 100% AcN) over 90 min followed by an isocratic step at 95% solvent B for 10 min.

3. The mass spectrometer is operated in the positive ion mode with a resolution of 4,000–6,000 full-width half-maximum (FWHM), using a source temperature of 80°C and a counter current nitrogen flow rate of 60 L/hr.

4. Employ data-dependent analysis (DDA) in order to detect the three most abundant ions in each cycle. The cycles are set up as follows: MS scans are acquired every 1 s (m/z 350–1,500) and MS/MS is performed on automatically selected peptide ions, also for 1 s (m/z 50–2,000, continuum mode) using the function switching in MassLynx 4.0 software (Waters).

3.5.3. Data Analysis

1. Convert raw MS/MS data using either MassLynx or Distiller (Matrix Science) software. Smooth (Savitzky Golay, two channels twice) and centroid (at 80%) spectra.

2. Perform database searching using Mascot (version 2.0.02) or similar to search and compare experimental masses with theoretical ones in the IPI human database.

3. Parameters for protein identification using the Mascot search engine are: enzyme (trypsin, bovine); miscleavages (2); charge of ions (+2 and +3); database (IPI human); species (*Homo sapiens*); mass tolerance of the precursor peptide ion (100 ppm) and mass tolerance for the MS/MS fragment ions (0.8 mmu). Carbamidomethylation of cysteines (alkylation with IAM, **Subheading 3.5.1**) is considered as a fixed modification, whilst oxidation of methionine, pyro-glutamic acid, N-acetylation and phosphorylation of tyrosine, serine and threonine are considered as variable modifications.

4. For protein identification from MS/MS data it is recommended to follow the guidelines provided by the Molecular Cell Proteomics journal *(21)*. Basically, protein identifications are accepted when at least two peptides match an entry with a Mascot score above 40. In addition, it is desirable to manually corroborate the selected identifications (particularly those derived from single peptide hits), by assigning the fragment ions in the MS/MS spectra to theoretical peptide fragmentations using Protein Prospector *(22)*. An example of a phosphorylated peptide identified after phosphotyrosine enrichment is shown in **Fig. 3**.

3.6. Immunoblotting

1. Separate a second set of samples by SDS-PAGE (*see* **Subheading 3.3**) and transfer electrophoretically to PVDF membrane using 1× transfer buffer and a Bio-Rad transfer tank (Transfer-Blot cell). Wet the PVDF membrane in 100% methanol for 1 min and place on top of the gel without air bubbles. Sandwich between two pieces of Whatman paper soaked in transfer buffer, role out air bubbles with a plastic pipette and secure in the tank cassette, as per the manufacture's instructions. Insert the cassette into the

GSTAENAEpYLR

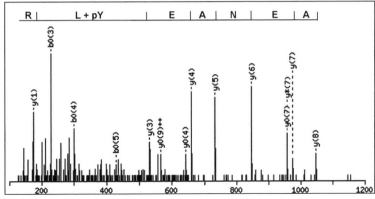

Epidermal growth factor receptor pY1197

Fig. 3. H_2O_2 induces tyrosine phosphorylation in EGFR. The figure shows the MS/MS spectrum and sequence (GSTAENAEpYLR) of a peptide derived from the epidermal growth factor receptor (EGFR), which appears phosphorylated at tyrosine 1,197 after stimulation with H_2O_2.

transfer tank containing transfer buffer such that the membrane is located between the gel and the anode. Connect the tank to an appropriate power supply and transfer at 350 mA for 5 h.

2. Once transfer is complete, remove the "sandwich" from cassette and disassemble. The coloured molecular weight markers should be clearly visible on the membrane if transfer has worked.

3. Incubate the membrane in 100 mL of blocking buffer for 1 h at RT on a rocking platform. The membrane can also be left in blocking buffer over night at 4°C.

4. Discard blocking buffer and rinse membrane with TBS-T prior to addition of a dilution of primary antibody. It is possible to cut the membrane with a razor blade into different molecular weight ranges for blotting with different antibodies.

5. Incubate with anti-phosphotyrosine antibody (PY99) for 1 h at RT on a rocking platform.

6. Remove primary antibody and wash membrane in TBS-T 3 × 15 min on a rocking platform.

7. Incubate with HRP-conjugated anti-mouse secondary antibody for 1 h at RT on a rocking platform.

8. Remove secondary antibody and wash membrane 3 × 15 min on a rocking platform.

9. Mix the two ECL reagents 1:1 according and incubate immediately with the membrane for 1–2 min ensuring full coverage of the membrane.

10. Remove membrane from ECL reagents, drain excess fluid, wrap in Saran wrap and tape into an X-ray cassette.

11. In a dark room, place X-ray film on the top of the membrane and close the cassette. Leave to expose for an appropriate time (20 s-1 min) to give a reasonable signal that is not saturated.

12. Scan film on a BioRad GS-800 densitometer and quantify intensity with ImageQuant software. The anti-phosphotyrosine immunoblotting results are presented in **Fig. 2b**.

4. Notes

1. Prepare the staining solution at least 1 h prior to use considering that it takes the ammonium sulphate a while to dissolve completely.

2. Primary antibody dilutions can be stored for re-use at 4°C in 0.01% sodium azide for several months.

3. It is recommended to pre-clear/pre-incubate the cell lysate with protein A-sepharose beads in order to reduce the level of non-specific binding to the beads.

4. After elution, the beads can be suspended in SDS sample buffer, boiled and run on a gel and immunoblotted to check that no tyrosine phosphorylated proteins remain bound.

5. The agarose-conjugated monoclonal antibody can be regenerated for re-use by washing with 20 bed volumes of 1.5 M NaCl followed by 20 bed volume of PBS. Store the conjugate in PBS containing 0.01% sodium azide at 4°C.

6. Protein recovery from H_2O_2-treated cells by anti-phosphotyrosine enrichment was ~6 μg from 18 mg total protein prior to concentration, whereas less than 0.5 μg was recovered from the untreated cells.

7. It is necessary to concentrate the sample for loading onto a 1D SDS-PAGE gel and would require desalting for subsequent analysis by 2D gel electrophoresis. Yields of protein were found to be poor using centrifuge concentrators, so a TCA/acetone precipitation method was employed.

8. Gels in containers should be placed on a rocking platform in all steps, in order to help staining. In addition, containers should be properly sealed to avoid evaporation.

9. The use of siliconised tubes avoids protein loss due to adsorption to the tube walls and also avoids contaminants derived from the tube plastic.

References

1. Ames, B. N., Gold, L. S., and Willett, W. C. (1995) The causes and prevention of cancer. *Proc Natl Acad Sci U S A* 92, 5258–65.

2. Olinski, R., Gackowski, D., Foksinski, M., Rozalski, R., Roszkowski, K., and Jaruga, P. (2002) Oxidative DNA damage: assessment of the role in carcinogenesis, atherosclerosis, and acquired immunodeficiency syndrome. *Free Radic Biol Med* 33, 192–200.

3. Taniyama, Y., and Griendling, K. K. (2003) Reactive oxygen species in the vasculature: molecular and cellular mechanisms. *Hypertension* 42, 1075–81.

4. Sbarra, A. J., and Karnovsky, M. L. (1959) The biochemical basis of phagocytosis. I. Metabolic changes during the ingestion of particles by polymorphonuclear leukocytes. *J Biol Chem* 234, 1355–62.

5. Rollet-Labelle, E., Grange, M. J., Elbim, C., Marquetty, C., Gougerot-Pocidalo, M. A., and Pasquier, C. (1998) Hydroxyl radical as a potential intracellular mediator of polymorphonuclear neutrophil apoptosis. *Free Radic Biol Med* 24, 563–72.

6. Jackson, A. L., and Loeb, L. A. (2001) The contribution of endogenous sources of DNA damage to the multiple mutations in cancer. *Mutat Res* 477, 7–21.

7. Suh, Y. A., Arnold, R. S., Lassegue, B., Shi, J., Xu, X., Sorescu, D., Chung, A. B., Griendling, K. K., and Lambeth, J. D. (1999) Cell transformation by the superoxide-generating oxidase Mox1. *Nature* 401, 79–82.

8. zatrowski, T. P., and Nathan, C. F. (1991) Production of large amounts of hydrogen peroxide by human tumor cells. *Cancer Res* 51, 794–8.

9. Rhee, S. G. (1999) Redox signaling: hydrogen peroxide as intracellular messenger. *Exp Mol Med* 31, 53–9.

10. Behrend, L., Henderson, G., and Zwacka, R. M. (2003) Reactive oxygen species in oncogenic transformation. *Biochem Soc Trans* 31, 1441–4.

11. Sundaresan, M., Yu, Z. X., Ferrans, V. J., Irani, K., and Finkel, T. (1995) Requirement for generation of H2O2 for platelet-derived growth factor signal transduction. *Science* 270, 296–9.

12. Rhee, S. G., Chang, T. S., Bae, Y. S., Lee, S. R., and Kang, S. W. (2003) Cellular regulation by hydrogen peroxide. *J Am Soc Nephrol* 14, S211-5.

13. Guyton, K. Z., Liu, Y., Gorospe, M., Xu, Q., and Holbrook, N. J. (1996) Activation of mitogen-activated protein kinase by H2O2. Role in cell survival following oxidant injury. *J Biol Chem* 271, 4138–42.

14. Bae, Y. S., Sung, J. Y., Kim, O. S., Kim, Y. J., Hur, K. C., Kazlauskas, A., and Rhee, S. G. (2000) Platelet-derived growth factor-induced H(2)O(2) production requires the activation of phosphatidylinositol 3-kinase. *J Biol Chem* 275, 10527–31.

15. Manna, S. K., Zhang, H. J., Yan, T., Oberley, L. W., and Aggarwal, B. B. (1998) Overexpression of manganese superoxide dismutase suppresses tumor necrosis factor-induced apoptosis and activation of nuclear transcription factor-kappaB and activated protein-1. *J Biol Chem* 273, 13245–54.

16. Rhee, S. G., Bae, Y. S., Lee, S. R., and Kwon, J. (2000) Hydrogen peroxide: a key messenger that modulates protein phosphorylation through cysteine oxidation. *Sci STKE* 2000, E1.

17. Wu, Y., Kwon, K. S., and Rhee, S. G. (1998) Probing cellular protein targets of H2O2 with fluorescein-conjugated iodoacetamide and antibodies to fluorescein. *FEBS Lett* 440, 111–5.

18. Stamps, A. C., Davies, S. C., Burman, J., and O'Hare, M. J. (1994) Analysis of proviral integration in human mammary epithelial cell lines immortalized by retroviral infection with a temperature-sensitive SV40 T-antigen construct. *Int J Cancer* 57, 865–74.

19. Laemmli, U. K. (1970) Cleavage of structural proteins during the assembly of the head of bacteriophage T4. *Nature* 227, 680–5.

20. Shevchenko, A., Wilm, M., Vorm, O., and Mann, M. (1996) Mass spectrometric sequencing of proteins silver-stained polyacrylamide gels. *Anal Chem* 68, 850–8.

21. Carr, S., Aebersold, R., Baldwin, M., Burlingame, A., Clauser, K., and Nesvizhskii, A. (2004) The need for guidelines in publication of peptide and protein identification data: Working Group on Publication Guidelines for Peptide and Protein Identification Data. *Mol Cell Proteomics* 3, 531–3.

22. Clauser, K. R., Baker, P., and Burlingame, A. L. (1999) Role of accurate mass measurement (+/– 10 ppm) in protein identification strategies employing MS or MS/MS and database searching. *Anal Chem* 71, 2871–82.

Chapter 4

Enrichment and Characterization of Phosphopeptides by Immobilized Metal Affinity Chromatography (IMAC) and Mass Spectrometry

Tine E. Thingholm and Ole N. Jensen

Summary

The combination of immobilized metal affinity chromatography (IMAC) and mass spectrometry is a widely used technique for enrichment and sequencing of phosphopeptides. In the IMAC method, negatively charged phosphate groups interact with positively charged metal ions (Fe^{3+}, Ga^{3+}, and Al^{3+}) and this interaction makes it possible to enrich phosphorylated peptides from rather complex peptide samples. Phosphopeptide enrichment by IMAC is sensitive and specific for peptide mixtures derived from pure proteins or simple protein mixtures. The selectivity of the IMAC method is, however, limited when working with peptide mixtures derived from highly complex samples, e.g., whole-cell extracts, where sample prefractionation is advisable. Furthermore, lowering the pH value of the sample loading buffer reduces nonspecific binding to the IMAC resin significantly, thereby improving the selectivity of IMAC for phosphopeptides. The retained phosphopeptides are released from the IMAC resin by using alkaline buffers (pH 10–11), EDTA, or phosphate-containing buffers. We have described a detailed and robust protocol for IMAC for phosphopeptide enrichment from semi-complex mixtures.

Key words: Protein phosphorylation, Phosphopeptide enrichment, Immobilized metal affinity chromatography (IMAC), Trifluoroacetic acid, Mass spectrometry.

1. Introduction

Immobilized metal affinity chromatography (IMAC) is a widely used affinity enrichment technique for the preparation of phosphopeptides for mass spectrometric analysis and sequencing. IMAC was initially developed for protein affinity-purification based on interactions of histidine and cysteine residues with

Marjo de Graauw (ed.), *Phospho-Proteomics, Methods and Protocols, vol. 527*
© 2009 Humana Press, a part of Springer Science + Business Media, New York, NY
Book DOI: 10.1007/978-1-60327-834-8_4

the IMAC resin *(1, 2)*, but the binding of phosphoproteins and phosphoamino acids to metal ions demonstrated by Andersson and Porath *(3)* gave the technique an extra dimension. Neville et al. extended the IMAC technique to the enrichment of phosphopeptides obtained from proteolytically digested proteins *(4)*. The IMAC technique has since then been used extensively for the enrichment of phosphorylated peptides prior to MS analysis and sequencing *(5–9)*.

The IMAC technique facilitates the recovery and identification of phosphopeptides in complex protein samples of biological origin *(7, 10, 11)*. However, nonphosphorylated peptides containing multiple acidic amino acid residues may co-purify with the phosphopeptides in IMAC, thereby reducing the selectivity of the method. In addition, nonphosphorylated peptides ionize much better in the mass spectrometer ion source than do phosphorylated peptides. Thus, the presence of nonphosphorylated peptides decreases the number of identified phosphorylated peptides in large-scale phosphoproteomic studies by mass spectrometry. *O*-methylesterification of carboxylic acid containing amino acid residues has been suggested to increase the specificity of the IMAC technique, but this step may introduce unwanted side reactions and loss of peptides due to extensive lyophilization *(12)*. Another possibility to enhance the performance of IMAC is to lower the complexity of the peptide samples by pre-fractionation using e.g. isoelectric focusing (IEF) *(13)* or ion exchange chromatography *(7, 11, 14)* prior to MS analysis. However, this requires significant amounts of starting material.

Saha et al. demonstrated that the pK_a value of phosphoric acid decreased to 1.1 upon methylation *(15)*. As the addition of an organic group decreases the pKa value of phosphoric acid, it is a reasonable assumption that the pK_a values of phosphopeptides is significantly lower than that of phosphoric acid due to the organic environment provided by the surrounding amino acid residues. Decreasing the pH of the IMAC loading conditions to below pH 1.9 means that more acidic peptides in a sample will become neutralized while phosphopeptides will retain their negative charges and their binding affinity towards the IMAC resin. Loading the peptide sample in 0.1% TFA, 50% acetonitrile reduces the level of nonspecific binding to the IMAC and increases the specificity of the method towards phosphopeptides *(16)*. The phosphopeptides are subsequently eluted from the IMAC material using alkaline buffers such as ammonia water (pH 10–11). Alternatively, phosphopeptides can be eluted with EDTA, highly acidic solutions or solutions including phosphate or phosphoric acid. It is often advantageous to desalt and concentrate the IMAC eluate by reversed phase chromatography prior to MS analysis and sequencing.

2. Materials

2.1. Model Proteins

1. Transferrin (human).
2. Serum albumin (bovine).
3. β-Lactoglobulin (bovine).
4. Carbonic anhydrase (bovine).
5. β-Casein (bovine).
6. α-Casein (bovine).
7. Ovalbumin (chicken).
8. Ribonuclease B (bovine pancreas).
9. Alcohol dehydrogenase (Baker's yeast).
10. Myoglobin (whale skeletal muscle).
11. Lysozyme (chicken).
12. α-Amylase (*Bacillus* species).

All proteins were from Sigma (St. Louis. MO).

2.2. Reduction, Alkylation, and Digestion of Model Proteins

1. Ammonium bicarbonate.
2. Dithiotreitol (DTT).
3. Iodoacetamide.
4. Modified trypsin.

2.3. Immobilized Metal-Ion Affinity Chromatography (IMAC)

1. Fe(III)-coated PHOS-select™ metal chelate beads (Sigma®), stored at –20°C. NTA-silica material from Qiagen is also an option.
2. IMAC Loading buffer: 0.1% trifluoroacetic acid (TFA), 50% acetonitrile, HPLC grade.
3. 200-μL GELoader tips from Alpha Laboratories (Hampshire, S050 4NU, UK).
4. IMAC elution buffer: 20 μL ammonia solution (25%), 980 μL UHQ water (pH ~11), prepare fresh when required.
5. Formic acid.
6. Water was obtained from a Milli-Q system (Millipore, Bedford, MA) (UHQ water). (*see* **Notes 1** and **2**).

2.4. Reversed Phase (RP) Chromatography

1. POROS Oligo R3 reversed phase material (Applied Biosystems, Foster City, CA).
2. GELoader tip (Eppendorf, Hamburg, Germany).
3. Syringe for HPLC loading (P/N 038250, N25/500-LC PKT 5, SGE, Ringwood, Victoria, Australia).
4. RP loading buffer: 0.1% TFA.

5. RP elution buffer (for LC-ESI MS/MS analysis): 70% acetonitrile, 0.1% TFA.

6. DHB elution buffer (for MALDI MS analysis): 20 mg/mL 2,5-dihydroxybenzoic acid (DHB) in 50% acetonitrile, 0.1% TFA, 1% *ortho*-phosphoric acid.

3. Methods

The principle of the IMAC protocol is demonstrated by using a semi-complex peptide mixture originating from tryptic digestions of 12 standard proteins (model proteins) that are commercially available (*see* **Notes 3** and **4**). The IMAC purification method is a simple and "easy-to-do" method; however, buffers used for batch incubation with IMAC material should contain 0.1% TFA, 50% acetonitrile to optimize adsorption of phosphopeptides to the IMAC beads, and to reduce nonspecific binding *(16)* (*see* **Note 5**).

3.1. Model Proteins and Peptide Mixture

1. Dissolve each protein in 50 mM ammonium bicarbonate, pH 7.8, 10 mM DTT and incubated at 37°C for 1 h. After reduction, add 20 mM iodoacetamide and incubate the samples at room temperature for 1 h in the dark. After incubation, quench each reaction with 10 mM DTT.

2. Digest each protein using trypsin (1–2% w/w) at 37°C for 12 h.

3.2. Batch Incubation with IMAC Beads

1. For 0.3 pmol tryptic digest use 5 μL of IMAC beads. When working with low amounts of sample use less IMAC beads to reduce the level of nonspecific binding from nonphosphorylated peptides. For more complex samples, where more material is available, more IMAC beads should be used.

2. Transfer 5 μL IMAC beads to a fresh 1-mL Eppedorf tube.

3. Wash the IMAC beads twice using 50 μL IMAC loading buffer.

4. Resuspend the beads in 40 μL of IMAC loading buffer and add the sample (*see* **Note 6**).

5. Incubate the sample with IMAC beads in a Thermomixer (Eppendorf) for 30 min at 20°C.

3.3. Packing the IMAC Micro-column

1. Squeeze the tip of a GELoader tip to prevent the IMAC beads from leaking, while still allowing for liquid to pass through.

2. Pack the beads in the constricted end of the GELoader tip by application of air pressure to generate an IMAC micro-column *(17)* (**Fig. 1**).

3. Wash the IMAC column using 40 μL of IMAC loading buffer.

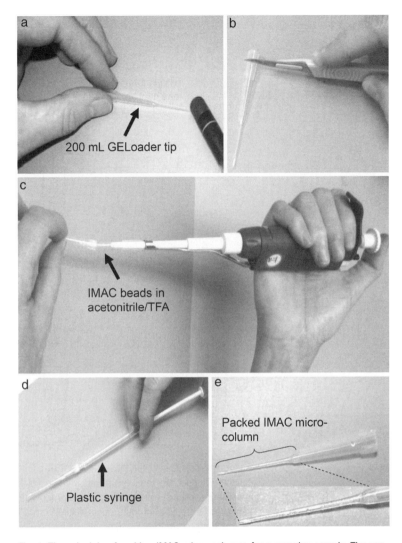

Fig. 1. The principle of making IMAC micro-columns for a complex sample. The constricted end of a 200-µL GELoader tip is squeezed to prevent the IMAC beads from leaking. This is done by pressing the tip of a pen onto the end of the GELoader tip (**a**). Cut the top of the GELoader tip to make a plastic syringe fit into the opening (**b**). Load the IMAC beads onto the inside of the GELoader tip (**c**). Pack the IMAC beads to form an IMAC micro-column by applying air pressure using a 1-mL plastic syringe (**d**). The packed IMAC micro-column (**e**). The principle for a simple sample is the same, however, using an Eppendorf GELoader tip instead of a 200 µL GELoader tip.

3.4. Elution of Phosphorylated Peptides from the IMAC Micro-column

1. Elute the phosphorylated peptides bound to the IMAC micro-column using 30 µL of IMAC elution buffer (**Fig. 2**). It is important that this step is performed slowly (~1 drop/s). (N.B. For MALDI MS analysis the peptides can be eluted off the IMAC micro-column directly onto the MALDI target using 1 µL of DHB solution. After crystallization the sample is ready for MALDI MS analysis) (*see* **Note 7**).

Flowchart for IMAC

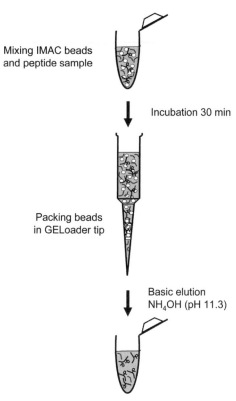

Mixing IMAC beads
and peptide sample

Incubation 30 min

Packing beads
in GELoader tip

Basic elution
NH₄OH (pH 11.3)

Enriched phosphopeptides

Fig. 2. The setup used for the enrichment of phosphopeptides using IMAC micro-columns. The peptide sample is mixed with the IMAC beads and incubated for 30 min in a Thermomixer at roomtemperature. After incubation the beads are packed into a GELoader tip forming an IMAC micro-column (*see* **Fig. 1**). The phosphopeptides are subsequently eluted from the IMAC micro-column using basic elution conditions.

2. For LC-ESI MS^n analysis the IMAC eluate should be acidified using 100% formic acid (pH should be ~2–3) and desalted/concentrated using reversed phase chromatography (*see* below).

3.5. Packing a Reversed Phase (RP) Micro-column for Desalting/Concentrating the Sample

1. Use RP micro-columns of ~3–6 mm (0.4–0.8 µg) depending on the amount of material.

2. Suspend ~2 mg POROS Oligo R3 reversed phase (RP) material in 200 µL of 50% acetonitrile.

3. Squeeze the tip of a GELoader tip to prevent the RP beads from leaking, while still allowing liquid to pass through (*see* **Note 8**).

4. Pack the beads in the constricted end of the GELoader tip by application of air pressure forming an micro-column (*17*) (*see* **Note 9**).

3.6. Loading the Sample Onto a the Reversed Phase (RP) Micro-column

1. Load the acidified sample slowly onto the RP micro-column (~1 drop/sec).
2. Wash the RP micro-column using 30 μL 0.1% TFA.

3.7. Elution of Phosphorylated Peptides from the RP Micro-column

1. Elute the phosphopeptides from the RP micro-column using 20 μL of RP elution buffer, followed by lyophilization of the phosphopeptides. (N.B. For MALDI MS analysis the peptides can be eluted off the RP micro-column directly onto the MALDI target using 1 μL DHB solution. After crystallization the sample is ready for MALDI MS analysis).

2. LC-ESI-MS^n analysis: Redissolve the dried phosphopeptides in 0.5 μL 100% formic acid and immediately dilute to 10 μL with UHQ water.

LC-ESI-MS^n analysis of multi-phosphorylated peptides is improved by re-dissolving the phosphopeptides by sonication in an EDTA containing buffer prior to LC-ESI-MS/MS analysis *(18)*. An example of the results produced is shown in **Fig. 3**. The figure

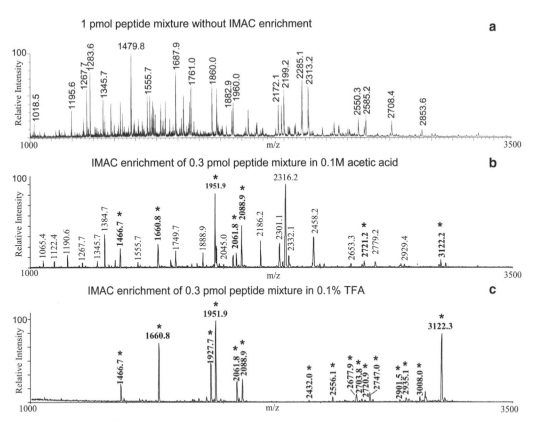

Fig. 3. MALDI MS spectra of 1 pmol peptide mixture before IMAC phosphopeptide enrichment. (**a**), MALDI MS spectra of 0.3 pmol peptide mixture after IMAC enrichment using 0.1 M acetic acid (**b**) or 0.1% TFA as loading buffer (**c**). The MALDI MS was performed using a Bruker Ultraflex (Bruker Daltonics, Bremen, Germany) instrument. The phosphopeptides are illustrated by *asterisks* and *bold* (*see* **Note 10**).

illustrates the complexity of 1 pmol peptide mixture analyzed on MALDI TOF MS as a dried droplet without IMAC enrichment (**a**), the MALDI MS peptide mass map obtained from 0.3 pmol peptide mixture after IMAC phosphopeptide enrichment using 0.2 M acetic acid (**b**) or 0.1% TFA (**c**) as loading buffer (*see* **Notes 10** and **11**).

4. Notes

1. It is important to obtain the highest purity of all chemicals used.

2. All solutions should be prepared in UHQ water.

3. Always start by testing the method using a model peptide mixture. It is important to freshly prepare the peptide mixture as peptides bind to the surface of the plastic tubes in which they are stored. In addition, avoid transferring the peptide sample between different tubes to minimize adsorptive losses of the sample.

4. The peptide mixture used for the experiment illustrated in this chapter contained peptides originating from tryptic digestions of 0.3 pmol of each of the 12 model proteins.

5. If the sample was pre-suspended in another buffer, check whether the buffer interferes with IMAC binding *(19)*. If the sample contains reagents that interfere with IMAC binding, dilute the samples sufficiently to reduce the concentration of these reagents. If EDTA is present in the sample, the peptides have to be purified using reversed phase chromatography prior to IMAC. Always test the pH value of the sample before IMAC batch incubation. The pH value should be approximately 1.7.

6. When working with larger amounts of IMAC beads, remember to increase the incubation, washing, and elution volumes.

7. When working with larger amounts of material, it is better to elute the phosphopeptides using 30 μL of IMAC elution buffer followed by reversed phase chromatography, as 1 μL of DHB buffer will not be sufficient to elute all phosphopeptides. In addition, the sample may be too concentrated for MALDI MS^n analysis and LC-ESI MS^n may be a better solution.

8. Use same principle for squeezing the GELoader tip as described in **Subheading 3.3**.

9. When working with more complex samples, the capacity of the RP column packed in a GELoader tip might be too low. Pack the RP micro-column in a P10 pipette tip by stamping out a small plug of C_{18} material from a 3 M Empore™ C_{18} extraction disk

and place it in the constricted end of a GELoader tip. Pack the RP material on top of the membrane plug as describe in **Subheading 3.5, steps 3–4**, Chapter "The Use of Titanium Dioxide Micro-Columns to Selectively Isolate Phosphopeptides from Proteolytic Digests."

10. The results obtained using this protocol will differ according to the mass spectrometer used for the analysis of the phosphopeptides, not only between MALDI MS and ESI MS but also within different MALDI MS instruments, depending on laser optics, laser frequency, instrumental configuration, sensitivity etc.

11. The exact binding affinity of the IMAC beads is not known; however, the amount of nonspecific binding from nonphosphorylated peptides are very dependent on the ratio between amount of sample and IMAC beads. It may be necessary to opmitize the ratio for different samples.

Acknowledgements

This work was supported by grants from the Danish Research Agency and the Lundbeck Foundation to O.N.J.

References

1. Chaga, G., Hopp, J., and Nelson, P. (1999) Immobilized metal ion affinity chromatography on Co2+ -carboxymethylaspartate–agarose Superflow, as demonstrated by one-step purification of lactate dehydrogenase from chicken breast muscle *Biotechnol Appl Biochem* 29, 19–24.

2. Hochuli, E., Dobeli, H., and Schacher, A. (1987) New metal chelate adsorbent selective for proteins and peptides containing neighbouring histidine residues *J Chromatogr* 411, 177–84.

3. Andersson, L., and Porath, J. (1986) Isolation of phosphoproteins by immobilized metal (Fe3+) affinity chromatography *Anal Biochem* 154, 250–4.

4. Neville, D. C., Rozanas, C. R., Price, E. M., Gruis, D. B., Verkman, A. S., and Townsend, R. R. (1997) Evidence for phosphorylation of serine 753 in CFTR using a novel metal-ion affinity resin and matrix-assisted laser desorption mass spectrometry *Protein Sci* 6, 2436–45.

5. Figeys, D., Gygi, S. P., McKinnon, G., and Aebersold, R. (1998) An integrated micro-fluidics-tandem mass spectrometry system for automated protein analysis *Anal Chem* 70, 3728–34.

6. Li, S., and Dass, C. (1999) Iron(III)-immobilized metal ion affinity chromatography and mass spectrometry for the purification and characterization of synthetic phosphopeptides *Anal Biochem* 270, 9–14.

7. Nuhse, T. S., Stensballe, A., Jensen, O. N., and Peck, S. C. (2003) Large-scale analysis of in vivo phosphorylated membrane proteins by immobilized metal ion affinity chromatography and mass spectrometry *Mol Cell Proteomics* 2, 1234–43.

8. Posewitz, M. C., and Tempst, P. (1999) Immobilized gallium(III) affinity chromatography of phosphopeptides *Anal Chem* 71, 2883–92.

9. Stensballe, A., Andersen, S., and Jensen, O. N. (2001) Characterization of phosphoproteins from electrophoretic gels by nanoscale Fe(III) affinity chromatography with off-line mass spectrometry analysis *Proteomics* 1, 207–22.

10. Ficarro, S. B., McCleland, M. L., Stukenberg, P. T., Burke, D. J., Ross, M. M., Shabanowitz, J., Hunt, D. F., and White, F. M. (2002) Phos-

phoproteome analysis by mass spectrometry and its application to Saccharomyces cerevisiae *Nat Biotechnol* 20, 301–5.

11. Gruhler, A., Olsen, J. V., Mohammed, S., Mortensen, P., Faergeman, N. J., Mann, M., and Jensen, O. N. (2005) Quantitative phosphoproteomics applied to the yeast pheromone signaling pathway *Mol Cell Proteomics* 4, 310–27.

12. Stewart, I. I., Thomson, T., and Figeys, D. (2001) O-18 Labeling: a tool for proteomics *Rapid Commun Mass Spectrom* 15, 2456–65.

13. Krijgsveld, J., Gauci, S., Dormeyer, W., and Heck, A. (2006) In-gel isoelectric focusing of peptides as a tool for improved protein identification *J Proteome Res* 5, 1721–30.

14. Beausoleil, S. A., Jedrychowski, M., Schwartz, D., Elias, J. E., Villen, J., Li, J., Cohn, M. A., Cantley, L. C., and Gygi, S. P. (2004) Large-scale characterization of HeLa cell nuclear phosphoproteins *Proc Natl Acad Sci U S A* 101, 12130–5.

15. Saha, A., Saha, N., Ji, L. N., Zhao, J., Gregan, F., Sajadi, S. A. A., Song, B., and Sigel, H. (1996) Stability of metal ion complexes formed with methyl phosphate and hydrogen phosphate *J Biol Inorg Chem* 1, 231–38.

16. Kokubu, M., Ishihama, Y., Sato, T., Nagasu, T., and Oda, Y. (2005) Specificity of immobilized metal affinity-based IMAC/C18 tip enrichment of phosphopeptides for protein phosphorylation analysis *Anal Chem* 77, 5144–54.

17. Gobom, J., Nordhoff, E., Mirgorodskaya, E., Ekman, R., and Roepstorff, P. (1999) Sample purification and preparation technique based on nano-scale reversed-phase columns for the sensitive analysis of complex peptide mixtures by matrix-assisted laser desorption/ionization mass spectrometry *J Mass Spectrom* 34, 105–16.

18. Liu, S., Zhang, C., Campbell, J. L., Zhang, H., Yeung, K. K., Han, V. K., and Lajoie, G. A. (2005) Formation of phosphopeptide-metal ion complexes in liquid chromatography/electrospray mass spectrometry and their influence on phosphopeptide detection *Rapid Commun Mass Spectrom* 19, 2747–56.

19. Jensen, S. S., and Larsen, M. R. (2007) Evaluation of the impact of some experimental procedures on different phosphopeptide enrichment techniques. *Rapid Commun Mass Spectrom* 21, 3635–45.

Chapter 5

The Use of Titanium Dioxide Micro-Columns to Selectively Isolate Phosphopeptides from Proteolytic Digests

Tine E. Thingholm and Martin R. Larsen

Summary

Titanium dioxide has very high affinity for phosphopeptides and it has become an efficient alternative to already existing methods for phosphopeptide enrichment from complex samples. Peptide loading in a highly acidic environment in the presence of 2,5-dihydroxybenzoic acid (DHB), phthalic acid, or glycolic acid has been shown to improve selectivity significantly by reducing unspecific binding from nonphosphorylated peptides. The enriched phosphopeptides bound to the titanium dioxide are subsequently eluted from the micro-column using an alkaline buffer. Titanium dioxide chromatography is extremely tolerant towards most buffers used in biological experiments. It is highly robust and as such it has become one of the methods of choice in large-scale phospho-proteomics. Here we describe the protocol for phosphopeptide enrichment using titanium dioxide chromatography followed by desalting and concentration of the sample by reversed phase chromatography prior to MS analysis.

Key words: Protein phosphorylation, Phosphopeptide enrichment, Titanium dioxide chromatography (TiO$_2$), Glycolic acid, Trifluoroacetic acid (TFA), Mass spectrometry.

1. Introduction

For many years the adsorption of proteins to titanium dioxide (TiO$_2$) films has been studied with the aim of determining a method for pursuing bioelectrochemical studies of protein functions (1). More interestingly for phosphoproteomic studies, TiO$_2$ has been shown to have affinity for phosphate ions from aqueous solutions (2, 3), and recently, TiO$_2$ chromatography has been adapted as an efficient alternative to already existing methods for phosphopeptide enrichment from complex samples (4–12). In 2004, Heck

Marjo de Graauw (ed.), *Phospho-Proteomics, Methods and Protocols, vol. 527*
© 2009 Humana Press, a part of Springer Science + Business Media, New York, NY
Book DOI: 10.1007/978-1-60327-834-8_5

and co-workers described the ability of TiO_2 to selectively bind to phophorylated peptides using an online two-dimensional liquid chromatography (LC) setup with spherical particles of TiO_2 (Titansphere) as the first dimension and reversed phase (RP) material as the second dimension. The sample was loaded onto a TiO_2 column under acidic conditions (pH 2.9) to promote the binding of phosphopeptides to the TiO_2 particles. The unbound non-phosphorylated peptides were trapped on the RP column. After elution from the RP column, the non-phosphorylated peptides were analyzed using nanoLC-ESI-MS/MS. The phosphopeptides were subsequently eluted from the TiO_2 column using an alkaline buffer (pH 9.0), concentrated on the RP precolumn, and analyzed using nanoLC-ESI-MS/MS *(9)*. However, they observed, as for IMAC phosphopeptide enrichment (*see* Chapter "Enrichment and Characterization of Phosphopeptides by Immobilized Metal Affinity Chromatography (IMAC) and Mass Spectrometry"), a significant amount of non-phosphorylated peptides co-purified with the phosphopeptides. Recently, we introduced an offline setup for TiO_2 chromatography using 2,5-dihydroxybenzoic acid (DHB), phthalic acid, or glycolic acid and a high concentration of trifluoroacetic acid (TFA) in the loading buffer, which have proven to reduce unspecific binding from non-phosphorylated peptides significantly *(8, 13, 14)*. In addition, ammonia solution at even higher pH (11.3) was shown to elute phosphorylated peptides from the TiO_2 column more efficiently than at pH 9 and thereby improve phosphopeptide recovery *(8)*.

The high selectivity of TiO_2 towards phosphorylated peptides makes it a powerful tool for phosphoproteomic studies. In addition, TiO_2 chromatography of phosphorylated peptides is extremely tolerant towards most buffers and salts used in biochemistry and cell biology laboratories *(13)*. The offline setup is simple and fast, and does not require expensive equipment.

2. Materials

2.1. Model Proteins

1. Transferrin (human).
2. Serum albumin (bovine).
3. β-Lactoglobulin (bovine).
4. Carbonic anhydrase (bovine).
5. β-Casein (bovine).
6. α-Casein (bovine).
7. Ovalbumin (chicken).

8. Ribonuclease B (bovine pancreas).

9. Alcohol dehydrogenase (Baker's yeast).

10. Myoglobin (whale skeletal muscle).

11. Lysozyme (chicken).

12. α-Amylase (*Bacillus* species).

All proteins were from Sigma (St. Louis. MO).

2.2. Titanium Dioxide (TiO₂) Chromatography

1. Titanium dioxide (TiO$_2$) beads (Titansphere, 5 µm, GL sciences Inc., WWW.inertsil.com).

2. GELoader tips (Eppendorf, Hamburg, Germany) or p10 pipette tips depending on the size of the column needed.

3. 3 M Empore C8 disk (3 M, Bioanalytical Technologies, St. Paul, MN).

4. Acetonitrile, HPLC Grade.

5. Syringe for HPLC loading (P/N 038250, N25/500-LC PKT 5, SGE, Ringwood, Victoria, Australia) to create a small plug of C$_8$ membrane material.

6. TiO$_2$ loading buffer: 1 M glycolic acid in 5% trifluoroacetic acid (TFA), 80% acetonitrile.

7. TiO$_2$ washing buffer: 1% TFA, 80% acetonitrile.

8. TiO$_2$ elution buffer 1: 20 µL ammonia solution (25%), 980 µL UHQ water (pH 11.3).

9. TiO$_2$ elution buffer 2: 30% acetonitrile.

10. Formic acid.

11. Water was obtained from a Milli-Q system (Millipore, Bedford, MA) (UHQ water) (*see* **Notes 1** and **2**).

2.3. Reversed Phase (RP) Chromatography

1. POROS Oligo R3 reversed phase material (Applied Biosystems, Foster City, CA).

2. GELoader tips (Eppendorf) or p10 pipette tips depending on the size of the column needed.

3. 3 M Empore C18 disk (3 M, Bioanalytical Technologies, St. Paul, MN).

4. Syringe for HPLC loading (P/N 038250, N25/500-LC PKT 5, SGE).

5. RP washing buffer: 0.1% TFA.

6. RP elution buffer (for LC-ESI MSMS analysis): 70% acetonitrile, 0.1% TFA.

7. DHB elution buffer (for MALDI MS analysis): 20 mg/mL DHB in 50% acetonitrile, 1% Ortho-phosphoric acid (Bie & Berntsen A/S, UK).

3. Methods

The principle of the TiO₂ method described in this chapter is illustrated using a peptide mixture originating from tryptic digestions of 12 standard proteins (Model proteins) (*see* **Notes 3** and **4**). The TiO₂ purification method is a simple and straight-forward method. It is fast and efficient for enrichment of phos-phopeptides even from highly complex samples *(8, 14)* (*see* **Note 5**). The experimental setup of the method is illustrated in **Fig. 1**.

3.1. Model Proteins and Peptide Mixture

1. Dissolve each protein in 50 mM ammonium bicarbonate, pH 7.8, 10 mM DTT and incubate at 37°C at 1 h. After reduction, add 20 mM iodoacetamide and incubate the samples at room temperature for 1 h in the dark. After incubation, quench each reaction with 10 mM DTT.

2. Digest each protein using trypsin (1–2% w/w) at 37°C for 12 h.

3.2. Packing the TiO₂ Micro-Column

1. When working with less complex samples use TiO₂ micro-columns of ~3 mm (~0.4 µg TiO₂ beads). Use longer columns

Flowchart for TiO₂ chromatography

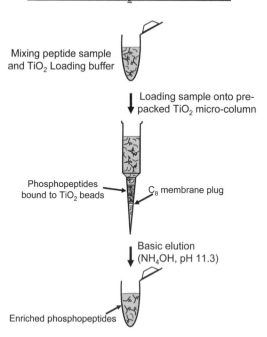

Mixing peptide sample and TiO₂ Loading buffer

Loading sample onto pre-packed TiO₂ micro-column

Phosphopeptides bound to TiO₂ beads

C₈ membrane plug

Basic elution (NH₄OH, pH 11.3)

Enriched phosphopeptides

Fig. 1. The setup used for the enrichment of phosphopeptides using TiO₂ micro-columns. The peptide sample is diluted with the TiO₂ buffer. The sample is subsequently loaded onto a pre-packed TiO₂ micro-column. The phosphopeptides are then eluted from the TiO₂ micro-column using basic elution conditions.

for more complex samples. The capacity of TiO_2 towards phosphorylated peptides is very high so even a small column will retain significant amount of phosphorylated peptide.

2. Suspend ~2 mg TiO_2 beads in 500 μL 100% acetonitrile.

3. Stamp out a small plug of C_8 material from a 3 M Empore™ C_8 extraction disk using a HPLC syringe needle, and place this plug in the constricted end of a GELoader tip using e.g., a LC capillary (**8, 14**). The C_8 material will prevent the TiO_2 beads from leaking, while still allowing for liquid to pass through.

4. Wash the membrane using 100% acetonitrile to reduce the level of polymers in the membrane plug: Load 20 μL 100% acetonitrile into the GELoader tip and apply gentle air pressure using a plastic syringe *(15)*.

5. Prepare a TiO_2 micro-column by packing the TiO_2 beads on top of the C_8 membrane disk in the GELoader tip. Load an aliquot of the TiO_2 suspension (depending on the amount of the sample) into the GELoader tip and apply gentle air pressure as in last step. **Figure 2** illustrates the principle for packing the TiO_2 micro-column.

3.3. Loading the Sample Onto the TiO_2 Micro-Column

1. It is important that buffers used for loading or washing of the TiO_2 micro-column contain 50–80% acetonitrile to abrogate adsorption of peptides to the C_8 membrane and the TiO_2 beads due to hydrophobic interactions.

2. Dilute the peptide sample at least five times in the TiO_2 Loading buffer (v/v) and mix well. Load the sample onto the TiO_2 micro-column by applying air pressure *(15)* (*see* **Note 6**).

3. Wash the TiO_2 micro-column using 5 μL TiO_2 Loading buffer and subsequently with 30 μL TiO_2 washing buffer.

3.4. Elution of Phosphorylated Peptides from the TiO_2 Micro-column

1. Elute the phosphopeptides bound to the TiO_2 micro-column using a minimum of 25 μL of TiO_2 elution buffer 1 (depending on the size of the column) (*see* **Note 7**).

2. Since peptides will have binding affinity towards the C_8 membrane plug, some phosphopeptides may have bound to the C_8 membrane during the previous elution step (**step 1**). Elute these using 1 μl of TiO_2 elution buffer 2 and pool this with the eluate from **step 1**. (*see* **Note 8**).

3. Acidify the pooled eluates using 1 μL of 100% formic acid per 10 μL eluate. Make sure that the eluate obtains a pH value of 2–3. At this stage the purified phosphopeptides can be analyzed by MS. It is, however, advisable to desalt/concentrate the samples first using reversed phase chromatography.

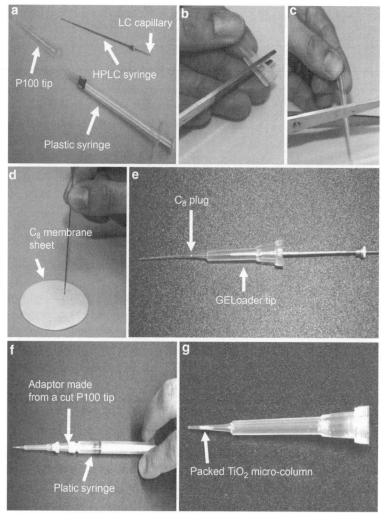

Fig. 2. The principle of packing a TiO$_2$ micro-column. A P100 tip, a plastic syringe, a HPLC syringe and a LC capillary is needed (**a**). An adaptor for the plastic syringe is made by cutting both ends of a P100 tip (**b** and **c**). A C$_8$ membrane plug is stamped out from a C$_8$ membrane sheet using a HPLC syringe tube (**d**). The membrane plug is placed in the constricted end of a GELoader tip by pushing it out from the HPLC syringe using a LC capillary (**e**). The cut P100 tip is placed on the constricted end of the plastic syringe to function as an adaptor for the GELoader tip. TiO$_2$ beads are resuspended in 100% acetonitrile and loaded into the GELoader tip and the TiO$_2$ micro-column is subsequently packed by applying air pressure using the plastic syringe (**f**). The packed TiO$_2$ micro-column (**g**).

3.5. Packing a Reversed Phase (RP) Micro-Column for Desalting/Concentrating the Sample

1. Use RP GELoader tip micro-columns of ~3–6 mm (0.4–0.8 μg) depending on the amount of material.

2. Suspend ~2 mg POROS Oligo R3 RP material in 500 μL 50% acetonitrile.

3. Prepare a RP micro-column by stamping out a small plug of C$_{18}$ material from a 3 M Empore™ C$_{18}$ extraction disk and place

it in the constricted end of a GELoader tip as describe in **Sub-heading 3.2, step 3**.

4. Pack the RP beads in the GELoader tip as described for the TiO$_2$ beads in **Subheading 3.2, step 5**.

5. Load the acidified sample slowly onto the RP micro-column (~1 drop/s).

6. Wash the RP micro-column using 30 µL RP washing buffer.

3.6. Elution of Phosphorylated Peptides from the RP Micro-Column

1. Elute the phosphopeptides from the RP micro-column using 20 µL of RP elution buffer, followed by lyophilization of the phosphopeptides. (N.B. For MALDI MS analysis the peptides can be eluted off the RP micro-column directly onto the MALDI target using 1 µL of DHB solution).

2. Redissolve the dried phosphopeptides in 0.5 µL 100% formic acid and dilute immediately to 10 µL with UHQ water. The sample is ready for LC-ESI-MS^n analysis.

LC-ESI-MS^n analysis of multi-phosphorylated peptides is improved by re-dissolving the phosphopeptides by sonication in an EDTA containing buffer prior to LC-ESI-MS^n analysis *(16)* (*see* **Note 9**). An example of the results obtained by TiO$_2$ chromatography is shown in **Fig. 3**. The figure shows the MALDI

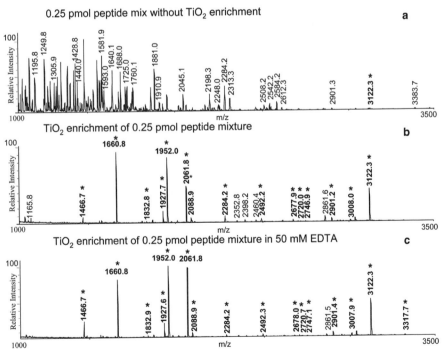

Fig. 3. MALDI MS spectra of 0.25 pmol peptide mixture before TiO$_2$ phosphopeptide enrichment. (**a**), after phosphopeptide enrichment using TiO$_2$ (**b**) and after phosphopeptide enrichment using TiO$_2$ in the presence of 50 mM EDTA. The MALDI MS was performed using a Bruker Ultraflex (Bruker Daltonics, Bremen, Germany) instrument. The phosphopeptides are illustrated by asterisks and bold (*see* **Note 10**).

MS results obtained on a Bruker Ultraflex (Bruker Daltonics, Bremen, Germany) from a dried droplet of 0.25 pmol peptide mixture without phosphopeptide enrichment (**a**). The MALDI MS peptide mass map of the TiO_2 enriched phosphopeptides from 0.25 pmol of peptide mixture is shown in (**b**). The unique tolerance towards biological buffers is illustrated by the enrichment of phosphopeptides from 0.25 pmol peptide mixture in the presence of 50 mM EDTA. The phosphopeptides are illustrated in by asterisks and bold (*see* **Note 10**).

4. Notes

1. It is important to obtain the highest purity of all chemicals used.

2. All solutions should be prepared in UHQ water.

3. Always start by testing the method using a model peptide mixture. It is important to freshly prepare the peptide mixture as peptides bind to the surface of the plastic tubes in which they are stored. In addition, avoid transferring the peptide sample between different tubes to minimize adsorptive losses of the sample.

4. The peptide mixture used for the experiment illustrated in this chapter contained peptides originating from tryptic digestions of 250 fmol of each of the 12 model proteins. Experiments have shown that the presented method is sensitive down to the low femtomole level *(14)*.

5. The capacity of the TiO_2 beads have been tested, however, more experiments need to be performed. The capacity of the beads appears to be very high, but when working with high amounts of material it is recommended to use P10 pipette tips instead of GELoader tips and to split the peptide sample and load it onto several TiO_2 micro-columns. The resulting eluates can then be pooled prior to MS analysis. For example for an experiments using 120 µg tryptic peptides, use three TiO_2 micro-columns packed in P10 tips (~4 mm).

6. It is very critical that the sample is mixed thoroughly with the TiO_2 Loading buffer before loading the sample onto the TiO_2 micro-column.

7. When eluting the phosphopeptides from the TiO_2 micro-column it is critical to use at least 25 µL TiO_2 elution buffer 1, since some phosphopeptides, especially multi-phosphorylated peptides bind strongly to the TiO_2.

8. It is important to follow this elution with an elution step using 1 µL TiO_2 elution buffer 2, to elute peptides bound to the C_8 membrane plug.

9. TiO$_2$ chromatography seems to have a preference for mono-phosphorylated peptides likely due to the fact that it is difficult to elute multi-phosphorylated peptides from the resin once they have bound *(13)*; however, the combination of sequential elution from immobilized metal affinity chromatography (IMAC) and TiO$_2$ has made it possible to enrich for mono- as well as multi-phosphorylated peptides as described in Chapter "Quantitative Phospho-proteomics Based on Soluble Nanopolymers" *(17)*.

10. The results obtained using this protocol will differ according to the mass spectrometer used for the analysis of the phosphopeptides, not only between MALDI MS and ESI MS but also within different MALDI MS instruments, depending on laser optics, laser frequency, instrumental configuration, sensitivity etc.

Acknowledgements

This work was supported by the Danish Natural Science Research Council (grant no. 21-03-0167 (M.R.L)).

References

1. Topoglidis, E., Cass, A.E.G., Gilardi, G., Sadeghi, S., Beaumont, N., Durrant, J.R. (1998) Protein adsorption on nanocrystalline TiO2 films: An immobilization strategy for bioanalytical devices. *Analytical Chemistry* 70, 5111–13.

2. Connor, P.A., Dobson, K.D., and McQuillan, J. (1999) Infrared spectroscopy of the TiO2/aqueous solution interface. *Langmuir* 15, 2402–08.

3. Connor, P. A., and McQuillan, A. J. (1999) Phosphate adsorption onto TiO2 from aqueous solutions: an in situ internal reflection infrared spectroscopic study. *Langmuir* 15, 2916–21.

4. Ikeguchi, Y., and Nakamura, H. (1997) Determination of organic phosphates by column-switching high performance anion-exchange chromatography using on-line preconcentration on titania. *Analytical Sciences* 13, 479–83.

5. Ikeguchi, Y., and Nakamura, H. (2000) Selective enrichment of phospholipids by titania. *Analytical Sciences* 16, 541.

6. Jiang, Z.-T., and Zuo, Y.-M. (2001) Synthesis of porous titania microspheres for HPLC packings by polymerization-induced colloid aggregation (PICA). *Analytical Chemistry* 73, 686–88.

7. Kawahara, M., Nakamura, H., and Nakajima, T. (1990) Titania and zirconia: possible new ceramic microparticulates for high-performance liquid chromatography. *Journal of Chromatography A* 515, 149–58.

8. Larsen, M. R., Thingholm, T. E., Jensen, O. N., Roepstorff, P., and Jorgensen, T. J. (2005) Highly selective enrichment of phosphorylated peptides from peptide mixtures using titanium dioxide microcolumns. *Molecular and Cellular Proteomics* 4, 873–86.

9. Pinkse, M. W., Uitto, P. M., Hilhorst, M. J., Ooms, B., and Heck, A. J. (2004) Selective isolation at the femtomole level of phosphopeptides from proteolytic digests using 2D-NanoLC-ESI-MS/MS and titanium oxide precolumns. *Analytical Chemistry* 76, 3 935–43.

10. Sano, A., and Nakamura, H. (2004) Chemoaffinity of titania for the column-switching HPLC analysis of phosphopeptides. *Analytical Sciences* 20, 565.

11. Tani, K., Sumizawa, T., Watanabe, M., Tachibana, M., Koizumi, H., and Kiba, T. (2002) Evaluation of titania as an ion-exchanger and as a ligand-exchanger in HPLC. *Chromatographia* 55, 33–37.

12. Tani, K., and Suzuki, Y. (1994) Syntheses of spherical silica and titania from alkoxides on a laboratory scale. *Chromatographia* 38, 291–94.

13. Jensen, S. S., and Larsen, M. R. (2007) Evaluation of the impact of some experimental procedures on different phosphopeptide enrichment techniques. *Rapid Communication in Mass Spectrometry* 21, 3635–45.

14. Thingholm, T. E., Jorgensen, T. J., Jensen, O. N., and Larsen, M. R. (2006) Highly selective enrichment of phosphorylated peptides using titanium dioxide. *Nature Protocols* 1, 1929–35.

15. Gobom, J., Nordhoff, E., Mirgorodskaya, E., Ekman, R., and Roepstorff, P. (1999) Sample purification and preparation technique based on nano-scale reversed-phase columns for the sensitive analysis of complex peptide mixtures by matrix-assisted laser desorption/ionization mass spectrometry. *Journal of Mass Spectrometry* 34, 105–16.

16. Liu, S., Zhang, C., Campbell, J. L., Zhang, H., Yeung, K. K., Han, V. K., and Lajoie, G. A. (2005) Formation of phosphopeptide-metal ion complexes in liquid chromatography/electrospray mass spectrometry and their influence on phosphopeptide detection. *Rapid Communication in Mass Spectrometry* 19, 2747–56.

17. Thingholm, T. E., Jensen, O. N., P.J, R., and Larsen, M. R. (2007) SIMAC - A phosphoproteomic strategy for the rapid separation of mono-phosphorylated from multiply phosphorylated peptides. *Published online by MCP on November 27th 2007 (M700362-MCP200).*

Chapter 6

Enrichment and Separation of Mono- and Multiply Phosphorylated Peptides Using Sequential Elution from IMAC Prior to Mass Spectrometric Analysis

Tine E. Thingholm, Ole N. Jensen, and Martin R. Larsen

Summary

Phospho-proteomics relies on methods for efficient purification and sequencing of phosphopeptides from highly complex biological systems using low amounts of starting material. Current methods for phosphopeptide enrichment, e.g., immobilized metal affinity chromatography and titanium dioxide chromatography, provide varying degrees of selectivity and specificity for phosphopeptide enrichment. Furthermore, the number of multiply phosphorylated peptides that are identified in most published studies is rather low. Here the protocol for a new strategy that separates mono-phosphorylated peptides from multiply phosphorylated peptides using sequential elution from immobilized metal affinity chromatography is described. The two separate phosphopeptide fractions are subsequently analyzed by mass spectrometric methods optimized for mono-phosphorylated and multiply phosphorylated peptides, respectively, resulting in improved identification of especially multiply phosphorylated peptides from a minimum amount of starting material. The new method increases the coverage of the phosphoproteome significantly.

Key words: Phosphopeptide enrichment, Multiply phosphorylated peptides, Immobilized metal affinity chromatography (IMAC), Sequential elution, SIMAC, Titanium dioxide (TiO_2) chromatography, Mass spectrometry.

1. Introduction

Several techniques exist for phosphopeptide enrichment prior to mass spectrometric analysis. Today the most commonly used methods are immobilized metal affinity chromatography (IMAC) and titanium dioxide (TiO_2) chromatography (*see* Chapters

Marjo de Graauw (ed.), *Phospho-Proteomics, Methods and Protocols, vol. 527*
© 2009 Humana Press, a part of Springer Science + Business Media, New York, NY
Book DOI: 10.1007/978-1-60327-834-8_6

"Enrichment and Characterization of Phosphopeptides by Immobilized Metal Affinity Chromatography (IMAC) and Mass Spectrometry" and " The Use of Titanium Dioxide Micro-Columns to Selectively Isolate Phosphopeptides from Proteolytic Digests"). Recent studies comparing three different phosphopeptide enrichment methods, including phosphoramidate chemistry (PAC) (1), IMAC, and TiO_2 chromatography, showed that each method isolated distinct, partially overlapping segments of a phosphoproteome, whereas none of the tested methods were able to provide a whole phosphoproteome (2).

One of the challenges in large-scale phospho-proteomics is the analysis of multiply phosphorylated peptides. Multiply phosphorylated peptides are in general suppressed in the ionization process in the mass spectrometric (MS) analysis in the presence of mono- or non-phosphorylated peptides and thereby the chance to detect and identify them decreases. In addition, most mass spectrometers are only able to perform a limited number of tandem MS experiments (MS/MS) in a given time period resulting in the negligence of the less abundant multiply phosphorylated peptides. Furthermore, collision-induced dissociation (CID) fragmentation of phosphopeptides usually results in the loss of phosphoric acid with poor peptide backbone fragmentation and consequently little sequence information is obtained. Phosphorylation-directed multistage tandem MS (pdMS^3), where an ion originating from a neutral loss signal detected in the MS^2 is selected for a second round of CID (MS^3), provides better sequence information for mono-phosphorylated peptide ions with relatively high abundance. However, multiply phosphorylated peptides will result in additional losses of phosphoric acid. Optimized pdMS^3 (3), multistage activation (MSA) (4), or electron capture/transfer dissociation (ECD/ETD) (5, 6) will most likely give better results for multiply phosphorylated peptides. Though, multiply phosphorylated peptides need to be separated from mono-phosphorylated peptides in order to analyze them using different mass spectrometric setups.

Recently, we developed a new method for the separation of mono-phosphorylated peptides from multiply phosphorylated peptides where we are using sequential elution from IMAC (SIMAC) (3). In this strategy the peptide mixture is incubated with IMAC beads, which have a stronger selectivity for multiply phosphorylated peptides than for mono-phosphorylated peptides (7). After incubation, the sample is split in three "elution" fractions (see **Fig. 1**), an IMAC flow-through fraction, an acidic (1% TFA) fraction, and a basic (pH 11.3) fraction. The IMAC flow-through and acidic fractions that contain mainly mono-phosphorylated peptides in addition to a significant number of nonphosphorylated peptides are further enriched by

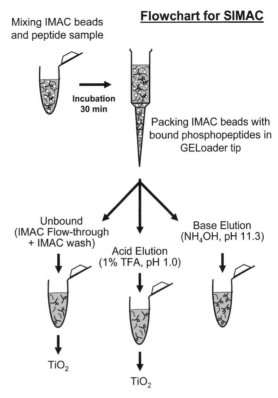

Fig. 1. The setup used for the enrichment and separation of mono- from multiply phosphorylate peptides using the SIMAC strategy. The peptide sample is mixed with the IMAC beads and incubated for 30 min in a Thermomixer at room temperature. After incubation the beads are packed into a GELoader tip forming an IMAC micro-column. The IMAC flow-through is collected and further enriched using TiO_2 chromatography. The mono-phosphorylated peptides are eluted from the IMAC micro-column using acidic elution conditions (1% TFA, pH 1.0) and for complex samples this eluate is also further enriched using TiO_2 chromatography. The multiply phosphorylated peptides are subsequently eluted from the IMAC micro-column using basic elution conditions (ammonia water, pH 11.3). All three fractions are then ready for mass spectrometric analysis.

TiO_2 chromatography to achieve pure mono-phosphorylated peptide fractions prior to tandem MS analysis. The basic fraction is analyzed directly by tandem MS analysis without further TiO_2 purification, as multiply phosphorylated peptides binds very hard to TiO_2 material and may be lost at this step (8). Furthermore, this fraction in general has a relatively low level of nonphosphorylated peptides, and thus no further enrichment is necessary.

SIMAC greatly improves the number of phosphorylation sites identified even from very low amounts of starting material and offers a way to identify and characterize multiply phosphorylated peptides at large-scale levels (3).

2. Materials

2.1. Model Proteins

1. Transferrin (human).
2. Serum albumin (bovine).
3. β-Lactoglobulin (bovine).
4. Carbonic anhydrase (bovine).
5. β-Casein (bovine).
6. α-Casein (bovine).
7. Ovalbumin (chicken).
8. Ribonuclease B (bovine pancreas).
9. Alcohol dehydrogenase (Baker's yeast).
10. Myoglobin (whale skeletal muscle).
11. Lysozyme (chicken).
12. α-Amylase (*Bacillus* species).

All proteins were from Sigma (St. Louis. MO).

2.2. Reduction, Alkylation, and Digestion of Model Proteins

1. Ammonium bicarbonate.
2. Dithiotreitol (DTT).
3. Iodoacetamide.
4. Modified trypsin.

(*see* **Note 1**)

2.3. Cell Culture and Lysis of Human Mesenchymal Stem Cells (hMSCs)

1. MEM (EARLES) media w/o phenol red, with glutamax-I (GibcoTM) containing 1% penicillin/streptomycin (GibcoTM) and 10% fetal bovine serum (FBS) (GibcoTM).
2. Phosphate saline buffer (PBS).
3. Phosphatase inhibitor cocktail 1 and 2 (P2850 and P5726, Sigma®).
4. Cell dissociation buffer (CDB) (GibcoTM).

2.4. Extraction, Reduction, Alkylation, and Digestion of hMSC Proteins

1. Lysis buffer: 7 M urea, 2 M thiourea, 1% *N*-octyl glycoside, 40 mM Tris-base, 300 U benzonase.
2. Dithiotreitol (DTT).
3. Iodoacetamide.
4. Acetone.
5. Endoproteinase Lys-C (lysyl endopeptidase®, WAKO).
6. Ammonium bicarbonate.
7. Modified trypsin (*see* **Note 1**).

2.5. Immobilized Metal Affinity Chromatography (IMAC)

1. Ironcoated PHOS-select™ metal chelate beads (Sigma®).

2. IMAC Loading buffer: 0.1% trifluoroacetic acid (TFA), 50% acetonitrile, HPLC Grade.

3. GELoader tips from Alpha Laboratories (Hampshire, S050 4NU, UK).

4. IMAC Elution buffer 1: 1.0% TFA, 20% acetonitrile.

5. IMAC Elution buffer 2: 20 μL ammonia solution (25%), 980 μL UHQ water (pH 11.3).

6. Formic acid.

7. Water was obtained from a Milli-Q system (Millipore, Bedford, MA) (UHQ water). (*see* **Notes 1** and **2**).

2.6. Titanium Dioxide (TiO₂) Chromatography

1. Titanium dioxide (TiO_2) beads (Titansphere, 5 μm, GL sciences Inc., www.inertsil.com).

2. GELoader tips (Eppendorf, Hamburg, Germany) or p10 pipette tips depending on the size of the column needed (*see* **Note 3**).

3. 3M Empore C8 disk (3M, Bioanalytical Technologies, St. Paul, MN).

4. Acetonitrile.

5. Syringe for HPLC loading (P/N 038250, N25/500-LC PKT 5, SGE, Ringwood, Victoria, Australia) to create a small plug of C_8 membrane material.

6. TiO_2 loading buffer: 1 M glycolic acid in 5% TFA, 80% acetonitrile.

7. TiO_2 washing buffer: 1% TFA, 80% acetonitrile.

8. TiO_2 elution buffer 1: 20 μL ammonia solution (25%), 980 μL UHQ water.

9. TiO_2 elution buffer 2: 30% acetonitrile.

10. Formic acid.

2.7. Reversed Phase (RP) Chromatography

1. POROS Oligo R3 reversed phase material (Applied Biosystems, Foster City, CA).

2. GELoader tips (Eppendorf, Hamburg, Germany) or p10 pipette tips depending on the size of the column needed (*see* **Note 4**).

3. 3M Empore C18 disk (3M, Bioanalytical Technologies, St. Paul, MN).

4. Syringe for HPLC loading (P/N 038250, N25/500-LC PKT 5, SGE, Ringwood, Victoria, Australia) to create a small plug of C_{18} membrane material.

5. RP Loading buffer: 0.1% TFA.

6. RP Elution buffer (for LC-ESI MSMS analysis): 70% acetonitrile, 0.1% TFA.

7. DHB Elution buffer (for MALDI MS analysis): 20 mg/mL DHB in 50% acetonitrile, 1% *ortho*-phosphoric acid (Bie & Berntsen A/S, UK).

3. Methods

The principle of the SIMAC method is illustrated in this chapter using a peptide mixture originating from tryptic digestions of 12 standard proteins (Model proteins) (*see* **Notes 5** and **6**). The protocol is then applied to enrich for phosphorylated peptides from whole cell lysates from 120 μg protein from human mesenchymal stem cells (hMSCs).

The SIMAC purification method is a simple and very straightforward method that can be adapted to already existing methods. It is fast and efficient for enrichment of phosphopeptides even from highly complex samples *(3, 8)*. The experimental setup of the method is illustrated in **Fig. 1.** For illustrating the anticipated results, a peptide mixture originating from tryptic digestions of 12 standard proteins was prepared.

3.1. Model Proteins and Peptide Mixture

See details in **Subheading 3.1**, Chapter "Enrichment and Characterization of Phosphopeptides by Immobilized Metal Affinity Chromatography (IMAC) and Mass Spectrometry."

3.2. Human Mesenchymal Stem Cell (hMSC) Proteins

The large-scale experiments shown in this article were performed using a peptide mixture originating from tryptic digestion of 120-μg total protein from cell extract from hMSCs.

1. Grow human mesenchymal stem cells (hMSC-TERT20) in T75 flasks in MEM media without phenol red, with glutamax-I containing penicillin/streptomycin and fetal bovine serum at 37°C until they reached 90% confluence. Wash the confluent cells once with PBS buffer (37°C) and add 5-mL media (37°C) to cover the cells. Add phosphatase inhibitor cocktails (50 μL of each) to the media. Incubate the cells with the phosphatase inhibitors for 30 min at 37°C. Wash the cells with ice-cold PBS buffer harvest them using cell dissociation buffer.

2. After harvesting the cells, resuspend the cell pellet in 1.5-mL lysis buffer. Sonicate the cells at 40% output with intervals three times 15 s on ice to disrupt the cells and incubate at −80°C for 30 min. Add dithiotreitol (DTT, 20 mM) and incubate the sample at room temperature for 35 min. After incubation, add

40 mM iodoacetamide and incubate for 35 min at room temperature in the dark.

3. Add 14-mL ice-cold acetone to the solution and incubate at −20°C for 20 min to precipitate the proteins. Pellet the proteins by centrifugation at $6,000 \times g$ for 10 min at −4°C. Lyophilize the pellet and store it at −20°C until further use.

4. Redissolve a total of 1 mg of the precipitated and lyophilized protein from hMSC (weight out) in 6 M urea, 2 M thiourea to a concentration of 50 μg protein/μL. Incubate the proteins with endoproteinase Lys-C (1–2% w/w) at room temperature for 3 h. After incubation, dilute the endoproteinase Lys-C digested sample five times with 50 mM NH_4HCO_3 and add 1–2% (w/w) chemically modified trypsin. Incubate the sample at room temperature for 18 h.

3.3. Batch Incubation with IMAC Beads

1. Always adjust the amount of IMAC beads to the amount of sample in order to reduce the level of nonspecific binding from nonphosphorylated peptides. For 1 pmol tryptic digest use 7-μL IMAC beads (*see* Chapter "Enrichment and Characterization of Phosphopeptides by Immobilized Metal Affinity Chromatography (IMAC) and Mass Spectrometry"). For more complex samples, where more material is available, use more IMAC beads. The following is describing the protocol when using 120-μg tryptic digest.

2. Transfer 40 μL of IMAC beads to a fresh 1-mL Eppedorf tube.

3. Wash the IMAC beads twice using 200 μL of IMAC loading buffer.

4. Resuspend the beads in 150 μL of IMAC loading buffer and add the sample (here 12 μL).

5. Incubate the sample with IMAC beads in a Thermomixer (Eppendorf) for 30 min at 20°C.

3.4. Packing the IMAC Micro-Column

1. Squeeze the end of a 200-μL GELoader tip to prevent the IMAC beads from leaking.

2. After incubation pack the beads in the constricted end of the GELoader tip by application of air pressure forming an IMAC micro-column essentially as described in Chapter "Enrichment and Characterization of Phosphopeptides by Immobilized Metal Affinity Chromatography (IMAC) and Mass Spectrometry" *(9)*.

3. It is critical to collect the IMAC flow-through in a fresh 1.5-mL Eppendorf tube for further analysis by TiO_2 chromatography.

4. Wash the IMAC column using 50-μL IMAC loading buffer and pool this with the IMAC flow-through.

5. Lyophilize the IMAC flow-through and wash and enrich for mono-phosphorylated peptides using TiO$_2$ chromatography (*see* **Subheading 3.7–3.9**).

3.5. Elution of Mono-phosphorylated Peptides from the IMAC Micro-Column

1. Elute the mono-phosphorylated peptides bound to the IMAC micro-column using 50 μL of IMAC elution buffer 1 (**Fig. 1**).

2. Dry the eluate by lyophilisation.

3. Redissolve the peptides in 5-μL 1% SDS and purify further by TiO$_2$ chromatography (*see* **Subheadings 3.7–3.9**).

3.6. Elution of Multiply phosphorylated Peptides from the IMAC Micro-Column

1. Elute the multiply phosphorylated peptides bound to the IMAC micro-column using 70 μL of IMAC elution buffer 2 (**Fig. 1**).

2. For LC-ESI MS^n analysis this IMAC eluate should be acidified using 100% formic acid (pH should be ~2–3) and desalted/concentrated using reversed phase chromatography.

3.7. Packing the TiO$_2$ Micro-Column

1. Prepare six TiO$_2$ micro-columns of ~4 mm by packing the TiO$_2$ beads on top of the C$_8$ membrane disks in P10 tips as described in **Subheading 3.2**, Chapter "The Use of Titanium Dioxide Micro-Columns to Selectively Isolate Phosphopeptides from Proteolytic Digests."

3.8. Loading the Sample onto the TiO$_2$ Micro-Column

1. Divide the IMAC flow-through + wash between three different TiO$_2$ micro-columns. Do the same for the IMAC eluate. See details for loading onto the TiO$_2$ micro-columns in **Subheading 3.3**, Chapter "SILAC for Global Phosphoproteomic Analysis."

3.9. Elution of Phosphorylated Peptides from the TiO$_2$ Micro-Column

1. Elute the phosphopeptides bound to each TiO$_2$ micro-columns using at least 35 μL of TiO$_2$ elution buffer 1 followed by an elution step using 1 μL of TiO$_2$ elution buffer 2 and pool the eluates as described in **Subheading 3.4**, Chapter "The Use of Titanium Dioxide Micro-Columns to Selectively Isolate Phosphopeptides from Proteolytic Digests" (*see* **Notes** 7 and **8**).

3.10. Packing a Reversed Phase (RP) Micro-Column for Desalting/Concentrating the Sample

1. Prepare a RP micro-column of ~5–6 mm by packing the R3 beads on top of the C$_{18}$ membrane disk in P10 tips as described for TiO$_2$ micro-columns in **Subheading 3.2**, Chapter "The Use of Titanium Dioxide Micro-Columns to Selectively Isolate Phosphopeptides from Proteolytic Digests.".

3.11. Loading the Enriched Phosphopeptides onto the RP Micro-Column

1. Load the acidified sample slowly onto the RP micro-column (~1 drop/s).

2. Wash the RP micro-column using 30 μL of 0.1% TFA.

3.12. Elution of Phosphorylated Peptides from the RP Micro-Column

1. Elute the phosphopeptides from the RP micro-column using 70 μL RP elution buffer, followed by lyophilization of the phosphopeptides. (N.B. For simple samples, the peptides can be eluted off the RP micro-column directly onto the MALDI target using 1 μL DHB solution).

2. LC-ESI-MS^n analysis: Redissolve the dried phosphopeptides in 0.5 μL 100% formic acid and immediately dilute to 10 μL with UHQ water.

LC-ESI-MS^n analysis of multiply phosphorylated peptides is improved by redissolving the phosphopeptides by sonication in an EDTA containing buffer prior to LC-ESI-MS^n analysis *(10)*.

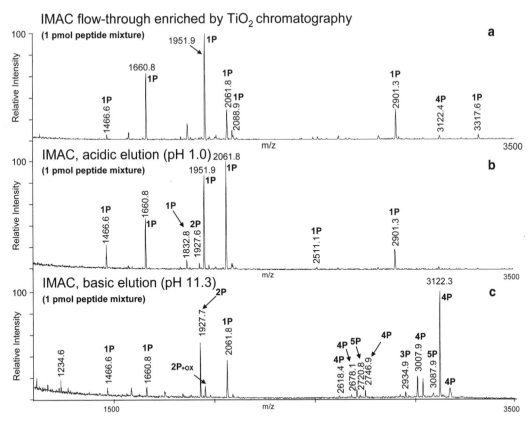

Fig. 2. Results obtained from 1 pmol peptide mixture using the SIMAC strategy. MALDI MS spectra of peptides identified from the IMAC flow-through after further enrichment using TiO$_2$ chromatography (**a**), MALDI MS spectra of peptides eluted from the IMAC micro-column using 1% TFA (**b**) and MALDI MS spectra of peptides eluted from the IMAC micro-column using ammonia water (pH 11.30) (**c**). The number of phosphate groups on the individual phosphopeptides are indicated by "#P". *Asterisk* indicates the loss of phosphoric acid. (*see* **Note 9**).

Figure 2 illustrates the results obtained by using the SIMAC strategy on 1 pmol standard peptide mixture. MALDI MS spectra of the IMAC flow-through after TiO_2 enrichment (a) and MALDI MS spectra of the phosphopeptides eluted from the IMAC material using acidic conditions (1% TFA, pH 1.0) (b) show that mainly mono-phosphorylated peptides are identified in these two peptide fractions. Mainly multiply phosphorylated peptides are identified when using basic elution conditions (ammonia water, pH 11.3) (c) (*see* **Note 9**).

Figure 3 compares the number of mono-phosphorylated and multiply phosphorylated as well as nonphosphorylated peptides identified from 120-µg hMSC tryptic digests using either TiO_2 enrichment or the SIMAC strategy. The results show that mainly mono-phosphorylated peptides are identified by TiO_2 chromatography, whereas only 17% (54 of 325) of the identified peptides are multiply phosphorylated. In the SIMAC experiment, the IMAC flow-through as well as the acidic elution fraction consist mainly of mono-phosphorylated peptides and the level of multiply phosphorylated peptides identified is even lower than for TiO_2 chromatography (3% and 4%, respectively). However, the level of multiply phosphorylated peptides identified is significantly increased by basic elution (68%). In total SIMAC gives rise to more than three times as many multiply phosphorylated peptides identified from 120-µg hMSC tryptic digest when compared with TiO_2 chromatography.

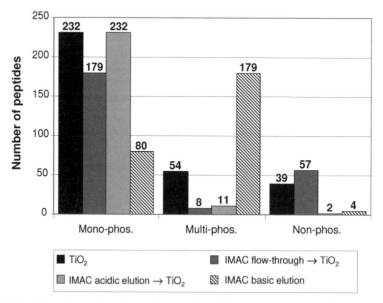

Fig. 3. Results obtained from the analysis of phosphorylated peptides from acetone precipitated proteins from human mesenchymal stem cells (hMSCs) using TiO_2 chromatography or SIMAC. The number of mono-phosphorylated peptides, multiply phosphorylated peptides and nonphosphorylated peptides identified from 120-µg hMSC tryptic digests using TiO_2 enrichment and the number of mono-phosphorylated peptides, multiply phosphorylated peptides and nonphosphorylated peptides identified in the three peptide fractions obtained from SIMAC analysis of 120-µg hMSC tryptic digests.

4. Notes

1. It is important to obtain the highest purity of all chemicals used.

2. All solutions should be prepared in UHQ water.

3. The capacity of the TiO_2 beads have been tested; however, more experiments need to be performed. The capacity of the beads appears to be very high, but when working with high amounts of material it is recommended to use P10 pipette tips instead of GELoader tips and to split the peptide sample and load it onto several TiO_2 micro-columns. The resulting eluates can then be pooled prior to MS analysis. For the experiments using 120-µg hMSC protein, six TiO_2 micro-columns of ~4 mm were packed in P10 tips.

4. For the experiments using 120-µg hMSC protein, a R3 micro-column of ~5–6 mm were packed in P10 tips.

5. Always start by testing the method using a model peptide mixture. It is important to freshly prepare the peptide mixture as peptides bind to the surface of the plastic tubes in which they are stored. In addition, avoid transferring the peptide sample between different tubes to minimize adsorptive losses of the sample.

6. The peptide mixture used for the experiment illustrated in this chapter contained peptides originating from tryptic digestions of 1 pmol of each of the 12 proteins.

7. Use at least 35 µL of TiO_2 elution buffer 1 due to the size of the TiO_2 micro-columns. It is important to follow the first elution step using 1 µL of TiO_2 elution buffer 2, to elute peptides bound to the C_8 membrane plug.

8. At this step, pool the eluates from the three TiO_2 micro-columns with IMAC flow-through + wash by eluting the phosphopeptides into the same Eppendorf tube. Do the same for the three TiO_2 micro-columns with IMAC eluates.

9. The results obtained using this protocol will differ according to the mass spectrometer used for the analysis of the phosphopeptides, not only between MALDI MS and ESI MS but also within different MALDI MS instruments, depending on laser optics, laser frequency, instrumental configuration, sensitivity etc.

Acknowledgements

This work was supported by the Danish Natural Science Research Council (grant no. 21-03-0167 (M.R.L)) and the Danish Strategic Research Council (O.N.J).

References

1. Zhou, H. L., Watts, J. D., and Aebersold, R. A. (2001) A systematic approach to the analysis of protein phosphorylation. *Nat. Biotechnol.* 19, 375–78.

2. Bodenmiller, B., Mueller, L. N., Mueller, M., Domon, B., and Aebersold, R. (2007) Reproducible isolation of distinct, overlapping segments of the phosphoproteome. *Nat. Methods* 4, 231–37.

3. Thingholm, T. E., Jensen, O. N., Robinson, P. J., and Larsen, M. R. (2007) SIMAC – A phosphoproteomic strategy for the rapid separation of mono-phosphorylated from multiply phosphorylated peptides. *Published online by MCP on November 27th 2007 (M700362-MCP200).*

4. Schroeder, M. J., Shabanowitz, J., Schwartz, J. C., Hunt, D. F., and Coon, J. J. (2004) A neutral loss activation method for improved phosphopeptide sequence analysis by quadrupole ion trap mass spectrometry. *Analytical Chemistry* 76, 3590–98.

5. Chalmers, M. J., Hakansson, K., Johnson, R., Smith, R., Shen, J., Emmett, M. R., and Marshall, A. G. (2004) Protein kinase A phosphorylation characterized by tandem Fourier transform ion cyclotron resonance mass spectrometry. *Proteomics* 4, 970–81.

6. Schroeder, M. J., Webb, D. J., Shabanowitz, J., Horwitz, A. F., and Hunt, D. F. (2005) Methods for the detection of paxillin post-translational modifications and interacting proteins by mass spectrometry. *J Proteome Res* 4, 1832–41.

7. Ficarro, S. B., McCleland, M. L., Stukenberg, P. T., Burke, D. J., Ross, M. M., Shabanowitz, J., Hunt, D. F., and White, F. M. (2002) Phosphoproteome analysis by mass spectrometry and its application to *Saccharomyces cerevisiae. Nat Biotechnol* 20, 301–5.

8. Jensen, S. S., and Larsen, M. R. (2007) Evaluation of the impact of some experimental procedures on different phosphopeptide enrichment techniques. *Rapid Commun. Mass Spectrom.* 21, 3635–45.

9. Gobom, J., Nordhoff, E., Mirgorodskaya, E., Ekman, R., and Roepstorff, P. (1999) Sample purification and preparation technique based on nano-scale reversed-phase columns for the sensitive analysis of complex peptide mixtures by matrix-assisted laser desorption/ionization mass spectrometry. *J. Mass Spectrom.* 34, 105–16.

10. Liu, S., Zhang, C., Campbell, J. L., Zhang, H., Yeung, K. K., Han, V. K., and Lajoie, G. A. (2005) Formation of phosphopeptide-metal ion complexes in liquid chromatography/electrospray mass spectrometry and their influence on phosphopeptide detection. *Rapid Commun Mass Spectrom.* 19, 2747–56.

Chapter 7

Use of Stable Isotope Labeling by Amino Acids in Cell Culture (SILAC) for Phosphotyrosine Protein Identification and Quantitation

Guoan Zhang and Thomas A. Neubert

Summary

In recent years, stable isotope labeling by amino acids in cell culture (SILAC) has become increasingly popular as a quantitative proteomic method. In SILAC experiments, proteins are metabolically labeled by culturing cells in media containing normal and heavy isotope amino acids. This makes proteins from the light and heavy cells distinguishable by mass spectrometry (MS) after the cell lysates are mixed and the proteins separated and/or enriched. SILAC is a powerful tool for the study of intracellular signal transduction. In particular, it has been very popular and successful in quantitative analysis of phosphotyrosine (pTyr) proteomes to characterize pTyr-dependent signaling pathways. In this chapter, we describe the SILAC procedure and use EphB signaling pathway as an example to illustrate the use of SILAC to investigate such pathways.

Key words: SILAC, Tyrosine phosphorylation, Quantitation, Identification, Mass spectrometry, HPLC, Immunoprecipitation, RTK, Phosphoproteome, Signal transduction.

1. Introduction

Phosphorylation is one of the most common posttranslational modifications of proteins. In eukaryotic cells, phosphorylation mainly occurs on serine, threonine, and tyrosine residues (1). Despite its infrequent occurrence compared with serine and threonine phosphorylation (pSer: pThr: pTyr=1,800:200:1), tyrosine phosphorylation plays critical roles in regulating intracellular signal transduction (1). In a typical receptor tyrosine kinase (RTK) pathway, upon stimulation by binding with ligand the receptor

Marjo de Graauw (ed.), *Phospho-Proteomics, Methods and Protocols, vol. 527*
© 2009 Humana Press, a part of Springer Science+Business Media, New York, NY
Book DOI: 10.1007/978-1-60327-834-8_7

becomes tyrosine phosphorylated and activated. The activated receptor then recruits and tyrosine phosphorylates other effectors, which triggers further signaling events down the pathway. Traditional approaches to study these pathways involve looking at one or a few effectors at a time, and it is difficult to get a complete picture of the signaling pathway in this way. Mass spectrometry (MS)-based quantitative proteomics has become a major tool for high throughput investigation of cell signal transduction, owing to rapid development of both quantitative methods and instrumentation in recent years *(2, 3)*. Quantitative proteomics allows the comparison of different cellular conditions to obtain a global view of changes in protein phosphorylation. These experiments provide valuable data to help us to screen for new signaling effector molecules and interpret the dynamic regulation of the pathway under study.

Stable isotope labeling by amino acids in cell culture (SILAC) is an excellent approach for high-accuracy quantitative proteomics *(4, 5)*. It involves culturing cells in a medium supplemented with amino acids containing either normal or heavy stable isotopes. The amino acids are metabolically incorporated into the proteins of the cells through protein synthesis. When the mixed light and heavy isotope labeled proteins are analyzed by MS, the source of the sample can be easily distinguished by the mass difference caused by the differential labeling. The relative abundance of the proteins is measured based on the intensities of the light and heavy peptides. Compared with other quantitative methods, SILAC has the following advantages:

(a) It allows combining the differentially labeled samples early on during sample preparation (usually directly after cell lysis and before any purification or fractionation steps). This feature is very important as it minimizes the possible quantitative error caused by handling different samples in parallel.

(b) It does not require chemical reactions to modify proteins or peptides.

(c) It is easy to obtain a high level of incorporation.

(d) It is simple to perform.

SILAC is not limited to mammalian cells. It can be applied to organisms that can be metabolically labeled with amino acids, which include bacteria, yeasts, and plants *(6–9)*.

Particularly in combination with anti-phosphotyrosine (pTyr) immunoprecipitation (IP), SILAC has proven to be a very powerful method to study RTK signaling *(10)*. It has been successfully used to study EGF, PDGF, FGF, EphB, insulin, and T-cell receptors *(11–19)*. The general workflow for such studies is shown in **Fig. 1**. Cells are first differentially labeled with SILAC. After labeling, in one cell population the RTK is activated to trigger tyrosine phosphorylation of downstream effectors. Then the lysates

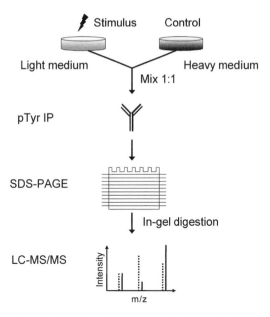

Fig. 1. Strategy for using SILAC to study RTK signaling. Two sets of cells are cultured in light and heavy SILAC medium, respectively. After complete SILAC labeling, one set of cells is treated to activate the RTK while the other is used as a control. The two cell lysates are combined at an equal ratio for anti-pTyr IP to pull down phosphotyrosine proteins and their binding partners. The immunoprecipitated proteins are fractionated by SDS-PAGE, digested in-gel and analyzed by LC-MS/MS. In MS spectra, light and heavy labeled peptides appear as peak doublets (light: *dashed line*; heavy: *solid line*). Their ratios can be used to indicate if the proteins participate in the RTK signaling.

of the stimulated and the control cells are combined for anti-pTyr IP to pull down pTyr proteins together with their tight binding partners. The precipitated proteins are then identified and quantified by MS, and the SILAC ratio of a protein (the abundance of proteins in the IP of stimulated cells/the abundance of proteins from the IP of unstimulated cells) is used to indicate whether the protein participates in the RTK pathway or not.

In this chapter, we describe the protocols to use SILAC to screen for downstream effector proteins in the EphB signaling pathway. These protocols can be easily adapted to study other similar RTK pathways.

2. Material

2.1. Cell Culture and SILAC Labeling

1. Cell line: NG108-15 cell line (mouse neuroblastoma and rat glioma hybrid) stably over-expressing EphB2 *(20)*.
2. Culture medium: Dulbecco's Modified Eagle Medium (DMEM) deficient in lysine and arginine (Specialty Media, Millipore).

3. Amino acids for labeling: Normal (light) L-lysine and L-arginine hydrochloride; Stable isotope-labeled (heavy) L-lysine $^{13}C_6$ and L-arginine $^{13}C_6$ hydrochloride (Cambridge Isotope Labs, Andover, MA). Heavy isotope amino acids (usually contain ^2H, ^{13}C, ^{15}N or ^{18}O) used in SILAC are not radioactive and therefore do not need special handling precautions.

4. Supplements: Dialyzed fetal bovine serum; post-fusion selective medium HAT (liquid mixture of sodium hypoxanthine, aminopterin and thymidine); penicillin/streptomycin; plasmid selecting antibiotic G418.

5. Other: Sterile phosphate buffered saline (PBS); trypsin-EDTA solution; 0.22-μm filter flasks.

6. Ligand stimulation: EphrinB1-Fc (Sigma-Aldrich, St. Louis, MO), recombinant human Fc and goat anti-human Fc IgG (Jackson ImmunoResearch Laboratories, West Grove, PA).

2.2. Cell Lysis, Immunoprecipitation, and SDS-PAGE

1. Lysis buffer: Contains 1% Triton X-100, 150 mM NaCl, 20 mM Tris–HCl, pH 8.0, 0.2 mM EDTA, 2 mM tyrosine phosphatase inhibitor sodium orthovanadate, 2 mM serine/threonine phosphatase inhibitor sodium fluoride, and protease inhibitors (Complete tablet; Roche, Mannheim, Germany). Fresh phosphatase and protease inhibitors should be added just before use.

2. Agarose conjugated anti-pTyr antibody PY99 (Santa Cruz Biotechnology, Inc., Santa Cruz, CA).

3. Elution buffer: 0.2% trifluoroacetic acid (TFA)/0.5% SDS.

4. Precast 7.5% Tris–HCl polyacrylamide gel.

5. Coomassie brilliant blue R250 (CBB) staining solution: 0.1% CBB/7% acetic acid/40% ethanol.

2.3. In-gel Digestion

1. Destaining buffer: 25 mM ammonium bicarbonate/50% acetonitrile.

2. Acetonitrile (HPLC grade).

3. Trypsin solution: 12.5 ng/μL sequencing grade trypsin (Promega Corporation, Madison, WI) in 25 mM ammonia bicarbonate. The trypsin stock solution (200 ng/μL in water) can be stored at –70°C for 6 months.

4. Extraction buffer: 5% formic acid (FA)/50% acetonitrile.

2.4. High Performance Liquid Chromatography–Tandem Mass Spectrometry (HPLC-MS/MS)

1. Nano-Aquity HPLC (Waters, Milford, MA) equipped with a Symmetry C18 trap column (180 μm × 2 cm, 5-μm particle diameter, Waters) and a Symmetry C18 analytical column (75 μm × 15 cm, 3.5-μm particle diameter, Waters).

2. LTQ-Orbitrap (Thermo Scientific, San Jose, CA), connected to the Nano-Aquity LC via a nanoelectrospray ion source.

3. Mobile phase A: 0.1% FA in water (HPLC grade).

4. Mobile phase B: 0.1% FA in acetonitrile (HPLC grade).

3. Methods

3.1. SILAC Medium Preparation

SILAC cell medium is the same as regular medium except for the following two components: (a) The amino acids that are used for labeling should be left out of the original formulation. Later the light and heavy amino acids are added to the base medium to make the light and heavy labeling media. (b) Dialyzed serum is used instead of regular serum because the latter contains free amino acids that can cause problems for labeling *(4)*.

1. Prepare 1,000× stock solutions for the light and heavy labeling amino acids by dissolving the amino acids in PBS. The use of stock solutions is generally recommended for ease of use. However this does not apply to some amino acids because of their poor solubility in water (*see* **Note 1**). In our case based on the DMEM formulation, the working concentrations of light arginine and lysine are 84 and 146 mg/L respectively. Note that because the molecular weights of heavy amino acids are different from their light counterparts, their concentrations (in mg/L) are different. In this case the concentrations of heavy arginine and lysine are 87.2 and 152.8 mg/L respectively. The use of lower concentrations of amino acids has been reported (*see* **Notes 2** and **3**) *(10)*. In addition to arginine and lysine, other amino acids can be used for labeling (*see* **Notes 4** and **5**) *(21)*.

2. Dilute the light and heavy amino acids 1,000-fold with DMEM deficient in arginine and lysine.

3. Add supplements to both media to their working concentrations: HAT (1×), penicilin/streptomycin (1×), and G418 (0.4 mg/mL).

4. Filter both media with 0.22-μm filter flasks.

5. Save some of the serum free media for the purpose of cell starvation prior to stimulation.

6. Add 10% dialyzed fetal bovine serum to light and heavy media. These media are now ready for cell culture.

3.2. SILAC Cell Culture

SILAC cell culture is nearly the same as regular cell culture.

1. Split one dish of NG108-EphB2 cells into two dishes: One with light medium and the other with heavy medium. When the cells are >80% confluent, rinse the cells with PBS, detach the cells with trypsin/EDTA and split the cells to new dishes.

2. To ensure complete labeling, cells should undergo at least five doublings in the SILAC media. Protein labeling is achieved through: (a) dilution of the non-labeled protein by newly synthesized (labeled) protein and (b) degradation of the non-labeled protein. Five cell doublings corresponds to a labeling incorporation of at least 96.9% when only the dilution effect is considered. Considering natural protein turnover, the actual incorporation should be even higher.

3. Prepare about 10^8 cells (10 cm I.D. dishes) for each labeling condition (*see* **Note 6**).

3.3. Cell Stimulation and Lysis

1. After SILAC labeling is complete, culture the cells in serum-free SILAC media (containing heavy or light amino acids consistent with previous labeling) for 24 h. The purposes of starvation are (a) to synchronize the cells and (b) to lower the basal tyrosine phosphorylation level to facilitate identification of pTyr proteins specific to the signaling pathway under study. The optimal starvation time is generally 12–24 h depending on cell lines.

2. Cluster the ligand by incubating ephrinB1-Fc (250 μg/mL) and anti-human Fc (65 μg/mL) at 4°C for 1.5 h. Non-clustered ephrin ligands are unable to activate the receptors efficiently *(20)*.

3. Add clustered ligand to one set of cells (either light or heavy) with a final ligand concentration of 2 μg/mL. To the other set of cells add 1 μg/mL anti-human Fc. These cells are used as a control. Incubate for 45 min or other desired time at 37°C.

4. Remove media from the dishes, wash twice with ice cold PBS. Tilt the dishes to facilitate complete removal of PBS. To each dish add 1 mL of ice cold lysis buffer. Scrape the cells and incubate the lysates on ice for 20 min. To remove cell debris, centrifuge the lysates at $14,000 \times g$ for 20 min. Carefully aspirate and save the supernatant fluids. Save 20 μL of the heavy lysate. This may be used to double check the labeling incorporation if necessary.

5. Mix equal amount of light and heavy lysates. The Bradford assay can be used to determine the lysate protein concentrations to ensure equal mixing. Save 20 μL of the mixed lysates. This will be analyzed by MS and used as a loading control.

3.4. Anti-pTyr IP and SDS-PAGE

The use of anti-pTyr IP is to isolate the pTyr proteome, which is critical because signaling proteins are usually of low abundance and difficult to detect by MS without enrichment. Care should be taken to minimize nonspecific binding in IP. Although SILAC is able to discriminate nonspecific binding proteins based on their ratios, if nonspecific binding proteins are too abundant they will dominate the analytical space of LC-MS/MS, leading to poor identification of target proteins.

1. Pre-clear the SILAC lysate: Add agarose beads to the lysate (10 µL of agarose beads per mL lysate) and incubate with rotation at 4°C for 1 h. Spin at $14,000 \times g$ for 20 min. Carefully aspirate and save the supernatant.

2. Add the agarose-conjugated anti-phosphotyrosine antibody PY99 to the lysate (10 µg in 20 µL of PY99 beads per mL lysate) and incubate with rotation at 4°C for 4 h. Remove the lysate and transfer the beads into 1.5-mL Handee™ spin cups with paper filters (Pierce, Rockford, IL).

3. Wash the beads four times with lysis buffer and once with water. Elute precipitated proteins by incubating the beads in equal volume of 0.2% TFA per 0.5% SDS for 10 min at room temperature. Repeat the elution once and combine eluates. Alternatively the proteins can be eluted using other approaches (*see* **Note 7**).

4. Neutralize the eluate with 1 M ammonium bicarbonate (1 µL ammonium bicarbonate per 100 µL elution buffer). Concentrate the eluate up to tenfold using a vacuum centrifuge (SpeedVac) to reduce the volume. Add Laemmli sample loading buffer. At this point the color of the sample should be blue. If the color is yellow it indicates the pH is acidic and should be further adjusted with ammonium bicarbonate. Heat the sample at 95°C for 5 min before loading the sample onto a Bio-rad 7.5% precast Tris–HCl PAGE gel (*see* **Note 8**).

5. After electrophoresis stain the gel with Coomassie blue, which can detect as little as 50–100 ng of a single protein in a gel. With sample loading of protein immunoprecipitated from five dishes of cells per PAGE lane, there should be hardly any bands visualized. Presence of dark bands often suggests considerable nonspecific binding in the IP.

3.5. In-Gel Digestion

Before digestion, horizontally cut each gel lane into 10–20 gel bands. This serves as a fractionation step to reduce sample complexity and improve the dynamic range of protein identification by LC-MS/MS. Wear gloves to prevent keratin contamination. Do the digestion in a laminar flow tissue culture hood if possible to minimize contamination by dust or other atmospheric particulates.

1. Cut each gel band into small pieces (~1-mm³ cubes) and destain in 25 mM ammonium bicarbonate in 50% acetonitrile. Repeat the destaining step until the gels are completely clear. Add acetonitrile to shrink the gels. Now the gels should be white in color. Any blue color indicates insufficient destaining, which may cause decreased sensitivity in LC-MS/MS analysis.

Remove the acetonitrile by aspiration and dry the gels with a SpeedVac.

2. Rehydrate the gel pieces with ice-cold 12.5 ng/μL trypsin solution in 25 mM ammonium bicarbonate and incubate on ice for 20 min or till the gels are fully swollen to allow trypsin to enter into the gel pieces. Remove surplus trypsin solution. Add 25 mM ammonium bicarbonate (about 25 μL for a 0.6-mL tube or 50 μL for a 1.5-mL tube) to cover the gels to prevent the gels from drying during digestion. Incubate overnight at 37°C.

3. Peptide extraction: Add the extraction buffer to the gels, mix for 10 min and sonicate for 10 min. Spin briefly and save the supernatant. Repeat the extraction once with the extraction buffer and once with acetonitrile. Concentrate the digests to almost dryness in a SpeedVac and reconstitute with 0.1% FA in 2% acetonitrile to about 5 μL. Now the digests are ready for LC-MS analysis.

3.6. HPLC-MS/MS

In this step the peptides from in-gel digestion are separated by HPLC and introduced into MS via online nanoelectrospray. In MS the peptide ions are analyzed in a data-dependent manner: First the m/z values of the intact peptides are measured. Then the most intense ions are selected for MS/MS through collision induced fragmentation *(22)*, in which a peptide is broken into fragment ions. The pattern of the fragment ions contains sequence information and is used to search protein databases for peptide sequence identification.

1. Load the peptides onto a trap column (180 μm × 2 cm Symmetry C18, Waters) with 100% mobile phase A for 4 min at 5 μL/min.

2. After sample loading, the peptides are eluted using a gradient of 6–40% mobile phase B over 120 min at 0.25 μL/min.

3. Mass spectra are acquired in data-dependent mode with one 60,000 resolution MS survey scan by the Orbitrap and up to five concurrent MS/MS scans in the LTQ for the most intense five peaks selected from each survey scan. Automatic gain control is set to 500,000 for Orbitrap survey scans and 10,000 for LTQ MS/MS scans. Survey scans are acquired in profile mode and MS/MS scans are acquired in centroid mode.

3.7. Data Processing

Peak list files containing peptide masses and their corresponding fragment ion masses and intensities are extracted from raw MS files. Then these peak lists are used by protein search tools to match the MS information to sequences in protein databases for protein identification. In this example Mascot is used for protein identification and MSQuant is used for quantitation. Alternatively there are other tools with similar functions available *(23)*.

1. Mascot generic format files are generated from the raw data using DTASuperCharge (version 1.01) and Bioworks (version 3.2, Thermo Fisher Scientific) for database searching.

2. Mascot software (version 2.1.0, Matrix Science, London, UK) is used for database searching. An IPI database containing mouse and rat protein sequences is used. Peptide mass tolerance is 10 ppm, fragment mass tolerance is 0.6 Da, trypsin specificity is applied with a maximum of one missed cleavage, and variable modifications were $^{13}C_6$ Lys, $^{13}C_6$ Arg, oxidation of methionine, and phosphorylation of serine, threonine, and tyrosine.

3. Proteins that are introduced during sample preparation should be excluded from the reported protein list. These proteins include: Keratins and trypsin (from in-gel digestion); immunoglobins (from the PY99 antibody and Fc fusion protein); ephrinB1 (the stimulating ligand); ferritin and serum albumin (from cell culture).

4. SILAC quantitation is carried out using the open source software MSQuant (version 1.4.2a13) developed by Peter Mortensen and Matthias Mann at the University of Southern Denmark. The SILAC ratio of a protein is calculated by comparing the summed MS intensities of all matched light peptides with those of the heavy peptides.

5. As a loading control, a small volume of the combined lysate was subjected to in-gel digestion, LC-MS/MS analysis and the identified proteins were also quantified. The average ratio for all quantified proteins was used as a correction for ratios of proteins identified from the IP.

6. The SILAC protein ratios are used to determine whether the proteins are involved in EphB signaling (*see* **Notes 9–11**). If after cell stimulation the SILAC ratio of a protein is "up regulated," i.e., more abundant in the pTyr IP, it indicates that the protein is very likely to be tyrosine phosphorylated after activation of EphB2 receptor. Another possibility is that the protein binds to another "up regulated" protein through non-pTyr dependent interactions (*see* **Note 9**). Conversely, the most likely explanations for a "down regulated" SILAC ratio are that the protein is dephosphorylated or binds to another "down regulated" effector. An unchanged SILAC ratio suggests either the tyrosine phosphorylation level of the protein is not affected by the stimulation or the protein is pulled down by pTyr IP through nonspecific binding. **Figure 2** shows an example of an MS spectrum corresponding to each scenario.

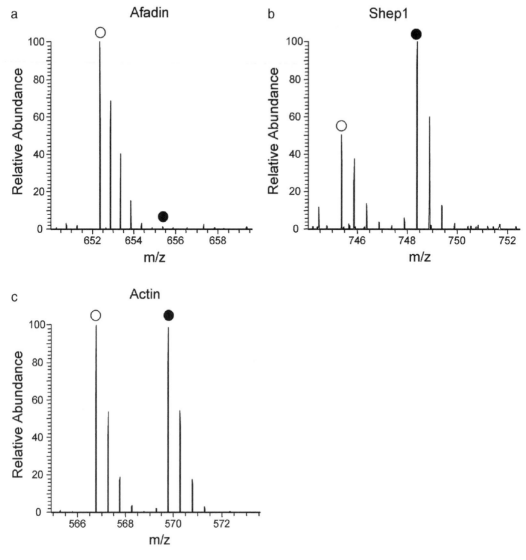

Fig. 2. Protein quantitation using SILAC. The lower-mass peak clusters (*open circles*) are from light Arg/Lys peptides from the stimulated cells while the higher-mass peak clusters (*black circles*) are from heavy Arg/Lys peptides from the control cells. All peptide ions shown are doubly charged and the m/z difference between the light and heavy peaks is three (mass difference is 6 Da). Ratios were determined by comparing the heights of the peaks from light and heavy peptides. (**a**) A SILAC peptide pair (LQLSVTEVGTEK) from Afadin, a downstream effector of EphB2 receptor, which was highly enriched in the anti-pTyr IP after ephrinB1 stimulation. (**b**) A peptide pair (YLEASYGLSQGSSK) from Shep1, a downstream effector of EphB2 receptor, which was less abundant in the anti-pTyr IP after ephrinB1 stimulation. (**c**) A peptide pair (GYSFTTTAER) from actin with a ratio of 1:1, indicating this protein does not participate in the signaling pathway.

4. Notes

1. Some hydrophobic amino acids, such as tyrosine, do not dissolve well in water and are therefore prepared directly at their working concentrations in medium.

2. Labeling amino acids can be used at concentrations lower than those specified in our standard formulations. For example, four times less arginine in the DMEM has been used for HeLa cell cultures (10). This significantly decreases the cost of SILAC experiments.

3. A known issue when using arginine labeling is that some cell lines can convert arginine metabolically into proline (5, 24). In MS, this will cause the signal of a proline containing peptide to split into multiple signals, depending on the number of prolines in the sequence. Decreasing the concentration of arginine in medium may prevent this conversion from happening (10). For cell lines in which the conversion cannot be inhibited, the options are: (a) exclude all proline containing peptides from quantitation; (b) for each heavy proline-containing peptide, add up all its split signals for quantitation; (c) use specially designed labeling strategies to cancel out the error (24).

4. In addition to arginine and lysine, some other amino acids can be used in SILAC (21). When choosing labeling amino acids, several important factors should be considered: (a) The mass difference between the light and heavy versions should be at least 4 Da to ensure sufficient separation of the naturally occurring isotopic envelopes of peptide doublets in MS. Mass differences smaller than 4 Da will cause overlaps of isotope peak clusters from light and heavy peptides (especially for large peptides) and complicate quantitation. (b) Peptides labeled with amino acids deuterated on side chains can elute earlier than their light counterparts in reverse-phase LC (25). The retention time difference depends on the number of ^{2}H and their positions within the molecule. This may affect the accuracy of quantitation. In contrast C^{13} and N^{15} labelings do not cause LC retention time shift and are considered better than ^{2}H labeling in this regard. (c) Arginine and lysine are the most commonly used labeling amino acids in SILAC. The reason is because trypsin, the most commonly used protease in proteomics, specifically cleaves at these amino acids. Therefore after digestion all peptides (except the one at the C-terminus of the protein) contain single labeling amino acids (more if trypsin misses one or more cleavages) and are thus quantifiable in MS.

5. In this chapter we describe a SILAC protocol to compare two cell conditions. Comparison of more than two conditions can be achieved by employing more labeling amino acids. For example, triple SILAC labeling with Arg0-Arg6 (^{13}C6)-Arg10 (^{13}C6^{15}N4) can be used for a three-condition experiment. Usually to conduct a time course study, two triple-labeling experiments can be conducted with one

common control condition in both experiments. Then the results of the two experiments are combined to obtain a five time point result *(12)*.

6. The number of cells needed for a SILAC study depends heavily on the cell line used, the expression level and tyrosine phosphorylation level of the proteins of interest, and the sensitivity of the LC-MS technology used. Based on our experiments, it seems that 50 mg total protein is a good starting point (This may translate into significantly different cell numbers depending on the cell lines used).

7. Another commonly used elution approach is to boil beads in Laemmli buffer. However, the use of DTT and heat will cause leaching of the crosslinked antibody from agarose beads. As a result, the light chain and heavy chain of the antibody will show up on the SDS-PAGE gel and affect identification of proteins with molecular weighs near 25 kDa (IgG light chain) and 50 kDa (IgG heavy chain).

8. Load the sample in as few lanes as possible to maximize the protein concentration in each lane. This will facilitate subsequent in-gel digestion and LC-MS/MS.

9. In general the anti-pTyr IP should pull down pTyr proteins together with their interacting proteins and thus these interacting proteins should also be detected by SILAC. However, it should be noted that the interactions that rely on phosphorylated tyrosines are generally not present in the IP due to competitive binding from the antibody.

10. The SILAC ratio of a protein is essentially a measure of the differences in the protein's affinity (directly or indirectly through a binding partner) towards anti-pTyr antibody before and after RTK activation. It generally reflects the activation/deactivation of the protein upon the RTK stimulation, but there are exceptions in rare occasions. The overall tyrosine phosphorylation level of a protein does not always reflect its level of activation. For example, Src family kinases are tyrosine phoshporylated at different sites both when inhibited and activated *(26)*.

11. The cutoff SILAC ratio to define significant changes depends on the accuracy of quantitation, which in turn can be affected by the type of MS instrument used, the quantitation software, the signal to noise ratio of the peptide in MS and the biological reproducibility. 1.5-fold is a commonly used cutoff when QTOF and Orbitrap instruments are used *(11, 12, 17)*. Ideally one should perform a control experiment using SILAC samples with predefined ratios to test the accuracy of the quantitation.

Acknowledgements

This work was supported by National Institutes of Health Grant P30 NS050276.

References

1. Hunter T. (1998) The croonian lecture 1997. The phosphorylation of proteins on tyrosine: its role in cell growth and disease *Philos Trans R Soc Lond B Biol Sci* 353(1368), 583–605.

2. Ong SE, Foster LJ, Mann M. (2003) Mass spectrometric-based approaches in quantitative proteomics *Methods* 29(2), 124–30.

3. Ong SE, Mann M. (2005) Mass spectrometry-based proteomics turns quantitative *Nat Chem Biol* 1(5), 252–62.

4. Ong SE, Blagoev B, Kratchmarova I, et al. (2002) Stable isotope labeling by amino acids in cell culture, SILAC, as a simple and accurate approach to expression proteomics *Mol Cell Proteomics* 1(5), 376–86.

5. Ong SE, Mann M. (2006) A practical recipe for stable isotope labeling by amino acids in cell culture (SILAC) *Nat Protoc* 1(6), 2650–60.

6. de Godoy LM, Olsen JV, de Souza GA, Li G, Mortensen P, Mann M. (2006) Status of complete proteome analysis by mass spectrometry: SILAC labeled yeast as a model system *Genome Biol* 7(6), R50.

7. Gruhler A, Olsen JV, Mohammed S, et al. (2005) Quantitative phosphoproteomics applied to the yeast pheromone signaling pathway *Mol Cell Proteomics* 4(3), 310–27.

8. Gruhler A, Schulze WX, Matthiesen R, Mann M, Jensen ON. (2005) Stable isotope labeling of Arabidopsis thaliana cells and quantitative proteomics by mass spectrometry *Mol Cell Proteomics* 4(11), 1697–709.

9. Neher SB, Villen J, Oakes EC, et al. (2006) Proteomic profiling of ClpXP substrates after DNA damage reveals extensive instability within SOS regulon *Mol Cell* 22(2), 193–204.

10. Blagoev B, Mann M. (2006) Quantitative proteomics to study mitogen-activated protein kinases *Methods* 40(3), 243–50.

11. Blagoev B, Kratchmarova I, Ong SE, Nielsen M, Foster LJ, Mann M. (2003) A proteomics strategy to elucidate functional protein-protein interactions applied to EGF signaling *Nat Biotechnol* 21(3), 315–8.

12. Blagoev B, Ong SE, Kratchmarova I, Mann M. (2004) Temporal analysis of phosphotyrosine-dependent signaling networks by quantitative proteomics *Nat Biotechnol* 22(9), 1139–45.

13. Hinsby AM, Olsen JV, Bennettt KL, Mann M. (2003) Signaling initiated by overexpression of the fibroblast growth factor receptor-1 investigated by mass spectrometry *Mol Cell Proteomics* 2(1), 29–36.

14. Hinsby AM, Olsen JV, Mann M. (2004) Tyrosine phosphoproteomics of fibroblast growth factor signaling: a role for insulin receptor substrate-4 *J Biol Chem* 279(45), 46438–47.

15. Ibarrola N, Molina H, Iwahori A, Pandey A. (2004) A novel proteomic approach for specific identification of tyrosine kinase substrates using [13C]tyrosine *J Biol Chem* 279(16), 15805–13.

16. Kim JE, White FM. (2006) Quantitative analysis of phosphotyrosine signaling networks triggered by CD3 and CD28 costimulation in Jurkat cells *J Immunol* 176(5), 2833–43.

17. Zhang G, Spellman DS, Skolnik EY, Neubert TA. (2006) Quantitative phosphotyrosine proteomics of EphB2 signaling by stable isotope labeling with amino acids in cell culture (SILAC) *J Proteome Res* 5(3), 581–8.

18. Zhang Y, Wolf-Yadlin A, Ross PL, et al. (2005) Time-resolved mass spectrometry of tyrosine phosphorylation sites in the epidermal growth factor receptor signaling network reveals dynamic modules *Mol Cell Proteomics* 4(9), 1240–50.

19. Kratchmarova I, Blagoev B, Haack-Sorensen M, Kassem M, Mann M. (2005) Mechanism of divergent growth factor effects in mesenchymal stem cell differentiation. *Science* 308(5727),1472–7.

20. Holland SJ, Gale NW, Gish GD, et al. (1997) Juxtamembrane tyrosine residues couple the Eph family receptor EphB2/Nuk to specific SH2 domain proteins in neuronal cells *Embo J* 16(13), 3877–88.

21. Beynon RJ, Pratt JM. (2005) Metabolic labeling of proteins for proteomics *Mol Cell Proteomics* 4(7), 857–72.

22. Aebersold R, Mann M. (2003) Mass spectrometry-based proteomics *Nature* **422**(6928), 198–207.

23. Nesvizhskii AI, Vitek O, Aebersold R. (2007) Analysis and validation of proteomic data generated by tandem mass spectrometry *Nat Methods* **4**(10), 787–97.

24. Van Hoof D, Pinkse MW, Oostwaard DW, Mummery CL, Heck AJ, Krijgsveld J. (2007) An experimental correction for arginine-to-proline conversion artifacts in SILAC-based quantitative proteomics *Nat Methods* **4**(9), 677–8.

25. Ong SE, Kratchmarova I, Mann M. (2003) Properties of 13C-substituted arginine in stable isotope labeling by amino acids in cell culture (SILAC) *J Proteome Res* **2**(2), 173–81.

26. Xu W, Doshi A, Lei M, Eck MJ, Harrison SC. (1999) Crystal structures of c-Src reveal features of its autoinhibitory mechanism *Mol Cell* **3**(5), 629–38.

Chapter 8

Hydrophilic Interaction Chromatography for Fractionation and Enrichment of the Phosphoproteome

Dean E. McNulty and Roland S. Annan

Summary

Mass spectrometry-based protein phosphorylation analysis on a proteome-wide scale remains a formidable challenge, hampered by the complexity and dynamic range of protein expression on the global level and multi-site phosphorylation at substoichiometric ratios at the individual protein level. It is recognized that reduction of sample complexity or enrichment of the phosphopeptide pool is a necessary prerequisite for global phospho-proteomics. Immobilized metal affinity chromatography (IMAC) and strong cation exchange chromatography, either alone or in tandem, have emerged as the most widely used chromatographic-based enrichment strategies. However, each is not without shortcomings. Both techniques provide little fractionation of phosphorylated species and are compromised by competition and co-elution of highly acidic peptides. Here, we describe a phosphopeptide prefractionation scheme using hydrophilic interaction chromatography, which both enriches the phosphopeptide pool and efficiently fractionates the remaining peptides. When used in front of IMAC, the selectivity of the metal affinity resin is improved to greater than 95%. The lack of significant numbers of nonphosphorylated peptides also allows for more efficient use of the mass spectrometer duty cycle in that the instrument spends nearly all of its time in sequencing the phosphopeptides.

Key words: Phosphorylation, Mass spectrometry, Hydrophilic interaction chromatrography, Proteomics, Phosphoproteome.

1. Introduction

Analysis of the phosphoproteome by mass spectrometry is hampered by the enormous complexity of the proteome and the large dynamic range of protein expression *(1)*. Many phosphorylated proteins will be less-abundant proteins involved in signaling and

Marjo de Graauw (ed.), *Phospho-Proteomics, Methods and Protocols, vol. 527*
© 2009 Humana Press, a part of Springer Science + Business Media, New York, NY
Book DOI: 10.1007/978-1-60327-834-8_8

regulation of cellular function. Most phosphoproteins will contain between 1 and 20 phosphorylation sites and these sites will largely be occupied at substoichiometric levels *(2–4)*. Many phosphorylation sites will occur in combination with other modifications, and multiple phosphorylation sites can exist in a single peptide. All of these factors suggest that the phosphoproteome is likely to be every bit as complex as the general proteome, but with a dynamic range spanning an additional two orders of magnitude and that the phosphopeptides will exist in the presence of a very large excess of nonphosphorylated peptides.

The most common strategy for enriching phosphopeptides in mixtures is the use of immobilized metal affinity chromatrography (IMAC) *(5–6)* or titanium dioxide (TiO_2) microcolumns *(7–8)*. Although phosphopeptides have a very high affinity for these resins, acidic peptides (containing aspartic acid and glutamic acid) and peptides containing histidine also have a high binding affinity. The conversion of peptide carboxylate groups to methyl ester derivatives has been shown to restore the selectivity of both IMAC *(9)* and TiO_2 *(7)*. Recently it was demonstrated that including dihydroxybenzoic acid (DHB) in the loading buffer can improve the selectivity of phosphopeptide binding to TiO_2 *(8)*.

Strong cation exchange chromatography (SCX) has been recently suggested as an alternative to metal chelating resins as a means of enriching phosphopeptides *(2)*. Although this appears to be highly reproducible, it does not provide a very high level of enrichment, and not all phosphopeptides elute in the enriched fractions. To provide a more comprehensive view of the phosphoproteome, a further reduction in sample complexity beyond affinity enrichment or partitioning by charge state is necessary. An additional high-resolution chromatographic separation at the peptide level, orthogonal to those currently in use would be highly desirable when coupled with an enrichment strategy such as IMAC or TiO_2.

Hydrophilic interaction chromatography (HILIC) *(10)* is a high-resolution separation technique where retention increases with increasing polarity (hydrophilicity) of the peptide, opposite to the trends observed in RPLC *(10–11)*. HILIC has been shown to have the highest degree of orthogonality to RPLC of all commonly used peptide separation modes *(12)*. This and the suitability of HILIC for the highly efficient separation of phosphopeptides prompted us to employ HILIC as part of a multidimensional separation strategy for phospho-proteomics. We show here how to use HILIC as a first dimension separation in combination with IMAC and reverse phase LC-MS/MS.

2. Materials

2.1. Cell Culture and Lysis

1. Dulbecco's Modified Eagle's Medium (DMEM) supplemented with 10% fetal bovine serum.
2. Serum-free Dulbecco's Modified Eagle's Medium (DMEM).
3. Solution of trypsin (0.25%) and ethylenediamine tetraacetic acid (EDTA) (1 mM).
4. 100×20 mm treated tissue culture dishes.
5. Dulbecco's phosphate buffered saline without calcium or magnesium.
6. Cell lysis buffer: 8 M Urea/20 mM Tris–HCl pH 8.0.
7. Teflon disposable cell scrapers.
8. Microson Ultrasonic Cell Disruptor (Misonix) with probe tip or equivalent.

2.2. Protein Digestion

1. 400 mM ammonium bicarbonate in water.
2. Reduction: 45 mM dithiothreitol (DTT) in water.
3. Alkylation: 100 mM iodoacetamide in water.
4. Digestion: Immobilized trypsin agarose (Pierce).
5. Stop: 100% trifluoroacetic acid.
6. Vacuum extraction manifold (Waters).
7. Desalt: 1 g Sep-Pak C18 6 cc/1 g cartridge (Waters).
8. Wash buffer: 0.1% TFA in water.
9. Elution buffer: 0.1% TFA containing 60% acetonitrile in water.
10. Speedvac concentrator (Savant).

2.3. Hydrophilic Interaction Chromatography

1. HILIC column: 4.6×250 mm TSKgel Amide-80 5 μm particle column (TOSOH Biosciences).
2. Solvent A: 98% water with 0.1% TFA.
3. Solvent B: 98% acetonitrile with 0.1% TFA.
4. HPLC grade acetonitrile.
5. HPLC grade water.
6. Trifluoroacetic acid (TFA), sequanal grade.

2.4. Immobilized Metal Affinity Chromatography

1. IMAC resin: PHOS-Select Iron Affinity Gel.
2. 0.22 μm Nylon Spin-X centrifuge tube filter.
3. Wash solution 1: 250 mM acetic acid with 30% acetonitrile.
4. Wash solution 2: water.

5. Elution buffer: 400 mM ammonium hydroxide.

6. Speedvac concentrator.

2.5. Reverse Phase Liquid Chromatography–Tandem Mass Spectrometry

1. Sample resuspension: 0.1% formic acid/0.02% TFA in water.

2. Trapping column: C18 PepMap100 300 μm × 5 mm 5-μm enrichment cartridge (LC Packings).

3. Analytical column: C18 PepMap100 75 μm × 15 mm 5-μm analytical column (LC Packings).

4. Load solution: 0.1% formic acid/0.05% TFA.

5. Solvent A: 0.1% formic acid in water.

6. Solvent B: 0.1% formic acid in acetonitrile.

2.6. Database Searching

1. Peak lists from individual LC-MS/MS runs are merged into one file using MasCat software (Agilent).

2. Database search software: Mascot version 2.2.

3. Database: IPI Human version 3.37 containing 69,164 entries; forward and reverse.

3. Methods

Hydrophilic interaction chromatography (HILIC) *(10)* is a high-resolution separation technique where the primary interaction between a peptide and the neutral, hydrophilic stationary phase is hydrogen bonding. In HILIC, retention increases with increasing polarity (hydrophilicity) of the peptide, opposite to the trends observed in RPLC *(10–11)*. Gilar et al. have shown that HILIC has the highest degree of orthogonality to RPLC of all commonly used peptide separation modes *(12)*. These characteristics prompted us to investigate the potential of HILIC as part of a multi-dimensional separation strategy for phospho-proteomics.

On HILIC, phosphopeptides exhibit increased retention relative to nonphosphorylated peptides. Therefore, we optimized the gradient with the intent to segregate that part of the digested lysate which does not contain phosphorylated peptides into the flow-through while providing optimal resolution of the phosphopeptide containing fractions during the gradient elution. This resulted in a separation that provides significant enrichment and fractionation of phosphopeptides (*see* **Fig. 1**).

We have reported that HILIC fractionation markedly improves the selectivity of IMAC *(13)*. The hydrophilicity-based separation yields fractions where all of the peptides in each fraction have a

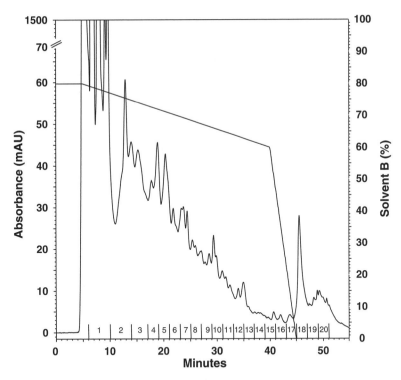

Fig. 1. Preparative HILIC fractionation of HeLa whole cell lysate (1 mg) tryptic digest. Tryptic peptides were separated using an optimized phosphopeptide gradient and pooled as indicated. A_{280} UV absorbance profile and gradient trace are shown.

similar polarity. This ensures fair competition between the phosphopeptides and the nonphospho-peptides in each fraction, and provides highly enriched phosphopeptide pools without the need for derivatization or additives. In the current study, HILIC prefractionation of phosphopeptide-containing pools prior to IMAC capture resulted in >99% selectivity based on MS/MS identification of phosphopeptides vs. nonphosphopeptides (**Fig. 2**). It is very likely that TiO_2 can be substituted just as effectively in this protocol with all the same advantages being observed *(14)*.

3.1. Cell Culture and Lysis

1. Maintain HeLa cells in DMEM/10% FBS. When they approach confluency, passage them with trypsin/EDTA to provide experimental cultures in 100×20 mm dishes for phosphoproteomic studies.

2. Grow the experimental cultures to near confluency and incubate them for 16–24 h in DMEM in the absence of serum. Stimulate the cells with 100 nM Calyculin A in serum-free DMEM for 30 min (*see* **Note 1**).

3. Wash cells twice with 10 mL of ice-cold DPBS.

4. Lyse cells in 1 mL of cold 8 M urea/20 mM Tris–HCl, pH 8.0. Scrape and pipette them into a 15-mL conical tube. (*see* **Note 2**). The sample will be highly viscous.

5. Disrupt the cellular lysate by sonication with three brief 15-s pulses using a microprobe tip set at 20 W output. Take care to keep the cell lysate ice-cold during this process. (*see* **Note 3**)

6. Clarify the sample by centrifugation at $10,000 \times g$ for 15 min at 4°C. The clarified supernatant may be processed further or stored at –80°C. (*see* **Note 4**)

3.2. Protein Digestion

1. Add 1:9 volume of 45 mM DTT in water to 1 mg of the clarified lysate and incubate the sample at 37°C for 30 min.

2. Allow the sample to reach room temperature and add 1:9 volume of 100 mM iodoacetamide in water. Allow the reaction to proceed at room temperature in dark for 30 min.

3. Dilute the lysate to 4 mL with 100 mM ammonium bicarbonate in water (*see* **Note 5**).

4. Add 250 µL of immobilized trypsin agarose slurry to the sample and digest at room temperature overnight on a rocking platform (*see* **Note 6**).

5. Add 40 µL of TFA to a final concentration of 1% to acidify and quench the digest reaction.

6. Desalt the digest by applying the sample to a prewetted (ethanol) and equilibrated (0.1% TFA) 1 g Waters Sep-Pak C18 6 cc/1 g cartridge using a vacuum manifold apparatus. Maintain a slow flow through the cartridge such that individual drops are visible in the effluent (*see* **Note 7**).

7. Wash the cartridge with 20 mL of 0.1% TFA, and elute the peptides with 4 mL of 60% CH3CN per 0.1% TFA.

8. Concentrate the eluent to dryness by Speedvac.

3.3. HILIC Chromatography

1. The protocol assumes access to high performance liquid chromatography equipment including an elution gradient programmer with data processing software, binary solvent delivery system, UV absorbance detector, and automated fraction collector. The present study was performed using Beckman 126 Solvent Module and Beckman 166 Detector controlled by 32 Karat software, and Gilson Model 203 fraction collector.

Preparative HILIC chromatography conditions:

Column: HILIC 4.6 × 250 mm TSKgel Amide-80 5 µm particle

Buffer- A: 98% water with 0.1% TFA.

Buffer-B: 98% acetonitrile with 0.1% TFA

Flow: 0.5 mL/min
Detection: A280 nm
Gradient:
80% B – 5 min
80–60% B – 40 min
60–0% B – 5 min

2. Dissolve the sample in 200 μL of 80% solvent B, and spin in a microcentifuge at maximum speed ($16,100 \times g$) for 5 min to remove particulates prior to injection (*see* **Note 8**).

3. Set the fraction collector to time-based mode and collect 1 min fractions throughout the gradient into 1.5-mL capless Eppendorf tubes (*see* **Note 9**).

4. Pool fractions appropriately based on experimental design (*see* **Note 9**, **Fig. 1**).

3.4. IMAC Enrichment

1. Equilibrate 500 μL of PHOS-Select Iron Affinity Gel slurry (250 μL gel) in 5 mL of 80% HILIC solvent B on a rocking platform for 15 min at room temperature. Centrifuge briefly and aspirate liquid. Resuspend the gel in 250 μL of 80% solvent B and add 20 μL into each fraction. (*see* **Notes 10** and **11**).

2. Rapidly vortex at room temperature for 30 min. Transfer the fractions to individual 0.22-μm Nylon Spin-X Centrifuge Tube Filters and centrifuge at $8,200 \times g$ in a microfuge for 30 s. (*see* **Note 12**)

3. Discard the unbound material. Wash the gel with 500 μL of 250 mM acetic acid containing 30% acetonitrile. Vortex rapidly for 5 min. Centrifuge and discard unbound wash.

4. Repeat as in **Subheading 3.3**, **step 3** using 500 μL of water.

5. Transfer the filter inserts to clean 1.5-mL Eppendorf tubes. Elute with 100 μL of 400 mM ammonium hydroxide with rapid vortexing for 10 min.

6. Collect the eluants and concentrate to dryness. These samples represent fractionated, highly enriched phosphopeptide pools.

3.5. Reverse Phase LC-MS/MS

1. The protocol assumes access to an on-line LC interfaced to a mass spectrometer capable of performing tandem MS/MS. The current study was conducted by injecting the sample using an 1100 series autosampler (Agilent) onto an enrichment trap through a 1100 Series binary pump, and back-flushing the trap via valve switching in-line to the analytical column which was gradient eluted using an 1100 Series nano pump. The nano pump was coupled to an 1100 Series LC/

MSD Trap XCT Ultra ion trap mass spectrometer (Agilent). (*see* **Note 13**)

LC conditions:

Trapping column: C18 PepMap100 300 μm × 5 mm 5-μm enrichment cartridge

Load buffer: 0.1% formic acid/0.05% TFA in water

Flow: 20 μL/min

Gradient: isocratic, 4-min wash following injection prior to valve switch

Analytical column: C18 PepMap100 75 μm × 15 mm 5 μm analytical column

Buffer A: 0.1% formic acid in water

Buffer B: 0.1% formic acid in acetonitrile

Flow: 300 nL/min

Gradient: 10% B – 4 min

10–30% B – 50 min

30–95% B – 5 min

Mass spectrometer parameters:

MS: drying gas flow: 4 L/min, 325°C; Vcap: 2,000 V; skim 1: 40 V; capillary exit: 128 V; trap drive: 70; averages: 2; ion charge control (ICC): on; maximum accumulation time: 200 ms; smart target: 250,000; scan range: 400–1,600; standard enhanced scan.

MS/MS: number of precursors, 2; averages, 4; fragmentation amplitude, 1.15 V; SmartFrag, on (50–200%); active exclusion, on, 2 spectra, 0.2 min; scan range: 100–2,000; ultra scan, on; ICC target, 500,000.

Two precursors per survey scan were selected for fragmentation. A neutral-loss triggered MS(3) scan was performed on ions detected in the MS(2) experiment which correspond to the loss of 49 Da from the precursor. This was performed in the fragmentation only mode in which no isolation is done between the MS(2) fragmentation and the "pseudo" MS(3) fragmentation. The two spectra were merged into a single spectrum for database searching.

2. Resuspend the lyophilized sample fractions from 3.4.6 in 15-μL load buffer, vortex vigorously, and microfuge for 2 min at 16,100 × *g* to remove particulates. Transfer the fractions to autosampler vials for injection. Inject 5 μL (33%) of each fraction for LC-MS/MS analysis.

3.6. Database Searching

1. The protocol assumes access to the MASCOT software for identifying proteins based on mass spectrometry data from primary sequence databases (*see* **Note 14**).

2. Merge peak lists from individual LC-MS/MS runs into one file. MasCat software from Agilent processes raw data into peak lists and concatenates individual files into a single MAS-COT compatible format (.mgf). The global charge setting is set to "NONE" to disable charge state determination by the instrument. In this way, peptide spectra are searched as both 2+ and 3+ charge states by MASCOT.

3. Search the .mgf file against a nonredundant protein database to which a reverse complement of itself has been appended *(15)*. Use the following restriction: *H. sapiens;* all peptides fully tryptic containing no more than one missed cleavage; carboxamidomethylated cysteine set as fixed modification; phosphorylation on serine/threonine/tyrosine set as variable modifications; mass tolerance ±0.3 Da for the precursor ions and ±0.5 Da for the product ions. Filter search results to yield a false-positive rate of approximately 1.0% in all cases. (*see* **Fig. 2**)

4. Manually validate sequence assignment of peptides with Mascot ion scores between the false-positive rate cut-off, and 25. Manually verify all phosphosite assignments where the difference in the score of the next most probable site is less than 10. Peptides with Mascot scores between the threshold for a 1.3% false-positive rate and 25 are examined for the following features: neutral loss of H_3PO_4 from the

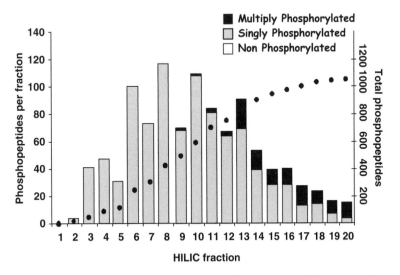

Fig. 2. IMAC enrichment and data-dependent LC-MS/MS analysis of HILIC fractionated HeLa cell lysate tryptic digest. Singly phosphorylated (*grey*) and multiply phosphorylated (*black*) peptides identified by MASCOT in each pool are shown. Nonphosphorylated peptides (*white*) are also plotted at the bottom of each bar, but are not visible due to the very high selectivity of the protocol for phosphopeptides. Summed cumulative total of unique phosphopeptides are indicated by *dotted line*.

MS/MS Fragmentation of **GGVTGpSPEASISGSK**
Found in IPI00021812, Tax_Id=9606 Neuroblast differentiation-associated protein AHNAK

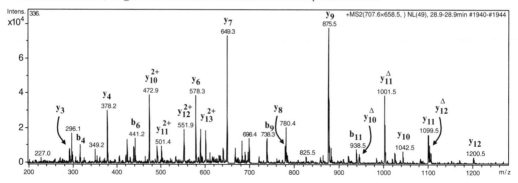

Monoisotopic mass of neutral peptide Mr (calc): 1412.6184
Variable modification: S6: phospho ST, with neutral losses of 97.9769 (shown in table)
Ions Score: 66 **Expect:** 6.5e-05
Matches: 30/198 fragment ions using 47 most intense peaks

	1	2	3	4	5	6	7	8	9	10	11	12	13	14	15
b++	29.5	58.0	107.6	158.1	186.6	221.1	269.6	334.2	369.7	413.2	469.7	513.3	541.8	585.3	
b	58.0	115.1	214.1	315.2	372.2	**441.2**	538.3	667.3	**738.5**	825.4	**938.5**	1025.5	1082.5	1169.5	
	G	G	V	T	G	S	P	E	A	S	I	S	G	S	K
y		1258.6	1201.6	1102.5	1001.5	944.5	875.5	778.4	649.4	578.3	491.1	378.2	291.2	234.1	147.1
y++		629.8	**601.3**	551.8	501.3	472.7	438.2	389.7	325.2	289.7	246.1	189.6	146.1	117.6	74.6
	15	14	13	12	11	10	9	8	7	6	5	4	3	2	1

Fig. 3. Typical identification data for phosphopeptide sequence assignment of tandem MS spectral data against the IPI protein database (extracted from Mascot). In this case, an abundant neutral loss of 49 Da from the initial precursor ion (m/z 707.6) was used to trigger the acquisition of the pseudo MS³ spectrum shown here from the [M + 2H-49]²⁺ at m/z 658.5.

precursor, contiguous series of three or more y or b ions, prominent y ion cleavage N-terminal to a proline, and no unexplained high mass fragment ions. (*see* **Fig. 3**)

4. Notes

1. Though HeLa cells were used in the present study, the choice of cell lines for phospho-proteomics studies are user dependent and the protocol described may be readily adapted to alternate adherent lines as well as suspension cultures. In addition, perturbation of phosphorylation-dependent pathways may be achieved using any variety of stimuli, for example extracellular growth factors (EGF), pathway specific activators (phorbol 12-myristate 13-acetate (PMA)), or general phosphatase inhibitors (Calyculin A). It may also

be desirable to employ subcellular prefractionation for the enrichment of specific organelles (e.g., nuclei) prior to sample processing for HILIC.

2. Although it is expected that 8 M urea will denature and irreversibly inactivate cellular phosphatases, the inclusion of 1 mM sodium pervanadate or alternate phosphatase inhibitor cocktails may be warranted.

3. Sonication is necessary to reduce the extreme viscosity of the lysate through shearing of chromosomal DNA released during 8-M urea lysis. Sonication by direct application of a microtip probe is more effective and preferable to immersion in a cup horn type sonicator.

4. Depending on the cell type and confluency, a yield of 1–5 mg of total protein is typically obtained from lysis of a single 10-cm culture dish. It may be desirable to reserve a small aliquot of clarified lysate in order to perform a protein assay for quantitation at this point. This can be accomplished with various colorimetric methods such as BCA, Bradford, Lowry, etc. Though optional, this may be useful in matching the desired peptide load to the appropriate HILIC column dimensions to maximize chromatographic resolution, or normalizing loads between multiple experiments. The method described has been optimized for 1 mg of starting cellular protein lysate.

5. Trypsin retains full enzymatic activity in the presence of urea if the concentration is maintained at <2 M. It is desirable to use sequencing grade trypsin to restrict cleavage to arginine and lysine residues; contaminating protease activity will produce peptides refractory to positive assignment against a search database.

6. If desired, an aliquot may be removed at this point for SDS-PAGE analysis to determine the efficiency of digest. Compared with the starting lysate, negligible Coomassie staining of protein bands should be visible following trypsin digest. If any stain is evident, it should be predominately <5 kDa apparent molecular weight, migrating near the dye front.

7. Trypsin agarose and any precipitates will be trapped on the head of the disposable solid phase extraction cartridge.

8. The sample may contain particulates and turbidity upon dissolution; solubilization may be aided by immersion into an ultrasonic water bath. However, the peptide content has been found to be highly soluble at acetonitrile concentrations as high as 90%.

9. In the current protocol, 20 fractions are collected throughout the gradient. The majority of the fractions are 1-mL

pools taken at 2 min intervals. Our experience using model peptides and literature reports *(12)* suggests that the peak capacity is much higher and that we are undersampling the resolving power of HILIC. Thus it may be preferable to extend the chromatographic separation and/or fractionate into more pools (e.g., 40 × 1-min pools) to achieve the full separation potential. This decision will be made based on specific experimental design and with the caveat that additional fractionation will require increased sample processing and significantly more dedicated mass spectrometer time.

10. Though not tested in this protocol, it should be possible to substitute other IMAC based media or alternative phosphopeptide enrichment resins such as TiO_2 and still obtain excellent results.

11. The iron affinity gel beads may clog or be excluded from a micropipettor tip. If it is difficult to dispense an accurate volume of slurry, cutting 1 mm off the end of the tip can aid in pipetting.

12. The spin filter is used as a bed for small-scale batch chromatography. Other filters may be substituted, though care should be taken to ensure the filter is resistant to high concentrations of acetonitrile. In addition, several manufacturers units are prone to high levels of leaching of polymeric materials, which can be severely detrimental to downstream mass spectrometric analysis.

13. Though the present study was conducted on a 3D quadrupole ion trap, any instrument capable of tandem MS may be employed including several current hybrid mass spectrometer platforms (e.g., qTOF, FT-ICR, LTQ-Orbitrap, etc.). The use of an instrument providing accurate mass determination of the precursor ion would be highly desirable for this experiment, as phosphorylated peptides are prone to neutral losses of H_3PO_4 and often produce much less analytically useful fragment ion spectra. As the IMAC enriched HILIC fractions still represent a highly complex and dynamically challenging mixture, the data acquisition regimen of the individual instrument chosen should be optimized for maximum sensitivity and duty cycle.

14. Several other algorithms for identifying proteins from primary sequence databases based on mass spectrometry data are available. Other packages in addition to MASCOT which might be considered include Sequest, X!Tandem, Spectrum Mill, and OMSSA. Search parameters will need to be optimized for each to maximize sensitivity and specificity of phosphopeptide identification. For example, some pertinent parameters to consider would include precursor

and fragment MS/MS mass accuracy tolerance, proteolytic enzyme specificity, variable post-translational modifications, instrument fragmentation type, etc.

References

1. de Godoy LM, Olsen JV, de Souza GA, Li G, Mortensen P, Mann M, (2006) Status of complete proteome analysis by mass spectrometry: SILAC labeled yeast as a model system. *Genome Biol* 6, R50.

2. Beausoleil SA, Jedrychowski M, Schwartz D, Elias JE, Villen J, Li J, Cohn MA, Cantley LC, Gygi SP, (2004) Large-scale characterization of HeLa cell nuclear phosphoproteins. *Proc Natl Acad Sci U S A* 101, 12130–12135.

3. Beausoleil SA, Villen J, Gerber SA, Rush J, Gygi SP, (2006) A probability-based approach for high-throughput protein phosphorylation analysis and site localization. *Nat Biotechnol* 24, 1285–1292.

4. Olsen JV, Blagoev B, Gnad F, Macek B, Kumar C, Mortensen P, Mann M, (2006) Global, in vivo, and site-specific phosphorylation dynamics in signaling networks. *Cell* 127, 635–648.

5. Nuwaysir LM, Stults JT, (1993) Electrospray ionization mass spectrometry of phosphopeptides isolated by on-line immobilized metal-ion affinity chromatography. *J Am Soc Mass Spectrom* 4, 662–667.

6. Posewitz MC, Tempst P (1999) Immobilized gallium(III) affinity chromatography of phosphopeptides. *Anal Chem* 71, 2883–2892.

7. Pinkse MW, Uitto PM, Hilhorst MJ, Ooms B, Heck AJ, (2006) Selective isolation at the femtomole level of phosphopeptides from proteolytic digests using 2D-NanoLC-ESI-MS/MS and titanium oxide precolumns. *Anal Chem* 76, 3935–3943.

8. Larsen MR, Thingholm TE, Jensen ON, Roepstorff P, Jorgensen TJ, (2005) Highly selective enrichment of phosphorylated peptides from peptide mixtures using titanium dioxide microcolumns. *Mol Cell Proteomics* 4, 873–886.

9. Ficarro SB, McCleland ML, Stukenberg PT, Burke DJ, Ross MM, Shabanowitz J, Hunt DF, White FM, (2002) Phosphoproteome analysis by mass spectrometry and its application to Saccharomyces cerevisiae. *Nat Biotechnol* 20, 301–305.

10. Alpert AJ. (1990) Hydrophilic-interaction chromatography for the separation of peptides, nucleic acids and other polar compounds. *J Chromatogr* 499, 177–196.

11. Yoshida T, Okada T, (1999) Peptide separation in normal-phase liquid chromatography: Study of selectivity and mobile phase effects on various columns. *J Chromatogr A* 840, 1–9.

12. Gilar M, Olivova P, Daly AE, Gebler JC, (2005) Orthogonality of separation in two-dimensional liquid chromatography. *Anal Chem* 77, 6426–6434.

13. McNulty DE and Annan RS, (2008) Hydrophilic-interaction chromatography reduces the complexity of the phosphoproteome and improves global phosphopeptide isolation and detection. *Mol Cell Proteomics* 7, 971–980.

14. Thingholm TE, Jensen ON, Robinson PJ, Larsen MR. (2008) SIMAC – A phosphoproteomics strategy for the rapid separation of monophosphorylated from multiply phosphorylated peptides. *Mol Cell Proteomics* 7, 661–671.

15. Elias JE, Gygi SP. (2007) Target-decoy search strategy for increased confidence in large-scale protein identifications by mass spectrometry. *Nat Methods* 4, 207–214.

Chapter 9

SILAC for Global Phosphoproteomic Analysis

Genaro Pimienta, Raghothama Chaerkady, and Akhilesh Pandey

Summary

Establishing the phosphorylation pattern of proteins in a comprehensive fashion is an important goal of a majority of cell signaling projects. Phosphoproteomic strategies should be designed in such a manner as to identify sites of phosphorylation as well as to provide quantitative information about the extent of phosphorylation at the sites. In this chapter, we describe an experimental strategy that outlines such an approach using stable isotope labeling with amino acids in cell culture (SILAC) coupled to LC-MS/MS. We highlight the importance of quantitative strategies in signal transduction as a platform for a systematic and global elucidation of biological processes.

Key words: Signal transduction, Phospho-proteomics, Quantitative proteomics, SILAC, Electron transfer dissociation.

1. Introduction

The identification of post-translational modifications such as serine, threonine, and tyrosine phosphorylation can be performed with liquid-chromatography (LC) systems coupled on-line to a tandem mass spectrometer. A global identification of phosphorylation sites is an essential first step towards establishing the functional role of proteins in signaling pathways. Examples include monitoring of phosphorylation changes in the proteome as a function of time (1–3).

A popular method of in vivo labeling is the use of heavy amino acids in the culture medium. In this SILAC strategy, a control sample is compared with at least one test sample (2 plex analysis). Up to five different samples can be routinely compared

Marjo de Graauw (ed.), *Phospho-Proteomics, Methods and Protocols, vol. 527*
© 2009 Humana Press, a part of Springer Science + Business Media, New York, NY
Book DOI: 10.1007/978-1-60327-834-8_9

Table 1
Recommended heavy amino acids for multiplexed SILAC experiments (4)

	Sample 1	Sample 2	Sample 3	Sample 4	Sample 5
2plex	^{12}C/^{14}N light amino acids	(^{13}C$_6$-Lys+ ^{13}C$_6$-Arg)			
3plex	^{12}C/^{14}N light amino acids	(D4-Lys+ ^{13}C$_6$-Arg)	(^{13}C$_6$,^{15}N$_2$-Lys+ ^{13}C$_6$,^{15}N$_4$-Arg)		
4plex	^{12}C/^{14}N light amino acids	(D4-Lys+ ^{13}C$_6$-Arg)	(^{13}C$_6$,^{15}N$_2$-Lys+ ^{13}C$_6$,^{15}N$_4$-Arg)	(^{13}C$_6$,^{15}N$_2$, D9-Lys+ ^{13}C$_6$,^{15}N$_4$, D7-Arg)	
5plex	^{12}C/^{14}N light amino acids	^{15}N$_4$-Arg	^{13}C$_6$-Arg	^{13}C$_6$,^{15}N$_4$-Arg	^{13}C$_6$,^{15}N$_4$, D7-Arg

in a SILAC experiment using the heavy amino acid combinations as shown in **Table 1**. The ratio of intensities of peptides labeled by light vs. heavy amino acids is used to establish relative quantities of the respective proteins in the different samples. The availability of this quantitative multiplex strategy allows for the simultaneous analysis of different cellular states in a model system.

2. Materials

1. Cell lysis buffer: 25 mM Tris–HCl, pH 7.5, 150 mM NaCl, 2 mM EDTA, 10% glycerol, 1% triton X-100.

2. Protein dissolution buffer: 8 M urea, 100 mM NH$_4$CO$_3$, pH 8.5, 1% SDS.

3. Reduction solution: 10 mM DTT in 50 mM NH$_4$CO$_3$, pH 8.5.

4. Alkylation solution: 50 mM iodoacetamide in 50 mM NH$_4$CO$_3$, pH 8.5.

5. TiO$_2$, Titanium (IV) nano-powder from Sigma-Aldrich (Cat. No. 634662).

6. TiO$_2$ loading solution: 90% acetonitrile, 2% TFA, 250 mg/mL lactic acid.

7. TiO$_2$ washing solution: 90% acetonitrile, 2% TFA.

8. TiO$_2$ elution solution: 25% NH$_4$OH, pH 10.5.

9. SCX solution A: 10 mM KH_2PO_4, pH 2.85, 25% acetonitrile.

10. SCX solution B: 10 mM KH_2PO_4, pH 2.85, 25% acetonitrile, 350 mM KCl.

11. RP-C loading and wash solution: 0.4% acetic acid.

12. RP-C elution solution: 0.4% acetic acid and 90% acetonitrile.

13. Trypsin: sequence grade from Promega (Cat. No. V5111).

14. SILAC medium: DMEM and RPMI media deficient in arginine and lysine (Invitrogen).

15. Light isotope amino acids to supplement SILAC media.

16. Heavy isotope-labeled amino acids (Cambridge Isotope Laboratories, Inc.).

17. Acetone, acetonitrile, and methanol should be HPLC grade.

18. Bio-Rad Protein Assay (Cat. No. 500–0006).

19. DL-dithiothreitol (DTT), ammonium bicarbonate (NH_4CO_3), and urea should be >99% purity and their solution stocks ought to be prepared fresh.

20. Phosphate buffered saline (PBS).

21. Phosphatase inhibitor cocktail 1 (Sigma-Aldrich).

22. Phosphatase inhibitor cocktail 2 (Sigma-Aldrich).

23. Protease inhibitor cocktail (Roche).

24. SCX column: polysulfoethyl A (100×2.1, 5 µm, 300-Å from PolyC).

25. SDS-PAGE: NuPAGE 4–12% Bis-Tris mini-Gel (Invitrogen).

26. SelfPack POROS 20 R2 (Reverse Phase Packing, RP-C).

27. 2-Amino-2-(hydroxymethyl)-1, 3-propanediol, hydrochloride (Tris–HCl).

3. Methods

A typical work flow is shown in **Fig. 1**.

3.1. SILAC Labeling

1. Propagate the control and stimulated cell line(s) to be compared in a particular medium (e.g., DMEM). One cell line should be grown in medium containing light amino acids, whereas the other(s) should be grown in media containing one of the heavy amino acid combinations listed in **Table 1**.

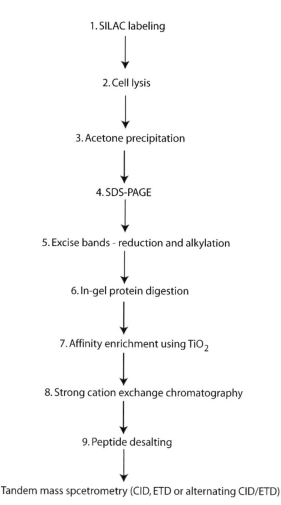

Fig. 1. A schematic of the steps involved in a representative SILAC-based phospho-proteomic analysis using TiO$_2$. The cells are lysed and the proteins precipitated with acetone. The precipitated proteins can be resolved by SDS-PAGE and the individual bands excised, proteins reduced, alkylated and digested with trypsin and extracted peptide from individual digests pooled. The phosphopeptides in the peptide mixture are enriched using TiO$_2$, and subjected to SCX-based fractionation followed by desalting. Finally, the samples can be analyzed by LC-MS/MS on a tandem mass spectrometer in the CID mode, ETD mode or alternating CID/ETD mode.

2. Each cell line must first be adapted to ensure complete incorporation prior to initiating the experiment (*see* **Note 1**). In general, five passages are sufficient for this purpose.

3.2. Cell Lysis

1. Replace the medium with serum free medium and leave the cells in culture for an additional 12–18 h.

2. Wash the cells twice with PBS and stimulate one of the cell populations with the required amount of the ligand (e.g., EGF).

3. Wash the cells three times with cold PBS.

4. In the case of 150-mm cell culture plates, place these on ice and rinse them with 15 mL of pre-chilled lysis buffer, which should be freshly prepared and with phosphatase inhibitors (*see* **Note 2**).

5. Dislodge the cells gently from the culture plate with a disposable cell scraper and place the liquid in a 50-mL centrifuge tube. Allow the tube to rock slowly for an hour at 4°C.

6. Centrifuge the cell lysate at $12,000 \times g$ at 4°C to remove the insoluble debris.

3.3. Acetone Precipitation

1. Precipitate the proteins from the lysate with a 5× volume of cold acetone, at –20°C, overnight.

2. Recover the precipitate by spinning the acetone/lysate mixture at $(12,000–15,000 \times g)$ for 15 min at 4°C (*see* **Note 3**).

3. Remove the supernatant carefully and leave the pellet to dry in a clean hood.

4. Recover the protein pellet with 200 μL of dissolution buffer and incubate the sample at 30°C with constant mixing and occasional vortexing.

5. Measure the total protein concentration for each sample with the Bio-Rad Protein assay following the manufacturer's instructions.

6. Prepare aliquots of 100–200 μg total protein in 1.5-mL micro centrifuge tubes.

7. Flash-freeze the samples in liquid nitrogen, and store them at –80°C if they are not to be used immediately.

8. Reduce, alkylate and use for in-solution trypsin digestion (*5*).

3.4. SDS-Polyacrylamide Gel Electrophoresis (SDS-PAGE)

1. For each sample, mix 100–200 μg of total protein in a volume of 10–15 with 5 μL of SDS-PAGE loading buffer.

2. Heat the samples at 95°C for 5 min.

3. Allow the samples to cool at room temperature for 5 min.

4. Load the samples onto a pre-cast mini-gel, assembled on a Novex mini-cell with SDS-PAGE running buffer.

5. Perform electrophoresis at 200 V until the loading buffer stain reaches the bottom of the gel.

6. Transfer the gel to a clean and covered container (*see* **Note 4**).

7. Wash the gel with H_2O twice briefly.

8. Stain the gel with colloidal Coomassie blue.

3.5. Reduction and Alkylation

1. Excise each gel lane in small pieces of about 1 mm.

2. Transfer the gel cubes in a clean 1.5-mL micro centrifuge tube.

3. Destain the gel pieces with 1:1 of a 50 mM NH_4CO_3/acetonitrile mixture for at least 2 hours.

4. Wash the samples step-wise with H_2O and H_2O/acetonitrile for 15 min each time. Repeat this step once.

5. Dehydrate the gel pieces with acetonitrile 100% for 15 min.

6. Swell the dehydrated gel pieces with 1 mL of reduction solution, and incubate for 60 min at 60°C.

7. Leave the reaction at room temperature to cool down.

8. Discard the reduction solution.

9. Add 1 mL of alkylation solution per tube, incubating the samples at room temperature and protected from light, for 30 min.

3.6. In-Gel Protein Digestion

1. Wash the samples step-wise with H_2O and H_2O/acetonitrile as in **Subheading 3.5, step 4**.

2. Dehydrate the gel pieces with acetonitrile 100% as in **Subheading 3.5, step5**.

3. Dissolve a lyophilized vial of trypsin with pre-chilled 50 mM NH_4CO_3 at a concentration of 10–15 ng/µL (*see* **Note 5**).

4. Swell the dehydrated gel pieces with freshly dissolved trypsin on ice, using a volume slightly beyond that required to cover the dry gel pieces in each tube.

5. Incubate the samples for 45 min on ice.

6. Remove the excess of liquid, and replace it with approximately the same volume of 50 mM NH_4CO_3, but without trypsin.

7. Allow the reaction to proceed at 37°C for 12–14 h.

8. Stop the trypsin digestion tube with a 1:5 v/v aliquot of 10% acetic acid.

9. Transfer the supernatant to a clean 1.5-mL micro centrifuge tube.

10. Extract the peptides remaining in the gel matrix three times by incubating with 0.1% acetic acid, 0.1% acetic acid in 50% acetonitrile and 100% acetonitrile, respectively.

11. For each sample, transfer the peptides from above wash and dehydration steps to the tube in which the supernatant was collected **Subheading 3.6, Step 8**.

12. Dry each sample using vacuum centrifugation.

3.7. Affinity Enrichment Using TiO₂

1. Redissolve the dried samples in 1 mL of TiO_2 loading solution.

2. Apply vigorous vortex to assure the recovery of the sample and spin briefly to remove insoluble particles.

3. Transfer 10 μL of a 1/1 H_2O/TiO_2 slurry per sample to a 1.5-mL clean micro centrifuge tube.

4. Make a suspension with the TiO_2 slurry in each tube, by adding 1 mL of TiO_2 loading solution.

5. Equilibrate the TiO_2 slurry for 15 min with slow rocking.

6. Spin the TiO_2 slurry briefly at low speed.

7. Discard the TiO_2 loading solution used for equilibration.

8. To each of the redissolved samples as in **Subheading 3.7, step 1**, add 10 μL of equilibrated TiO_2 slurry.

9. Allow the phosphopeptides to bind to the TiO_2 material for 30 min in slow rocking (*see* **Note 6**).

10. Spin the samples briefly.

11. Transfer the supernatant, which is the nonbound peptide solution to a 1.5-mL clean micro centrifuge tube.

12. Carefully redissolve the TiO_2 pellet in 100 μL of TiO_2 washing solution.

13. Incubate the above slurry for 15 min with slow rocking.

14. Repeat the previous step three times.

15. Carefully dissolve the TiO_2 pellet with 100 μL of 25% NH_4OH at pH 10.5.

16. Incubate the slurry for 15 min with slow rocking.

17. Repeat the previous step once more to assure complete elution of the bound peptides.

18. Bring the pH of the 200 μL of phosphopeptide elution to pH 7.5 by adding appropriate amount of 10% acetic acid.

19. Dry the samples obtained in **Subheading 3.7, step 11** (nonbound peptide solution) and **Subheading** 3.7, **step 18** (phosphopeptide elution) in a vacuum centrifuge.

20. Dissolve peptides in LC-MS solvent and use for phosphopeptide analysis.

3.8. Strong Cation Exchange (SCX) Chromatography

1. The in-solution digested (**Subheading 3.3, Step 8**) samples enriched with TiO_2 are redissolved in 1 mL of SCX-loading solution.

2. Inject the sample onto the SCX column and start the chromatographic experiment.

3. For an off-line LC-system coupled to an auto-sampler, the SCX gradient is as follows:

   – 70 min run starting with 0% solution B

   – 0–20 min, isocratic 0% solution B

   – 20–22 min, 0–8% solution B

   – 22–48 min, 8–35% solution B

– 48–58 min, 35–100% solution B

– 58–68 min, isocratic 100% solution B

– 68–69.5 min, 100–0% solution B

– 69.5–70 min, 0% solution B End

Peptides can be fractionated using strong cation exchange chromatography on PolySULFOETHYL A column (PolyLC, Columbia, MD) (100 × 2.1 mm, 5-μm particles with 300 Å pores). SCX chromatography can be performed using Ultimate HPLC system (LC Packings) connected to a Probot fraction collector. SCX fractions (0.5 mL) can be collected using a non-linear gradient of 0–350 of SCX solution B. First 20 min, SCX solution A is injected for cleaning SCX bound peptides and subsequently fractionation is done using 0–35% SCX solution B for 30 min, increasing solution B to 100% for further 15 min duration.).

4. Collect the elution aliquots in several fractions, depending on the chromatographic profile.

5. Dry each SCX fraction using vacuum centrifugation.

3.9. Peptide Desalting   The final step before LC-MS/MS is to desalt the SCX fractions by batchwise reversed-phase chromatography with Poros 20 R2 material.

1. Redissolve the dried samples obtained above in **Subheading 3.8**, **step 5** with 1 mL of loading RP-C solution.

2. Apply vigorous vortex to assure the recovery of the sample.

3. Transfer a small amount of Poros 20 R2 per sample to a 1.5 mL clean micro centrifuge tube.

4. Make a suspension with the Poros 20 R2 in each tube, by adding 1 mL of RP-C loading solution.

5. Equilibrate the Poros 20 R2 material for 15 min with slow rocking.

6. Spin the Poros 20 R2 slurry briefly at low speed.

7. Discard the loading solution used for equilibration.

8. Add each of the redissolved samples to an aliquot of equilibrated Poros 20 R2.

9. Incubate for 1 h and slow rocking.

10. Spin the samples briefly.

11. Discard the supernatant.

12. Carefully redissolve the Poros 20 R2 pellets in 100 μL of RP-C washing solution.

13. Incubate for 15 min with slow rocking.

14. Repeat the previous step three times.

15. Carefully dissolve the Poros 20 R2 pellet with 100 μL of RP-C elution solution.

16. Incubate the slurry for 15 min with slow rocking.

17. Repeat the previous step once more to assure complete elution of the bound peptides.

18. Dry each desalted sample with a vacuum centrifuge.

19. When needed, redissolve the samples in an appropriate LC-MS/MS loading solution.

4. Notes

1. The appropriate labeling should be monitored by analyzing an aliquot of each cell line by LC-MS/MS.

2. To each 50 mL of this buffered solution, 100 μL of cocktail 1 and 2 phosphatase inhibitors are used. Also a tablet of EDTA-free protease inhibitors is added. An excess of inhibitors is important to diminish protease degradation and in particular dephosphorylation reactions during sample processing.

3. Glass or plastic acetone resistant centrifuge tubes should be used.

4. It is important to proceed with all of the following steps in a clean hood and to use sterile material. Powder-free gloves and a coat should be worn. These precautions will diminish the chances of contaminating the samples with traces of keratin.

5. Use fresh lyophilized trypsin dissolved in cold buffer and keep it on ice at all times, before adding an aliquot to the dry gel pieces. This will prevent unwanted auto-proteolysis of the enzyme.

6. To enrich the phosphopeptides use a batchwise protocol with TiO_2 self-packing material. Both the nonbound eluate and the concentrated phosphopeptide elution are analyzed by LC-MS/MS. The classic approach of self-made micro-columns may also be used, although it is not described in this chapter (6).

References

1. Glavy, J.S., et al. (2007) Cell-cycle-dependent phosphorylation of the nuclear pore Nup107–160 subcomplex. *Proc Natl Acad Sci U S A.* 104, 3811–6.

2. Molina, H., et al. (2007) Global proteomic profiling of phosphopeptides using electron transfer dissociation tandem mass spectrometry. *Proc Natl Acad Sci U S A.* 104, 2199–204.

3. Olsen, J.V., et al. (2006) Global, in vivo, and site-specific phosphorylation dynamics in signaling networks. *Cell.* 127, 635–48.

4. Molina, H., et al. (2009) Temporal profiling of the adipocyte proteome during differentiation

using a 5-plex SILAC based strategy. *J Proteome Res.* 8, 48–55.

5. Harsha, H.C., Molina, H., Pandey, A. (2008) Quantitative proteomics using stable isotope labeling with amino acids in cell culture. *Nat Protoc.* 3, 505–16.

6. Rappsilber, J., Ishihama, Y., Mann, M. (2003) Stop and go extraction tips for matrix-assisted laser desorption/ionization, nanoelectrospray, and LC/MS sample pre-treatment in proteomics. *Anal Chem.* 75, 663–70.

Chapter 10

Quantitative Phospho-proteomics Based on Soluble Nanopolymers

Anton Iliuk and W. Andy Tao

Summary

Phospho-proteomics, the global analysis of protein phosphorylation, holds great promise for the discovery of cell signaling events that link changes in dynamics of protein phosphorylation to the progression of various diseases, particularly cancer and diabetes. Mass spectrometry has become a powerful tool for identification and global profiling of protein phosphorylation. However, even with continuously improving sensitivity of mass spectrometers, sub-stoichiometric nature of phosphorylation poses enormous challenges for phosphoprotein identification and, particularly, mapping phosphosites. Therefore, a successful mass spectrometry-based phosphoproteomic experiment depends largely on an effective method of phosphopeptide enrichment.

We present in this chapter two robust methods based on soluble nanopolymers functionalized for phosphopeptide enrichment. The first method describes the formation of reversible phosphoramidate bonds between amines on the nanopolymer and phosphate groups on peptides, thus enabling selective isolation of phosphopeptides using a molecule size-based filtering device. The second technique is based on the selective chelation of phosphopeptides to zirconia or titania functionalized nanopolymer, which can be isolated from the complex peptide mixture by binding the nanopolymer to solid-phase support through efficient hydrazide chemistry. Combined with stable isotope labeling approaches, both strategies provide reproducible and efficient meanings for quantitative phospho-proteomics.

Key words: Phospho-proteomics, Phosphopeptide enrichment, Dendrimer, Quantitation, Phosphoramidate, Zirconia.

1. Introduction

Proteomic approaches to phosphorylation usually involve selective isolation of phosphopeptides and subsequent fragmentation in a mass analyzer to identify the peptide sequence and phosphosites. Efficient isolation of phosphopeptides has been considered as a

Marjo de Graauw (ed.), *Phospho-Proteomics, Methods and Protocols, vol. 527*
© 2009 Humana Press, a part of Springer Science+Business Media, New York, NY
Book DOI: 10.1007/978-1-60327-834-8_10

crucial step toward low false identification and high sensitivity. There are three main strategies to enrich phosphopeptides: antibody affinity *(1)*, chemical derivatization *(2)*, and metal-ion affinity *(3)*. Among these approaches, immobilized metal-ion affinity chromatography (IMAC) based on Fe (III) *(4)* and Ga (III) *(5)*, and more recently on metal oxides, such as TiO_2 *(6)* and ZrO_2 *(7)*, is the most commonly used enrichment method for its simplicity. However, the efficiency and specificity of IMAC appears to be highly dependent on the solid phase support and buffer condition for binding and elution. Chemical derivatization, on the other hand, has the potential to be very specific but is limited by relatively long process. Therefore, there is a need to develop more reproducible and robust techniques to aid in phosphoproteomic studies.

We propose two novel strategies for phosphopeptide enrichment, which attempt to address the above problems with increasing specificity and reproducibility with relatively simple procedures (**Fig. 1**). The binding/reaction occurs in homogeneous solution, thus improving the linear kinetics and distribution of phosphopeptide isolation. Both methods are based on soluble nanopolymers (e.g., polyamidoamine (PAMAM) dendrimer), which can be functionalized to selectively target phosphate groups on peptides. Soluble nanopolymers, such as dendrimer, have high structural and chemical homogeneity, compact spherical shape, high branching, and controlled surface functionalities *(8)*. They also possess a valuable ability to cross cell membranes with low cytotoxicity, which has been utilized to deliver genomic materials *(9)*, drugs *(10)*, and vaccines *(11)* into cells. Therefore, soluble nanopolymers possess unique qualities necessary for biochemical tools.

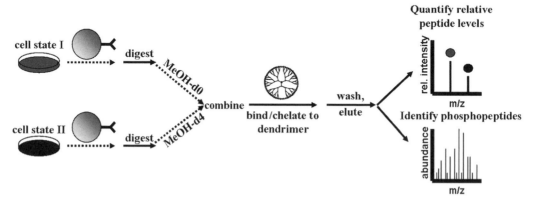

Fig. 1. General strategy for soluble nanopolymer-based phospho-proteomics. Cells from two different states are lysed (as an option, phosphotyrosine-containing proteins can be isolated by immuno-affinity using phosphotyrosine antibodies attached to beads). Resulting proteins are digested, and peptides are optionally methylated with either d_0- or d_4-methanol for selectivity and quantitation (alternatively, stable isotope labeling can be introduced during cell culture (e.g., SILAC), or on the peptide stage (e.g., iTRAQ)). Peptide mixtures are combined and incubated with soluble nanopolymer for chemical derivatization or chelation of the phosphorylated peptides on the polymer, thus, facilitating the isolation of phosphopeptides from complex peptide mixtures. The retained phosphopeptides are eluted off the nanopolymer, followed by MS analysis for the identification and quantification of phosphopeptides. (*Dashed arrows* represent optional steps).

The first technique utilizes a well-described derivatization of phosphates to phosphoramidates by coupling phosphopeptides to amine groups on the nanopolymer (PAMAM dendrimer) *(12)*. The dendrimer with bound phosphopeptides can be then isolated by using a size-selective membrane-based filter device, thus removing the unbound peptides. Mild acid hydrolysis detaches phosphopeptides from the dendrimer for mass spectrometry (MS) analysis.

The second technique is based on a strong and selective chelation of phosphate group to zirconium or titanium oxide immobilized on soluble nanopolymers in a homogeneous solution. Soluble nanopolymer (PAMAM dendrimer G4) is functionalized with zirconia or titania to form stable bidentate with (usually singly-) phosphorylated peptides. The isolation of phosphopeptide-bound nanopolymer is achieved using an efficient bioconjugation between the coupling groups on the dendrimer and solid-phase beads. After stingy washing steps, phosphopeptides are eluted under mild basic condition and analyzed with MS.

Both methods can be coupled with stable isotope labeling steps *(13)* for quantitative phospho-proteomics or with a phosphotyrosine antibody-based immuno-affinity step to enable more robust evaluation of less frequent tyrosine phosphorylation *(12)*.

2. Materials

2.1. Cell Culture and Lysis

1. RPMI medium 1640, supplemented with 10% fetal bovine serum (FBS), 2 µM L-glutamine, 100 µg/mL streptomycin sulfate, and 100 units/mL penicillin G.

2. 50 µM pervanadate solution, prepare freshly from sodium orthovanadate, okadaic acid and H_2O_2 solutions.

3. 10 mL lysis buffer: 1% Triton, 150 mM NaCl, 50 mM Tris–HCl (pH 7.8); store at 4°C.

4. Tyrosine enrichment (if necessary): agarose-conjugated 4G10 phosphotyrosine monoclonal antibodies, mixture of 50 mM phenylphosphate and 200 mM NaCl.

5. Microsep (10 K) centrifugal filter device (Pall Filtron Co.).

2.2. Protein Denaturation and Isotope Tagging

1. 20 mM ammonium bicarbonate (pH 8.0) containing 0.1% RapiGest (Waters Co.). RapiGest is freshly prepared in water and used immediately (*see* **Note 1**).

2. 200 mM dithiothreitol is dissolved in water and stored in single-use aliquots at –20°C.

3. 300 mM iodoacetamide is freshly prepared in water and used immediately.

4. 1 μg trypsin is dissolved in trypsin resuspension buffer (or 50 mM acetic acid) and stored at –20°C.

5. MCX desalting column (Waters Co.).

6. 75 μL methanolic HCl is prepared at 12°C by adding 100 μL of acetyl chloride to 500 μL of anhydrous methanol-d_0 or -d_4 (Cambridge Isotope Laboratories); use immediately.

2.3. Phosphopeptide Enrichment (Method 1)

1. 40-μL reaction solution: 50 mM 1-ethyl-3-(3-dimethylaminopropyl) carbodiimide hydrochloride EDC (*see* **Note 2**), 100 mM imidazole, 200 mM 2-(N-morpholino)ethanesulfonic acid (MES, pH 6.0), and 9 mg of PAMAM (polyamidoamine) dendrimer generation 5 (supplied by Sigma-Aldrich as 10% (wt/vol) solution in methanol; methanol is removed in vacuo); use immediately.

2. Biomax filter device (5 kDa cutoff; Millipore).

3. Washing solution 1: 500 μL of 3 M NaCl in water; store at room temperature.

4. Washing solution 2: 500 μL of 30% methanol in water; store at room temperature.

5. Elution solution: 10% trifluroacetic acid (TFA) in water; store at room temperature.

6. MS analysis solution: 0.1% formic acid.

2.4. Synthesis of Zirconia-Functionalized Dendrimer

1. PAMAM dendrimer generation 4 (Sigma-Aldrich; 200 μL of 10% (w/v) solution; methanol is removed in vacuo) dry and redissolve in 2 mL of dimethyl sulfoxide (DMSO); use on the same day.

2. "Handle" coupling mixture: dissolve 3.1 mg of Boc-aminooxyacetic acid (Sigma-Aldrich), 1.5 mg of hydroxybenzotriazole (HOBT), and 2 μL of 1,3-diisopropylcarbodiimide (DIPCI) in 1 mL DMSO; use on the same day.

3. Phosphonic acid coupling mixture: dissolve 16 mg of 2-carboxyethyl-phosphonic acid (Sigma-Aldrich), 16 mg N-hydroxy-succinamide (NHS; dissolved in 100 μL of water), and 160 mg EDC in 2 mL MES (pH 5.8); use immediately.

4. TFA (10%) in water; store at room temperature.

5. Zirconium oxychloride solution ($ZrOCl_2$; Alfa Aezar) – 100 mM in water; store at room temperature.

6. 95% TFA in water; store at room temperature.

7. Dialysis solution 1: 0.01% HCl in water; store at room temperature.

8. Dialysis solution 2: 0.1% HCl in 1:4 mixture of DMSO in water; store at room temperature.

9. Amicon Ultra centrifugal filter device (5 kDa MW cutoff; Millipore).

10. Snakeskin® pleated dialysis tubing (3,500 MWCO, 22 mm dry diameter, Pierce).

2.5. Synthesis of Aldehyde Beads

1. Reaction mixture: mix 200 mg of amine controlled pore glass (CPG) beads (NH2: 142 µM/g; Primer Synthesis) with 92 mg of Fmoc-serine-OH (dissolve in 500 µL *N,N*-dimehylformamide (DMF); Novabiochem), 38 mg of HOBT (hydroxybenzotriazole; dissolve in 500 µL DMF), and 90 µL DIPCI (1,3-diisopropylcarbodiimide).

2. Washing solutions: DMF (dimethylformamide) and dichloromethane (CH_2Cl_2)

3. Acetylation solution: mix 250 µL of DMF, 250 µL of CH_2Cl_2, 300 µL of pyridine, and 200 µL of acetic anhydride; use immediately.

4. De-protection solution: Mix 800 µL of DMF and 200 µL of piperidine; use immediately.

5. Oxidation solution: dissolve 8.5 mg sodium (meta)periodate ($NaIO_4$) in 200 µL of 40 mM mixture of acetic acid with sodium acetate; use immediately.

6. Oxidized beads washing solution: 0.1% TFA; store at room temperature.

7. Bio-spin® disposable chromatography column (Bio-Rad).

8. Micro-filter spin column (Bocascientific).

2.6. Phosphopeptide Enrichment (Method 2)

1. Loading buffer: mixture of 180 mM acetic acid, 180 mM sodium acetate, and 30% ethanol (adjust to pH 4.65 with sodium hydroxide); store at room temperature.

2. Washing solution: 1% acetic acid in 80% acetonitrile; store at room temperature.

3. Elution solution: 400 mM ammonium hydroxide (NH_4OH) in water; store at room temperature.

4. Spin columns with frit (Boca Scientific).

5. MS analysis solution: 0.1% formic acid.

3. Methods

3.1. Method 1: Phosphoramidate Chemistry on Soluble Nanopolymer (12)

In this method, phosphopeptide enrichment is based on a potent covalent derivatization between the phosphate groups and amines on the soluble nanopolymer (**Fig. 2**). The formation of phosphoramidate bond is catalyzed by EDC and imidazole. After the derivatization on the dendrimer, the mixture is transferred to a 5 kDa MWCO membrane-based spin column on which phosphopeptides bound on polymer are isolated from unbound peptides based

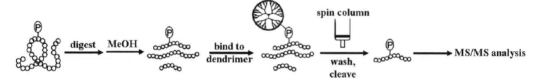

Fig. 2. Schematic illustration of phosphopeptide enrichment through phosphoramidate chemistry with PAMAM dendrimer. Proteins are digested and the carboxylic groups are methyl-esterified. Methylated peptides are subjected to a dendrimer conjugation through the phosphoramidate bond formation. Using a size-selective filtering device, the bound phosphopeptides are isolated from nonspecific peptides. Finally, methylated phosphopeptides are released from the dendrimer via acid hydrolysis, followed by mass spectrometry analysis.

on their size. Phosphopeptides can then be cleaved off the nano-polymer under acidic condition and collected using the same type of filtration device for mass spectrometry analysis.

The high specificity to isolate phosphopeptides is achieved through two steps. In the first step, the majority of nonphosphopeptides are methyl esterified to prevent the reaction between carboxylate and amine groups. A stable isotopic tag can also be introduced concurrently with either light methanol (CH_3OH) or deuterated methanol (CD_3OD). In the second step, only phosphopeptides can be detached from the dendrimer through the acid-labile phosphoramidate bond. Examples of phosphopeptide isolation from β-casein, as well as identification and quantification of a doubly tyrosine-phosphorylated peptide, are illustrated in **Fig. 3**.

3.1.1. Sample Preparation

1. Jurkat cells (two sets) are grown in RPMI medium 1640 in a 5% CO_2 incubator at 37°C.

2. Cells are stimulated with fresh sodium pervanadate and lysed in 10-mL lysis buffer (*see* **Note 3**).

3. Lysed cells are centrifuged down at $16,000 \times g$ and supernatant is collected (this step gets rid of insoluble fractions of the cell).

4. If necessary, at this point phosphotyrosine-containing proteins can be isolated by subjecting cellular supernatant to anti-phosphotyrosine immunoprecipitation with 2 mL of agarose-conjugated 4G10 monoclonal antibodies at 37°C for 2 h. The beads are then washed with lysis buffer, and bound phosphotyrosine proteins are eluted off five times with 50 mM phenylphosphate, 200 mM NaCl; the eluents are combined.

5. Proteins are concentrated with Microsep centrifugal filter device, collected and dried.

3.1.2. Protein Digestion and Isotopic Tagging

1. Resolubilize the protein mixture in 20 mM ammonium bicarbonate (pH 8.0) with 0.1% RapiGest and heat the solution at 95°C for 5 min to denature the proteins.

2. Reduce and alkylate proteins with 5 mM dithiothreitol for 30 min at 37°C, and with 15 mM iodoacetamide for 30 min at room temperature in the darkness, respectively.

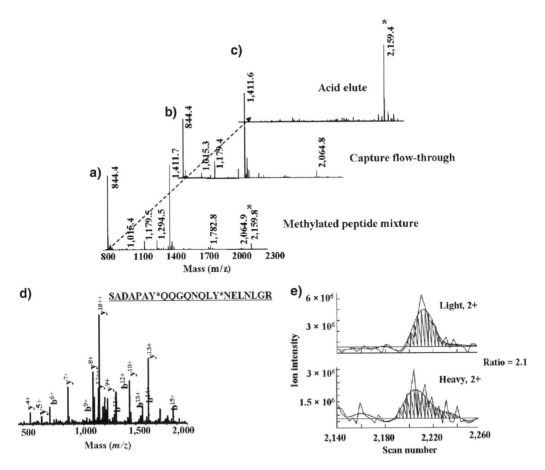

Fig. 3. Isolation via nanopolymer coupling and MS/MS analysis of phosphopeptides. (a–c) Enrichment of phosphopeptide FQS*EEQQQTEDELQDK from digested β-casein. Spectra were obtained using MALDI-TOF/TOF spectrometer. (a) Starting amount of material for isolation is 100 pmol of β-casein digests, of which 0.5 pmol was used for MS analyses. Ions from β-casein are labeled; *asterisk* indicates a phosphopeptide. (b) Flowthrough collected, indicating efficient binding and isolation of the phosphopeptide. (c) Cleaved methylated phosphopeptide, indicating clean phosphopeptide enrichment with little detectable nonspecific peptide ions. (d) MS/MS analysis of the CD3ζ-drived phosphopeptide SADAPAY*QQGQNQLY*NELNLGR after immuno-affinity-based isolation of tyrosine-phosphorylated proteins, followed by nanopolymer-based phosphopeptide enrichment. Its characteristic peptide bond fragment ions, type b and type y ions, are labeled. (e) Quantitation of the same phosphotyrosine peptide. Reconstructed ion chromatograms are shown of the precursor ion in its light and heavy versions using the ASAPRatio program. Smoothed chromatograms are depicted, which are used for peak ratio quantitation. (Reproduced from ref. *12*).

3. Digest proteins with 1 μg trypsin (~1:100 ratio of trypsin to proteins) overnight at 37°C.

4. Desalt the peptides using the MCX column (following the manufacturer's protocol) and lyophilize the peptides.

5. 500 μL anhydrous methanol, -d$_0$ or -d$_4$, is first cooled down to 12°C, and 50 μL of acetic chloride is slowly added and the solution is incubated for 5 min at 12°C (*see* **Note 4**).

6. Dried peptides (from two samples to be quantified) are dissolved in 75 μL of methanolic HCl -d0 and -d4, respectively,

and methyl esterification is allowed to proceed at 12°C for 90 min.

7. Dry methyl-esterified peptides using SpeedVac concentrator (Thermo Savant).

3.1.3. Phosphopeptide Enrichment

1. Resolubilize dried peptides in 40 μL of reaction solution, containing 50 mM EDC, 100 mM imidazole, 200 mM MES (pH 6.0), and 9 mg of PAMAM dendrimer generation 5; the reaction is allowed to proceed at room temperature for 10 h with vigorous shaking to couple phosphorylated residues to amines on the dendrimer.

2. Transfer the reaction mixtures into a membrane-based spin column (5 kDa MWCO) and wash with 500 μL of 3 M NaCl, 30% methanol and water several times to remove unbound peptides (discard filtrates).

3. Cleave the bound phosphopeptides off the dendrimers with 10% TFA at room temperature for 30 min and collect the filtrate.

4. Wash the dendrimer twice with 100 μL of 30% methanol and combine all filtrates.

5. Dry the phosphopeptides using SpeedVac concentrator.

3.1.4. Mass Spectrometry-Based Phosphopeptide Analysis

1. Resolubilize the dried phosphopeptides in 0.1% formic acid and pressure-load them on a capillary reverse-phase C_{18} column (75-μm inner diameter and 12 cm of bed length).

2. Analyze the peptides by micro flow LC-MS. Peptides are eluted at a flow rate of 200–300 nL/min to the mass spectrometer through an integrated electrospray emitter tip.

3. Instruments are operated in data-dependant mode of one MS scan followed by several (3–10) MS/MS scans.

4. MS data are searched against International Protein Index (IPI) database using the SEQUEST algorithm.

5. Static modifications are put into search parameters to include light or heavy modifications on Asp, Glu, and C-terminal (+14 and +17 accordingly), as well as alkylated Cys residue (+57).

6. Variable modifications of +80 are allowed on Ser, Thr, and Tyr during the database search to identify phosphorylated peptides.

7. Protein probability score is used to filter the research results with additional manual validation.

3.2. Method 2: Immobilized Metal Oxide Affinity Isolation on Soluble Nanopolymers

In this method, phosphopeptide enrichment is based on the strong affinity of zirconium oxide to the phosphate groups (**Fig. 4**). PAMAM dendrimer G4 (generation 4) is bi-functionalized by ZrO_2-phosphonate group and hydrazide ("handle") group. The affinity binding in the homogeneous solution can be completed in less than 5 min. Following the brief reaction of the reagent with

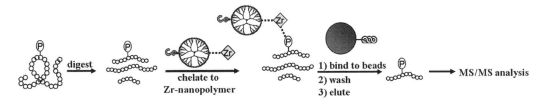

Fig. 4. Schematic illustration of affinity-based phosphopeptide enrichment with ZrO_2-functionalized nanopolymer. Proteins are digested, and the resulting peptides undergo short incubation with zirconia-functionalized nanopolymer, resulting in strong and selective chelation between zirconia and phosphopeptides. The nanopolymer is then covalently bound to the aldehyde-functionalized beads using the hydroxylamine "handle," facilitating the isolation of phosphopeptides. The retained phosphopeptides are eluted off under basic condition for MS analysis.

the peptide mixture, the bound peptides are isolated by binding the dendrimer onto functionalized beads using a specific "handle" and nonphosphopeptides are removed with stingy washing steps. Finally, the bound phosphopeptides are eluted under basic conditions and analyzed using mass spectrometry. This enrichment strategy has high recovery yield with relatively low contamination. A stable isotope labeling step (e.g., iTRAQ (**14**)) can be incorporated in the procedure, allowing quantitative analysis of two or more complex samples.

3.2.1. Synthesis of Zirconia-Functionalized Dendrimer

1. Dry 200 µL of PAMAM dendrimer generation 4 solution (provided as 10% (wt/vol) in methanol) in a microfuge tube.

2. Resolubilize dried dendrimer in 2-mL DMSO and transfer into a 10-mL round-bottom flask with a magnetic stir bar (*see* **Note 5**).

3. In a microfuge tube, add 3 mg of Boc-amino-oxyacetic acid, 1.5 mg of HOBT, and 2 µL of DIPCI; dissolve in 1 mL of DMSO and incubate for 30 min at room temperature. Add the mixture into the round-bottom flask containing dendrimer and stir overnight (*see* **Note 6**).

4. Dialyze the solution against water for 7–8 h to remove any remaining unreacted reagents (replace water periodically).

5. Transfer the solution into an Amicon Ultra centrifugal filter device and concentrate it to a volume under 2 mL (*see* **Note 7**). Bring up the mixture volume to 2 mL with water and transfer into a clean 10-mL round-bottom flask with a stir bar.

6. Add 1 mL of 250 mM MES (pH 5.8), 16 mg 2-carboxyethylphosphonic acid, 16 mg NHS and 160 mg EDC into the flask, and stir overnight to functionalize the dendrimer with phosphonic acid. (*see* **Note 8**).

7. Acidify the solution with 3 mL of 10% TFA and stir for 1 h to remove any phosphonic acid groups bound to amines on the dendrimer.

8. Dialyze the solution against water to remove any unreacted reagents (replace water periodically).

9. Concentrate again using Amicon filter tubes to the final volume of 1.5 mL (*see* **Note 9**). At this point, the mixture should be stored at 4°C and the remaining synthesis steps can be done multiple times using 300 μL of the mixture at one time.

10. Mix 300 μL of the above solution and 300 μL of 100 mM zirconium oxychloride ($ZrOCl_2$) and incubate for 1 h with agitation at room temperature to chelate zirconia with phosphonic acid groups on the dendrimer.

11. Dialyze the solution against 0.01% HCl to remove any unbound zirconia (replace dialysis solution periodically), and dry the solution under vacuum using SpeedVac concentrator.

12. Add 200 μL of 95% TFA into the dried sample and incubate for 1 hour with agitation at room temperature to remove the Boc group.

13. Dry the sample using nitrogen positive flow, and resolubilize in 1 mL of 0.1% HCl in 4:1 mixture of water in DMSO; incubate sample at room temperature until the precipitate dissolves (*see* **Note 10**). Dialysis the final product against 0.1% HCl in 4:1 mixture of water in DMSO. Dry the solution under vacuum and resolubilize in 300 μL of 0.01% HCl and stored at 4°C for further use.

3.2.2. Synthesis of Aldehyde Beads

1. Transfer 200 mg of controlled pore glass (CPG) beads into a Bio-spin® disposable chromatography column.

2. Dissolve 92 mg of Fmoc-serine-OH, 38 mg of HoBT (hydroxybenzotriazole), and 90 μL of DIPCI (1,3-diisopropylcarbodiimide) in 500 μL of DMF (dimethylformamide). Add the solution into the above-mentioned column and rotate the column end-over-end overnight at room temperature to couple Fmoc-serine group to the beads.

3. Wash the beads with 4 mL of DMF. Add sequentially 250 μL of DMF, 250 μL of dichloromethane, 300 μL of pyridine, and 200 μL of acetic anhydride to the beads to block the remaining amines; rotate end-over-end for 1 h at room temperature.

4. Wash the beads three times with 1 mL of dichloromethane and three times with 1 mL of DMF. Add a mixture of 800 μL of DMF and 200 μL of piperidine to remove the Fmoc group; rotate end-over-end for 1 h at room temperature.

5. Wash the beads three times with 1 mL of DMF and three times with 1 mL of dichloromethane. The resulting serine beads should be dried completely using SpeedVac concentrator and stored at 4°C.

6. Oxidation: transfer 5–7 mg of serine beads into a frit-based spin column and add 200 μL of oxidation solution (8.5 mg of sodium (meta)periodate in 200 μL of 40 mM acetic acid/sodium acetate solution) and incubate for 30 min with agitation in the dark at room temperature. The final aldehyde beads should be used on the same day.

3.2.3. Sample Preparation

1. Peptide samples are prepared using the protocols described in **Subheading 3.1.1, steps 1–3** and **Subheading 3.1.2, steps 1–3**. (*see* **Note 11**); for quantitative analysis, peptides can be isotopically tagged using standard iTRAQ reagents and procedures *(14)*.

3.2.4. Phosphopeptide Enrichment

1. Peptide mixtures are resuspended in 50 μL of 180 mM acetic acid/sodium acetate buffer in 30% ethanol (pH 4.65).

2. Dry 20 μL of synthesized zirconia-functionalized dendrimer solution (*see* **Subheading 3.2.1**) and resolubilize in 50 μL of 180 mM acetic acid/sodium acetate buffer in 30% ethanol (pH 4.65). (*see* **Note 12**).

3. Mix the functionalized dendrimer solution with the peptide sample and incubate for 3 min with agitation at room temperature to allow chelation of phosphopeptides to zirconia-functionalized nanopolymer.

4. Wash the aldehyde beads (oxidized on the same day) with 200 μL of 0.1% TFA. Add the above reaction mixture into the beads and incubate for 1 h with vigorous agitation at room temperature.

5. Wash the beads with 100 μL of loading buffer, 100 μL of 1% acetic acid in 80% acetonitrile solution, and 100 μL of water.

6. Elute the bound phosphopeptides by incubating the beads with 100 μL of 400 mM ammonium hydroxide (NH_4OH) for 5 min (twice) with agitation at room temperature. Dry the elute under vacuum and resolubilize in 0.1% formic acid (*see* **Note 13**) for mass spectrometry analysis.

3.2.5. Mass Spectrometry Analysis of Phosphopeptides

Mass spectrometry identification of the phosphopeptides can be achieved using the protocol in **Subheading 3.1.4**.

4. Notes

1. RapiGest should be stored dry in single-use aliquots at −20°C.

2. EDC is moisture-sensitive and its solution should be prepared in fresh.

3. This lysis procedure can be applied to other cell cultures.

4. Special care is needed when adding acetyl chloride to anhydrous methanol due to extreme exothermic reaction. Slowly add acetyl chloride into cooled methanol with agitation.

5. It might be necessary to incubate dendrimer with DMSO for 20–30 min with agitation at room temperature to fully dissolve dendrimer in DMSO.

6. Ninhydrin test may be used to monitor the existence of free amine groups on dendrimers: 10 µL of functionalized dendrimer solution is mixed with 50 µL of 1% ninhydrin in ethanol solution. Heat the mixture at 90°C for ~2 min and check solution color (change to purple or blue color indicates presence of free primary amines).

7. Amicon filter device's volume capacity is only 4 mL; therefore, it is necessary to concentrate all of the solution multiple times on the same filter to avoid loss of reagent.

8. At this point, there should be no free amine groups left (ninhydrin test should be negative – no color change).

9. A small portion of amine groups on dendrimer are modified with phosphonic acid and the reaction is reversible under acidic condition. Ninhydrin test should now show light purple (for ninhydrin test in this step, 10 µL of sample is resuspended in 50 µL of 180 mM acetic acid/sodium acetate buffer (pH 4.6–6.6) before adding ninhydrin). Ninhydrin test may not be effective under low pH condition.

10. Do not proceed to the next step until all the precipitate is dissolved (adding more DMSO might potentially help dissolve the solid).

11. Desalting peptides is unnecessary since beads-washing steps (**Subheading 3.2.4**, steps **4–5**) will remove most of the salt.

12. It is critical to maintain the right pH for the nanopolymer solution.

13. Formic acid is used here as an excellent ion pairing agent for LC-MS analysis.

Acknowledgments

This work was supported in part by Purdue University (funding through Purdue Doctoral Fellowship to A.I.), the National Scientific Foundation CAREER award, and American Society for Mass Spectrometry (ASMS). We acknowledge the use of the software in the Institute for Systems Biology developed using Federal funds from the National Heart, Lung, and Blood Institute, National Institutes of Health, under contract No. N01-HV-28179.

References

1. Zolodz, M.D., Wood, K.V., Regnier, F.E., and Geahlen, R.L. (2004) New approach for analysis of the phosphotyrosine proteome and its application to the chicken B cell line, DT40. *J Proteome Res.* 3, 743–750.

2. Zhou, H., Watts, J.D., and Aebersold, R. (2001) A systematic approach to the analysis of protein phosphorylation. *Nat Biotechnol.* 19, 375–378.

3. Nuhse, T.S., Stensballe, A., Jensen, O.N., and Peck, S.C. (2003) Large-scale analysis of in vivo phosphorylated membrane proteins by immobilized metal ion affinity chromatography and mass spectrometry. *Mol Cell Proteomics.* 2, 1234–1243.

4. Stensballe, A., Andersen, S., and Jensen, O.N. (2001) Characterization of phosphoproteins from electrophoretic gels by nanoscale Fe(III) affinity chromatography with off-line mass spectrometry analysis. *Proteomics.* 1, 207–222.

5. Posewitz, M.C. and Tempst, P. (1999) Immobilized gallium(III) affinity chromatography of phosphopeptides. *Anal Chem.* 71, 2883–2892.

6. Larsen, M.R., Thingholm, T.E., Jensen, O.N., Roepstorff, P., and Jorgensen, T.J. (2005) Highly selective enrichment of phosphorylated peptides from peptide mixtures using titanium dioxide microcolumns. *Mol Cell Proteomics.* 4, 873–886.

7. Zhou, H., Tian, R., Ye, M., Xu, S., Feng, S., Pan, C., Jiang, X., Li, X., and Zou, H. (2007) Highly specific enrichment of phosphopeptides by zirconium dioxide nanoparticles for phosphoproteome analysis. *Electrophoresis.* 28, 2201–2215.

8. Boas, U. and Heegaard, P.M. (2004) Dendrimers in drug research. *Chem Soc Rev.* 33, 43–63.

9. Kukowska-Latallo, J.F., Bielinska, A.U., Johnson, J., Spindler, R., Tomalia, D.A., and Baker, J.R., Jr. (1996) Efficient transfer of genetic material into mammalian cells using Starburst polyamidoamine dendrimers. *Proc Natl Acad Sci U S A.* **93**, 4897–4902.

10. Sakharov, D.V., Jie, A.F., Bekkers, M.E., Emeis, J.J., and Rijken, D.C. (2001) Polylysine as a vehicle for extracellular matrix-targeted local drug delivery, providing high accumulation and long-term retention within the vascular wall. *Arterioscler Thromb Vasc Biol.* 21, 943–948.

11. Patri, A.K., Myc, A., Beals, J., Thomas, T.P., Bander, N.H., and Baker, J.R., Jr. (2004) Synthesis and in vitro testing of J591 antibody-dendrimer conjugates for targeted prostate cancer therapy. *Bioconjug Chem.* 15, 1174–1181.

12. Tao, W.A., Wollscheid, B., O'Brien, R., Eng, J.K., Li, X.J., Bodenmiller, B., Watts, J.D., Hood, L., and Aebersold, R. (2005) Quantitative phosphoproteome analysis using a dendrimer conjugation chemistry and tandem mass spectrometry. *Nat Methods.* 2, 591–598.

13. Ficarro, S.B., McCleland, M.L., Stukenberg, P.T., Burke, D.J., Ross, M.M., Shabanowitz, J., Hunt, D.F., and White, F.M. (2002) Phosphoproteome analysis by mass spectrometry and its application to Saccharomyces cerevisiae. *Nat Biotechnol.* 20, 301–305.

14. Wiese, S., Reidegeld, K.A., Meyer, H.E., and Warscheid, B. (2007) Protein labeling by iTRAQ: a new tool for quantitative mass spectrometry in proteome research. *Proteomics.* 7, 340–350.

Chapter 11

Profiling the Tyrosine Phosphorylation State Using SH2 Domains

Kevin Dierck, Kazuya Machida, Bruce J. Mayer, and Peter Nollau

Summary

Global monitoring of cellular signaling activity is of great importance for the understanding of the regulation of complex signaling networks and the characterization of signaling pathways deregulated in diseases. Tyrosine phosphorylation of intracellular signaling proteins followed by the recognition and binding of Src homology 2 (SH2) domains are key mechanisms in the downstream transmission of many important biological signals. SH2 domains, comprising 120 members in humans, are small modular protein binding domains that recognize tyrosine phosphorylated signaling proteins with high specificity. Based on these binding properties, the large number of naturally occurring and currently available SH2 domains serve as excellent probes for the comprehensive profiling of the cellular state of signaling activity. Here we have described different experimental strategies for global SH2 profiling: high-resolution phosphoproteomic scanning by far-Western Blot analysis and high-throughput profiling by our recently developed oligonucleotide-tagged multiplex assay (OTM) and Rosette assay.

Key words: Signal transduction, Tyrosine phosphorylation, Phospho-proteomics, Src homology 2 (SH2) domain, Oligonucleotide-tagged multiplex assay, SH2 Rosette assay, Far-Western blot, High-throughput, PCR–ELISA.

1. Introduction

Tyrosine phosphorylation, a posttranslational modification tightly regulated by the balanced action of tyrosine kinases and phosphatases, plays a central role in signal transduction controlling many important biological processes such as proliferation, differentiation, apoptosis, cell adhesion, and motility (1–4).

Marjo de Graauw (ed.), *Phospho-Proteomics, Methods and Protocols, vol. 527*
© 2009 Humana Press, a part of Springer Science + Business Media, New York, NY
Book DOI: 10.1007/978-1-60327-834-8_11

In addition to its importance in physiological processes of signal transduction, aberrant activation of tyrosine phosphorylation has been observed in various pathological conditions, particularly in cancer development and progression (5–7). Global characterization of the cellular state of tyrosine phosphorylation is therefore of great interest for a deeper understanding of physiological processes of cell signaling and the identification of deregulated signaling pathways in diseases.

During signal transduction, the activation of tyrosine kinases leads to tyrosine phosphorylation of signaling proteins, providing docking sites for the binding of Src homology 2 (SH2) domains, a prototypic modular binding domain of ~100 amino acids present in many signaling proteins (8, 9). SH2 domains recognize tyrosine-phosphorylated peptides with high specificity, thereby acting as phospho-specific sensors in the cytosol, transmitting signals by monitoring the state of tyrosine phosphorylation of receptors and other signaling molecules (10). Based on screens of phosphopeptide libraries, the binding specificity of SH2 domains is defined by a sequence of amino acids, typically three residues C-terminal to the phosphotyrosine residue, within the target protein (11, 12). Binding affinities of SH2 domains are moderate, with affinity constants in the nanomolar to low micromolar range (13).

Because of their modular nature, SH2 domains can be expressed as isolated protein binding modules without losing their specific binding characteristics. Currently, all 120 SH2 domains present in the human genome have been cloned and expressed as bacterial glutathione S-transferase (GST) fusion proteins (14). Approximately two-thirds of the SH2 domains are soluble and bind to their target proteins in a phosphorylation-dependent manner (15). Thus, the large panel of SH2 domains presently available provide an ideal tool for the comprehensive and biologically relevant profiling of the cellular state of tyrosine phosphorylation (16–18). For phosphotyrosine profiling, SH2 domains of different binding specificities are generated and used as phosphotyrosine-specific probes for whole cellular extracts immobilized on membranes (19). Here we have described in detail three reverse-phase assay platforms, far-Western blot analysis, the OTM, and the Rosette assay, which have been developed for qualitative or quantitative SH2 profiling of the global state of tyrosine phosphorylation (15, 19, 20). All these platforms are suited for the analysis of whole cellular extracts (e.g., tumor samples) without the need for enrichment of tyrosine phosphorylated proteins. Thus cell type specific phosphotyrosine profiles can be obtained for rapid and comprehensive analysis of alterations in cell signaling activity in normal and pathological conditions.

2. Materials

2.1. Expression and Purification of Glutathione S-Transferase (GST)-SH2 Domains

1. For the generation of nonbiotinylated GST fusion proteins use SH2 domains cloned in the bacterial expression vector pGEX6P1 (Pharmacia) and transformed in BL21 DE3 cells (*E. coli* K12 strain; Invitrogen). For monobiotinylated probes, use SH2 domains cloned in pAC4 Avi-tag vector (Avidity) transformed in AVB100 bacteria (MC1061; Avidity) expressing the *birA* gene stably integrated into the bacterial genome (*see* **Note 1**).

2. Lauria-Bertani (LB) bacterial growth medium: Prepare 1 l of medium containing 20 g Bacto Trypton, 10 g Bacto yeast extract, 20 g NaCl in H_2O. Sterilize by autoclaving at 120°C and 1 bar. Store media at room temperature.

3. 1 M isopropyl-β-D-thiogalactopyranoside (IPTG). Dissolve in water and store at −20°C.

4. L-arabinose (Sigma).

5. 25 mM α-biotin solution (Avidity). Dissolve in water and adjust pH to 7.0 with 2 M NaOH and store at −20°C.

6. Bacteria lysis buffer: Prepare 10 ml of lysis buffer by adding 100 μl of aprotinin solution (Sigma), 100 μl of 100 mM PMSF (dissolved in isopropanol and stored at −20°C), 10 μl of 100 mM DTT to PBS containing 1% Triton X-100 (w/v) and 100 mM EDTA. Always make fresh buffer and cool to 4°C prior to use.

7. Sonicator (Bandelin, Berlin, Germany).

8. 20-ml PolyPrep columns (BioRad).

9. Gluthathione (GSH)-sepharose 4B (GE Healthcare).

10. Phosphate-buffered saline (PBS).

11. TN buffer: Prepare by combining 1.5 ml 1 M Tris–HCl, pH 7.5, with 15 ml 300 mM Nacl, and adjust to a final volume of 30 ml with H_2O.

12. GSH elution buffer: Prepare 50 ml of buffer with 0.31 g gluthathione and 5 ml 1 M Tris–HCl, pH 8.0, in H_2O. Set up buffer freshly prior to use or store at −20°C avoiding repeated freezing and thawing.

13. PD10 sephadex columns (GE Healthcare).

14. PBS/20% (w/v) glycerol for equilibration and elution of PD10 columns.

2.2. Preparation of Whole Cellular Extracts

1. Standard tissue culture media (e.g., DMEM or RPMI 1640) supplemented by 10% fetal calf serum containing penicillin and streptomycin.

2. Phosphate-buffered saline (PBS).

3. Cell lysis buffer: 25 mM Tris–HCl, pH 7.4, 150 mM NaCl, 5 mM EDTA, pH 8.0, 10% glycerol, 1% Triton X-100, 10 mM sodium pyrophosphate, 10 mM β-glycerol phosphate, 1 mM sodium orthovanadate; store as a stock solution at 4°C. Prepare 50 mM orthovanadate solution by dissolving 1.84 g Na-orthovanadate in 150 ml of H_2O and adjust the pH to 10 with conc. HCl (solution gets yellow). Boil the solution for 2 min in a microwave until the solution is clear again. Adjust to pH 10 again and repeat the procedure until the pH is stable at 10. Prepare sodium pervanadate solution by incubating 16 μl of H_2O_2 with 100 μl of sodium orthovanadate (50 mM) for 30 min at room temperature. Prior to cell lysis, add 10 μl of 1 M DTT, 100 μl of 100 mM PMSF, 100 μl of aprotinin (Sigma), 100 μl of 1 M sodium fluoride, and 20 μl of sodium pervanadate to 10 ml of cell lysis stock solution. Keep cell lysis buffer on ice.

4. Cell scraper.

5. Bradford reagent (BioRad).

2.3. SDS Polyacrylamide Gel Electrophoresis and Far-Western Blot Analysis

1. NuPAGE precast 4–12% gradient gels (Invitrogen) (see **Note 2**).

2. NuPAGE LDS sample buffer (4×) (Invitrogen).

3. PVDF membranes (0.45-μm pore size, Millipore).

4. Tank-blot TE22 (Hoefer).

5. Transfer buffer: 10 mM 3-(cyclohexylamino)-1-propanesulfonic acid (CAPS), pH 11, 20% (v/v) methanol. Store at 4°C.

6. Tris-buffered saline with Tween (TBST): 150 mM NaCl, 10 mM Tris–HCl, pH 8.0, 0.05% Tween 20 (v/v).

7. Blocking buffer: 10% (w/v) nonfat dry milk in TBST supplemented with sodium orthovanadate and EDTA at a final concentration of 1 mM each. Prepare fresh.

8. Streptavidin-conjugated horseradish peroxidase (streptavidin-HRP) (Pierce). Dissolve in water at a concentration of 1 μg/μl. Store at 4°C in the dark.

9. Monobiotinylated affinity-purified GST-SH2 domains (see **Note 3**).

10. ECL Western blotting detection reagent (GE Healthcare).

2.4. Oligonucleotide-Tagged Multiplex Assay (OTM)

1. Double-stranded biotinylated oligonucleotides (80 bp) for labeling of SH2 domains (see **Note 4**).

2. Primer pairs for PCR amplification of oligonucleotide tags (see **Note 4**).

3. Novex precast urea acrylamide gels (10%) (Invitrogen).

4. Gel extraction buffer: 500 mM ammonium acetate, 10 mM magnesium acetate in H_2O.

5. NAP-5 Sephadex columns (GE Healthcare).

6. Monobiotinylated affinity-purified GST-SH2 domains.

7. ImmunoPure Streptavidin (Pierce) dissolved in water at a final concentration of 10 μg/μl.

8. 25 mM α-biotin solution (Avidity).

9. PVDF membranes (0.45-μm pore size, Millipore).

10. TBST: 150 mM NaCl, 10 mM Tris–HCl, pH 8.0, 0.05% Tween 20 (w/v).

11. Blocking solution I: 5% (w/v) blocking reagent (Roche), 1 mM sodium orthovanadate, 1 mM EDTA. Prepare 5% (w/v) blocking reagent and store aliquots at –20°C. Add sodium orthovanadate and EDTA prior to use.

12. Blocking solution II: 2× Denhardt's solution, 1 mM sodium orthovanadate, 100 μg/ml of herring sperm DNA (sheared by sonication), and 500 μg/ml of heparin (Sigma) in TBST. Always prepare blocking solution freshly.

13. Plastic petri dishes (diameter: ~8 cm).

14. 24-well cell-culture plates.

15. Whatman 3 MM paper.

16. Denaturation buffer: 300 mM Tris–HCl, pH 6.8, 10% (w/v) SDS, 25% (v/v) β-mercaptoethanol.

17. Elution buffer: 10 mM phenyl phosphate in 10 mM Tris–HCl, pH 7.4. Store aliquots at –20°C. Avoid repeated freezing and thawing.

18. Hot start Taq master mix kit (Qiagen).

19. LightCycler fast start DNA master SYBR green I kit (Roche).

20. LightCycler instrument (Roche).

21. PCR–ELISA (Dig Detection) (Roche).

22. Biotinylated capture oligonucleotides for quantitation of PCR products by PCR–ELISA (*see* **Note 5**).

2.5. SH2 Rosette Assay

1. Spotting solution: 180 mM Tris–HCl, pH 6.8, 30% glycerol, 6% sodium dodecyl sulfate (SDS), 15% β-mercaptoethanol, and 0.03% bromophenol blue.

2. Transparency film (3 M, CG3300).

3. Nitrocellulose membrane: 120 × 75-mm² size (Schleicher & Schuell BioScience, BA83).

4. Light panel (Gagne Inc., LPA4).

5. Gel loading pipette tip (USA Scientific, 1022–0000).

6. Wetting solution: 10 mM 3-(cyclohexylamino)-1-propanesulfonic acid (CAPS), pH 11, 20% (v/v) methanol. Store at 4°C.

7. Blocking buffer: 10% (w/v) nonfat dry milk in TBST supplemented with sodium orthovanadate and EDTA to a final concentration of 1 mM each. Prepare fresh.

8. MBA96 chemotaxis chambers set (Neuroprobe, Gaithersburg, MD).

9. Membrane support plate: 96-well flat-bottom plate.

10. Probe reservoir plate: 96-well round-bottom plate.

11. Filter paper: 125 × 85–mm² size X2 (Schleicher & Schuell BioScience, 3 MM Chr, Cat #3030917).

12. Parafilm-M (American National Can).

13. Labeling reagent: Glutathione (GSH)-HRP conjugate (Sigma G6400, currently discontinued) or anti-GST-HRP conjugate (Sigma, A7340).

14. Repeater pipette.

15. TBST: 150 mM NaCl, 10 mM Tris–HCl, pH 8.0, 0.05% Tween 20 (v/v).

16. Chemiluminescence kits (*see* **Note 6**). We routinely use NEL103 for the Rosette assay

3. Methods

The setup of assay components differs between the different SH2 profiling platforms. For OTM, biotinylated SH2 domains are required, while the Rosette assay is currently performed with unbiotinylated SH2 domains. Far-Western blot analysis can either be performed with biotinylated or unbiotinylated SH2 domains leading to similar binding profiles. For the generation of bacterial GST fusion proteins *(19)*, SH2 domains are cloned into appropriate restriction sites in the vector pGEX6P1. When biotinylated probes are required, GST-SH2 domains are further subcloned in the pAC4 Avi-tag vector (Avidity) containing a biotinylation sequences C-terminal to the SH2 domain. For unbiotinylated probes, bacterial expression is performed in BL21 cells. The bacterial strain AVB100 is used for protein expression compatible with the biotinylation of GST-SH2 fusion proteins during bacterial expression by induction of the BirA system (Avidity).

Among the three SH2 profiling platforms described here, far-Western blot analysis is a commonly used technique for profiling of the cellular state of tyrosine phosphorylation *(19)*. Protein extracts are separated by 1D SDS-PAGE, transferred to membranes and tyrosine phosphorylated proteins are detected by a panel of SH2 domains applied in separate binding reactions. To enhance sensitivity, precomplexing between the SH2 domain and the detection reagent is required, increasing the binding avidity of the probe. Detection is performed by chemiluminescence. Replicate filter membranes are probed with different SH2 domains marked by different binding specificities. The resulting binding patterns are characterized by intensity and molecular weight of the detected phosphoproteins (**Fig. 1**). The combination of binding patterns of different SH2 domains leads to the identification of SH2 profiles specific for the type of cells or tissue and the conditions (e.g., stimulated vs. unstimulated cells) under which the cells are investigated. In general, SH2 profiling by far-Western blot analysis provides semiquantitative binding information.

For quantitative SH2 profiling, we have developed the OTM and the Rosette assay, two SH2-profiling platforms suitable for

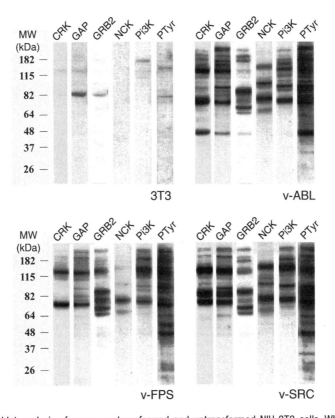

Fig. 1. Far-Western blot analysis of oncogene-transformed and untransformed NIH-3T3 cells. Whole cellular lysates (50 μg/lane) were separated by 4–12% gradient SDS-PAGE and transferred to PVDF membranes. Monobiotinylated GST-SH2 domains of CRK, GAP, GRB2, NCK and Pi3K were complexed with streptavidin-HRP for subsequent detection by chemiluminescence. As a control, total tyrosine phosphorylation (PTyr) was analyzed by the phosphotyrosine-specific, monoclonal antibody 4G10.

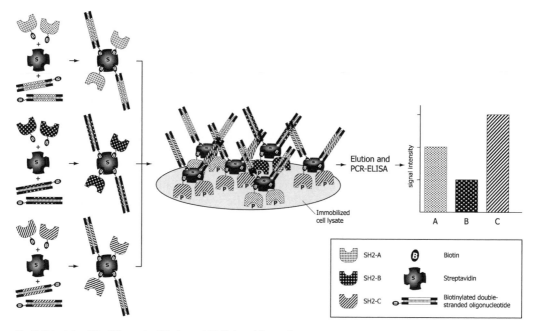

Fig. 2. Principle of the Oligonucleotide-tagged Multiplexed Assay. For quantitative multiplexed profiling of the cellular state of tyrosine phosphorylation OTM probes are generated by sequentially complexing of monobiotinylated SH2 domains (SH2-A, SH2-B, SH2-C) with streptavidin (S) and domain-specific biotinylated double-stranded oligonucleotides. The double-stranded oligonucleotides consist of a specific internal recognition sequence flanked by nonvariable primer-binding sites for PCR amplification. Labeled OTM probes are combined at equimolar concentrations and incubated with denatured proteins immobilized on PVDF membranes. After binding to tyrosine phosphorylated proteins and washing specifically bound probes are eluted and analyzed by quantitative PCR assay such as PCR–ELISA or quantitative real time PCR.

fast and high-throughput profiling of small amounts of whole cellular extracts (15, 20). For OTM, different SH2 domains are labeled by domain-specific oligonucleotide tags and applied as probes in a multiplex reaction. Monobiotinylated SH2 domains are coupled to biotinylated double-stranded oligonucleotide tags via streptavidin (Fig. 2). Up to ten SH2 domains are combined in a multiplex reaction and incubated with 10 µg of whole cellular extract (representing ~10^4 cells) spotted on circular pieces of PVDF membrane. After binding and washing, specifically bound complexes are eluted by phenyl phosphate competing for the binding to phosphotyrosine. Subsequently, the different oligonucleotide tags are amplified by PCR using primers complementary to common primer binding sites present at the 5′- and 3′-end of the oligonucleotide tags. PCR products are quantified by PCR–ELISA using capture oligonucleotides complementary to the SH2 domain-specific internal recognition sequences of the corresponding oligonucleotide tags.

In contrast to OTM, in which several SH2 domains are simultaneously competing for the binding to target sequences in a single binding reaction, the Rosette assay is based on the binding of a single SH2 domain to a large number of target samples (15) (Fig. 3). Small amounts of cellular extracts (~500 ng) are spotted in small

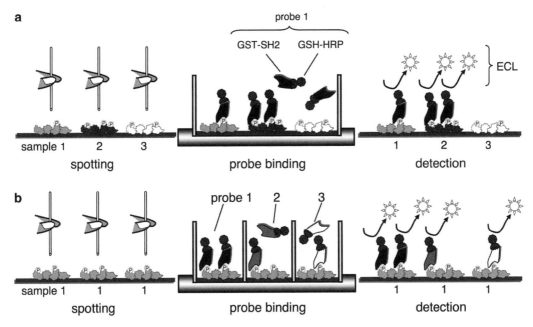

Fig. 3. Feature of Rosette assay. Dot blotting assays can be grouped into two categories. (a) In a protein macroarray, an array of sample spots is incubated with a single probe one at a time. (b) In a chamber assay, duplicate sample spots are incubated with different probes using a multichamber apparatus. The Rosette assay incorporates features of both: multiple samples (up to 19 spots per well) are incubated with multiple probes (up to 96 GST fusion proteins per plate). Immobilized proteins are incubated with GSH-HRP-labeled GST-SH2 probes, and bound probes are detected by enhanced chemiluminescence (ECL).

volumes (~100 nl) on nitrocellulose membranes in register with a well of a 96-well chamber plate allowing the immobilization of up to 12 or more different samples per well. Subsequently, lysates are probed with SH2 domains precomplexed with conjugates of GSH-HRP or anti-GSH-HRP and bound probe is detected by chemiluminescence. Binding reactions are performed in separate reactions with different SH2 domains, providing quantitative binding data for a large number of samples and SH2 domains.

Both high-throughput formats (OTM and Rosette assay) provide quantitative binding data for multiple samples and SH2 domains, and results are comparable. The choice of which platform to use is largely dependent on the number of specimens and the number of SH2 domains under investigation, the availability of instrumentation, and the degree of automation desired.

3.1. Expression and Purification of Glutathione S-Transferase (GST)-SH2 Domains

1. Depending on the generation of nonbiotinylated or biotinylated probes, use frozen stocks of BL21 or AVB100 bacteria, respectively, transformed by the corresponding GST-SH2 domain expression vectors (*see* **Note 1**).

2. Remove bacteria from frozen stocks with a sterile loop, inoculate 50 ml of prewarmed LB medium supplemented with 100 µg/ml ampicillin, and incubate overnight at 37°C with constant shaking.

3. The next day, inoculate 200 ml of prewarmed LB medium containing 100 µg/ml ampicillin with 10 ml of the overnight culture and shake at 37°C until OD (595 nm) reaches 0.5–0.6.

4. Add 20 µl of 1 M IPTG.

5. After 15 min add 0.6 g L-arabinose and 400 µl α-biotin (25 mM). This step is omitted when nonbiotinylated GST-SH2 domains are generated from BL21 cells.

6. Incubate at 37°C for 3 h under constant shaking.

7. Centrifuge bacteria with $5,000 \times g$ at 4°C for 10 min.

8. Discard the supernatant and resuspend the cell pellet in 8 ml of ice-cold bacteria lysis buffer. Sonicate three times for 2 min each with five cycles at a power of 40%. Always keep samples on ice and prevent heating of the suspension.

9. Centrifuge with $2,000 \times g$ at 4°C for 10 min to remove cell debris.

10. Load a 20-ml PolyPrep column (BioRad) with 1.5 ml of glutathione-sepharose 4B (GE Healthcare), let column drain and wash beads with 5 ml PBS.

11. Tightly close the bottom of the column, load the supernatant containing the GST-SH2 fusion protein on the affinity-column, tightly close the top and incubate at 4°C for 2 h with rotation.

12. Open the column and let the column drain and wash beads with 10 ml of TN buffer.

13. Elute the SH2 domain by applying 2.5 ml GSH buffer to the column. Repeat elution step once.

14. For buffer exchange, equilibrate a PD10 Sephadex column (GE Healthcare) with 15 ml PBS/20% (w/v) glycerol.

15. Load 2.5 ml of affinity-purified SH2 domains to the column and elute with 3.5 ml PBS/20% (w/v) glycerol. Use multiple columns for volumes larger than 2.5 ml.

16. Check yield, integrity, and functionality of SH2 domain (*see* **Note 7**).

17. Store SH2 domains in small aliquots at –80°C. Avoid repeated freeze and thaw cycles. SH2 domains stored for longer time periods (≥2 weeks) at 4°C may lose binding activity.

3.2. Preparation of Whole Cellular Extracts

1. Grow tissue culture cells to 70–80% confluency in tissue culture dishes or flasks.

2. Remove medium and wash cells three times with ice-cold PBS. Place culture dishes or flasks on ice and add cell lysis buffer. Use 1 ml of cell lysis buffer for ~10^6 cells and a surface area of ~50 cm². For adherent cells, scrape cells with a

cell scraper; cells grown in suspension are directly lysed after washing and centrifugation by resuspending the cell pellet in lysis buffer. For extraction of tissue samples, completely pulverize frozen tissue specimen by mortar and pestle under liquid nitrogen. Add 1 ml of lysis buffer to ~25–50 mg of frozen tissue specimen.

3. Incubate on ice for 30 min.

4. Transfer cell suspension to a 1.5-ml Eppendorf tube and centrifuge with $16,000 \times g$ at 4°C for 10 min.

5. Transfer the supernatant to a new tube and store aliquots of cell lysate at –80°C.

6. Determine protein concentration by the Bradford assay. Depending on the cell line or cell type, yields of lysates are typically in the range of 2–10 mg/ml.

3.3. SDS Polyacrylamide Gel Electrophoresis and Far-Western Blot Analysis

1. For far-Western blot analysis, separate 50 µg per lane of whole cellular extracts by SDS PAGE using precast 4–12% gradient gels (*see* **Note 2**).

2. Transfer proteins to a methanol-activated PVDF membrane by tank blotting at 4°C in CAPS transfer buffer at a constant current of 400 mA for 2 h.

3. Remove the membrane from the tank blot, immediately place the membrane in blocking buffer and incubate with gentle shaking at 4°C overnight.

4. The next day wash membranes three times in TBST for 1 min each to remove excess blocking buffer.

5. Preincubate SH2 domains (1 µg) with 5 µl of a 1:100 dilution of streptavidin-HRP conjugate for 1 h at room temperature. Prior to the binding reaction, add TBST to a final volume of 1 ml and incubate the shrink-wrapped membrane with the precomplexed, labeled SH2 domain (final domain probe concentration 1 µg/ml) for 1 h at room temperature (*see* **Note 8**).

6. Wash membranes in TBST for 2 h, changing washing buffer every 15 min.

7. Remove excess buffer and detect binding of SH2 domains by chemiluminescence using the ECL Kit (GE Healthcare)

3.4. Oligonucleotide-Tagged Multiplex Assay (OTM)

3.4.1. Generation of SH2 Domain-Specific Oligonucleotide Tags

1. For double-stranded oligonucleotide tags synthesize single-stranded sense and complementary antisense oligonucleotides (*see* **Note 4**). Lyophilized oligonucleotides should be dissolved in water at a concentration of 100 pmol/µl.

2. For purification, separate the single-stranded oligonucleotides by denaturing polyacrylamide gel electrophoresis using Novex precast urea gels (Invitrogen). Load 7.5 µg of oligonucleotide per lane denatured by boiling at 70°C for 3 min in

sample buffer (Novex). To prevent potential contamination, use separate gels for the different oligonucleotides.

3. Stain the gel by ethidium bromide (add 20 μl 1% ethidium bromide solution (Sigma) to 100 ml H_2O, incubate the gel for 15 min at room temperature and wash twice in 100 ml H_2O). Cut out the band corresponding to the 80 mer oligonucleotide from the gel, combine the gel fragments from several lanes, add the same volume of gel extraction buffer and freeze the gel fragments (*see* **Note 9**).

4. Crush the frozen gel fragments with a pipette tip.

5. Repeat **step 4** until the gel slice is completely crushed (*see* **Note 10**).

6. Centrifuge with $16,000 \times g$ at room temperature for 10 min and transfer the supernatant to a new tube.

7. Reextract the pellet once by adding 100 μl of gel extraction buffer, centrifuge and combine the supernatants. Avoid carrying over acrylamide residues from the pellet.

8. Equilibrate NAP-5 columns with 10 mM Tris–HCl, pH 7.4 and apply a total volume of 0.5 ml of the supernatant to the column.

9. Elute the size-purified oligonucleotides with 1 ml of 10 mM Tris–HCl, pH 7.4.

10. The purity of the extracted oligonucleotides should be checked by denaturing polyacrylamide gel electrophoresis followed by ethidium bromide staining as described earlier.

11. Determine yield and concentration of oligonucleotides by OD at 260 nm.

12. To generate double-stranded oligonucleotide tags, hybridize complementary oligonucleotides at equimolar concentrations in the presence of 10 mM KCl by heating to 95°C and cooling to room temperature followed by incubation overnight. Store double-stranded oligonucleotide tags at –20°C or use tags immediately for generation of OTM probes.

3.4.2. Complex Formation of SH2 Domains and Oligonucleotide Tags

1. For the generation of OTM probes, monobiotinylated SH2 domains should be coupled in separate reactions with domain-specific biotinylated oligonucleotide tags via strepavidin. Complexes should be prepared freshly on the day of use.

2. Sequentially complex the biotinylated SH2 domain, the corresponding biotinylated, double-stranded oligonucleotide tag and streptavidin at a molar ration of 2:2:1. First, incubate the biotinylated SH2 domain with streptavidin at a molecular ratio of 2:1 for 1 h at room temperature, then incubate the complex with the oligonucleotide tag at a concentration equimolar to the SH2 domain for 1 h at room temperature (*see* **Note 11**).

3. After complex formation, block unoccupied biotin binding sites of streptavidin by incubating the complex for 15 min at room temperature with a 100-fold excess of free biotin using 3 μl of a 1:100 dilution of α-biotin solution (25 mM).

4. After complex formation add TBST to a total volume of 1 ml per SH2 domain and pool the differentially labeled OTM probes in a 50-ml falcon tube. Adjust volume with TBST to a final concentration of 0.1 μg/ml for each SH2 domain (*see* **Note 12**).

3.4.3. Protein Immobilization and Blocking

1. Protein immobilization is performed on circular pieces of PVDF membrane. Punch out circular pieces of PVDF membrane with a commercial hole puncher (diameter of punch hole: 5 mm). One piece of PVDF membrane is needed per binding reaction (*see* **Note 13**).

2. Activate membranes in a culture dish in methanol for 1 min. Remove methanol from dish by pipetting. Float the circular membrane pieces on sterile H₂O and carefully shake the dish for 1 min. Membranes should settle to the bottom of the dish. Remove wash solution and keep membranes in fresh H₂O until further use.

3. Heat protein extracts in denaturation buffer at 95°C for 3 min.

4. For each sample place two layers of Whatman 3 MM paper in a petri dish and completely wet filter paper with H₂O. Carefully remove activated PVDF circles from H₂O and place on top of the Whatman filters (*see* **Note 14**).

5. Remove excess liquid before spotting lysate. Spot 10 μg of lysate in a volume of ~4 μl on top of each membrane circle and incubate for 10 min at room temperature. Cover the petri dish during incubation (*see* **Note 15**).

6. Transfer membrane pieces into blocking solution I in a petri dish. Membranes should be completely submersed in blocking solution. Incubate at 4°C overnight with slight shaking.

7. Remove blocking solution by pipetting or aspiration and wash membranes three times in TBST. Avoid transferring the membrane pieces.

8. Transfer membranes from TBST to wells of a 24-well cell-culture plate (up to five membrane circles per well) filled with 500 μl blocking solution II and incubate for 1 h at room temperature.

9. Remove blocking solution by pipetting or aspiration and wash membranes three times in TBST (*see* **Note 16**).

3.4.4. OTM Binding Reaction

1. Add 500 μl of the diluted OTM probes to wells of a 24-well cell-culture plate and carefully transfer the membrane pieces

from TBST buffer to the appropriate well (up to five membrane circles per well). Incubate for 1 h at room temperature with slight shaking.

2. Remove solution by aspiration and wash the membranes three times with TBST.

3. Transfer the membranes from the wells of the 24-well plate to individual petri dishes and wash three additional times with 25 ml of TBST. Always remove the wash buffer by pipetting.

3.4.5. Elution of Bound OTM Probes and PCR–ELISA

1. Transfer each membrane piece to individual wells of a 24-well cell-culture plate (one membrane circle per well) filled with 200 µl of 10 mM phenyl phosphate and incubate for 30 min at room temperature with slight shaking (*see* **Note 17**).

2. After elution, rinse membranes in the well by carefully pipetting several times up and down and transfer 190 µl of eluate to new tubes. Keep eluates on ice or store at –20°C.

3. Quantify 2 µl of eluates by quantitative real time PCR using the LightCycler fast start DNA master SYBR green I kit and a Light-Cycler instrument (Roche). Determine the optimal cycle conditions and linear range of PCR amplification (*see* **Note 18**).

4. Amplify 2 µl of eluates using the hot start Taq master mix kit (Qiagen).

5. Perform the PCR–ELISA with 10–20 µl of PCR products under conditions recommended by the supplier. Hybridization is performed for 2 h at 45°C (*see* **Note 19**)

6. For normalization and comparison of different binding reactions, OD readings obtained by PCR–ELISA are converted to relative binding intensities on the basis of internal standard curves generated from serial dilutions of double-stranded oligonucleotides with defined concentration. For each measurement, standards need to be included in the assay and amplified and quantified by PCR–ELISA in parallel.

3.5. SH2 Rosette Assay

This section gives a protocol for the Rosette assay, a high-throughput method based on far-Western blotting. An array of protein samples such as lysates and purified proteins is immobilized on a membrane in 96 replicates and incubated with labeled probes (**Fig. 3**). A 96-well chamber apparatus allows simultaneous binding of arrays to a large number of different probes for high throughput, but an equivalent experiment can be performed using standard 6-well culture dishes and smaller size membranes, especially when the number of probes being used is limited (*see* below). Although the method was developed for SH2 domain profiling analysis using GST-SH2 domains *(15)*, other modular protein domains, antibodies, etc., are easily incorporated into the system.

1. Using a standard office printer, print an array pattern on a transparent guide film indicating the positions of all spots and wells (**Fig. 4**) (*see* **Note 20**).

2. Position nitrocellulose membranes on the template and fix them to respective positions at each corner using a piece of adhesive tape. Always wear gloves when handling blotting membranes (*see* **Note 21**).

3. Place the template with membranes on a light panel and fix it at each corner using a piece of adhesive tape (**Fig. 2a**).

4. Dissolve lysate samples in spotting solution at a concentration of $4\,\mu g/\mu l$ (*see* **Note 22**).

5. Fill a gel loading pipette tip with the sample solution by capillary action, and manually spot by quickly touching the membrane surface at the spot positions indicated by the template underneath (*see* **Note 23**).

6. When spotting is done, draw a rectangular line around the spotted area with a ballpoint pen, which serves as a marker for aligning the membrane to 96-well chambers (**step 13**).

7. Allow the spotted membranes to dry at room temperature (RT) for at least 2 h (*see* **Note 24**).

8. Prewet the membrane in wetting solution and soak for 30 min with gentle agitation.

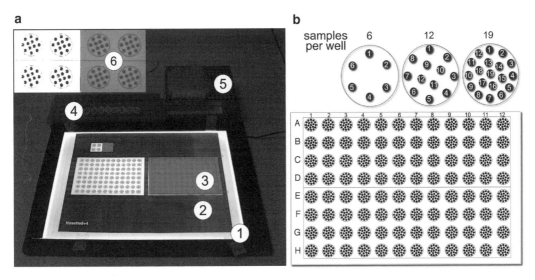

Fig. 4. Immobilization of samples for Rosette assay. A, setup for spotting. (**a**) Nitrocellulose membrane (3) is placed on a guide film (2) that is fixed on a light panel (1) allowing template spots to be visible through the membrane (6). Protein samples (4) are manually spotted in 96 replicates using gel loading tips (5). (**b**) An example of the template printed on a transparency film. Samples can be spotted in up to 19 spots per single well area.

9. Briefly rinse the membrane twice in TBST and block in blocking buffer for 1–2 h at RT until applying probes (*see* **Note 25**).

10. Prepare probe dilutions at the appropriate concentrations in the blocking buffer and reserve them in corresponding position of a 96-well round-bottom plate. The recommended loading volume of probe per well is 100 μl or more (*see* **Note 26**).

11. Assemble the MBA96 chemotaxis chambers. Briefly, flip the top plate of the chamber up and set a gasket. Place a membrane support plate into position in lower plate. Place two sheets of filter papers covered with parafilm on top of the membrane support plate. Always wear gloves when handling the plate.

12. Insert the blocked membrane. Hold the edge of membrane with tweezers allowing the excess block solution to drip off Kimwipe papers, and place it on the parafilm surface with a protein side up. Position the membrane with tweezers making sure all the spots are aligned within the holes of the upper plate (**Fig. 5**).

13. Close the top plate and press down firmly on it as you turn the knobs about one-third turn clockwise to seal the plates together.

14. Carefully apply 100 μl probe solutions from the reservoir plate (**step 10**) into each well using an 8-channel pipette. Cover the chambers with aluminum foil and incubate for 1–2 h (*see* **Note 27**).

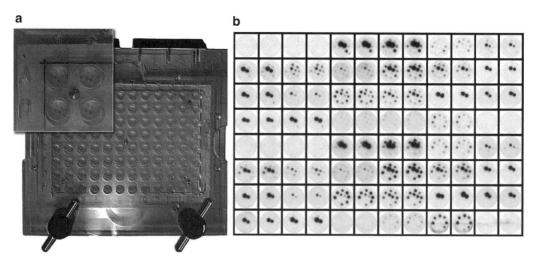

Fig. 5. Rosette assay using a 96 well chamber apparatus. (**a**) A blocked membrane is inserted in register with a 96 well chamber apparatus making sure that all spots are aligned inside the chambers (*see* inset). 100 μl of probe solutions are applied into wells using an 8-channel pipette and incubated 1–2 h before washing wells separately. (**b**) Representative Rosette assay result. Signal detected by ECL is quantifiable using densitometry, e.g., integrated density measurement by ImageJ.

15. To wash the wells, aspirate fluid from the wells and fill with 150 µl TBST solution per well using a repeater pipette. Repeat the washing three times (*see* **Note 28**).

16. After washing, aspirate fluid from the wells. Disassemble the MBA96 apparatus to remove the membrane and transfer it to a tray containing TBST. Rinse briefly and then wash for 15 min with gentle agitation, occasionally replacing the buffer (*see* **Note 29**).

17. Perform chemiluminescence detection according to the manufacturer's instructions (*see* **Note 30**).

4. Notes

1. For general information on SH2 domains refer to: http://proteoscape.uchicago.edu/sh2/. For detailed information on cloning and availability of SH2 domains, solubility, and binding data refer to: http://genetics.uchc.edu/faculty/mayer_linkpage.htm. Depending on the assay and method of detection used, either unlabeled or monobiotinylated SH2 domains are generated by bacterial expression. Unlabeled GST-SH2 domains are detected by glutathione-horseradish peroxidase (GSH-HRP) conjugate; for biotinylated SH2-domains, detection is performed with streptavidin-HRP or streptavidin-DNA conjugates.

SH2 domain cDNAs can be obtained from following sources.

   (a) ATCC: Mammalian Gene Collection (MGC) http://www.atcc.org/common/catalog/molecular/index.cfm

   (b) Open Biosystems: pET28 based SH2 constructs http://www.openbiosystems.com/GeneExpression/Mammalian/Human/SH2Domains/

   (c) Mayer Lab: pGEX-based SH2 constructs http://155.37.79.139/fmi/iwp/cgi?-db = Mayer%20Lab%20SH2%20Database&-loadframes

2. Alternatively, self-cast gels can be used. However, we observed a substantial improvement in reproducibility of protein separation with low day-to-day variability using commercially available precast gels.

3. For far-Western blot analysis, either monobiotinylated or unbiotinylated SH2 domains can be used. Detection of unbiotinylated SH2 domains is performed by gluthathione-horseradish peroxidase (HRP)-conjugate (Sigma) (for details *see* **Note 8**).

4. For multiplex reactions, different SH2 domains are labeled with domain-specific oligonucleotide tags. Use double-stranded tags instead of single-stranded tags to prevent cross-hybridization. Oligonucleotide tags should consist of a domain-specific internal recognition sequence flanked by non-variable primer binding sites for PCR amplification. Recognition sequences should be different in the order of nucleotides, but identical in composition to ensure PCR amplification with comparable efficiencies. The antisense oligonucleotide should be labeled with biotin at the 5′-end. The typical length of the oligonucleotide tags is 80-mers. A list of validated oligonucleotide tags along with PCR primers allowing the detection of PCR products by PCR–ELISA is given in **Table 1**. Oligonucleotides should be synthesized at a scale of 1 μmol, yielding a minimum amount equivalent to 10 OD (260 nm). Purification by HPLC is recommended followed by additional gel purification to remove incomplete synthesis products.

5. For quantitation by PCR–ELISA, biotinylated capture oligonucleotides complementary to the internal recognition sequence of the oligonucleotide tag are used. A selection of capture oligonucleotides complementary to the oligonucleotide tags is given in **Table 1**.

6. Comparison of chemiluminescence kits for Rosette assay:

Product	Company	Sensitivity	Background
RPN2106	GE Healthcare	Medium	Low
NEL103	PerkinElmer	High	Medium
#34079	Pierce	Very high	High
RPN2132	GE Healthcare	Very high	High

7. Determine protein yield by the Bradford assay. Depending on the SH2 domain expressed, yields vary between 5 and 25 mg of GST-fusion protein per liter of bacterial culture. Low yields indicate domain insolubility, which may be improved by the expression of SH2 domains at low temperatures (e.g., 16°C). Purity and quality of the SH2 domain should be assessed by SDS-PAGE followed by Coomassie blue staining. Integrity of the domain can further be checked by Western blot analysis using anti-GST antibodies. In comparison with the biotinylated standards (Avidity), efficiency of biotinylation should be analyzed by dot blot or Western blot analysis using streptavidin-HRP for detection. Far-Western blot analysis of cell lysates with different levels of tyrosine phosphorylation is required to evaluate the binding activity of the SH2

Table 1
List of oligonucleotides

Oligonucleotide tag 1	
Sense	5'-AAAAAAAACTGATTAACCCTCACTAAAGTACGCTAGATCGGACTATACACAGTGCTAAACTCGATT-TAAAAAAAAAAAAA-3'
Antisense	5'-TTTTTTTTTTTTTAAATCGAGTTTAGCACTGTGTATAGTCCGATCTAGCGTACTTTAGTGAGGGT-TAATCAGTTTTTTTT-3'
Oligonucleotide tag 2	
Sense	5'-AAAAAAAACTGATTAACCCTCACTAAAGTATGTGGTCAACCCTATACACAGTGCTAAACTCGATT-TAAAAAAAAAAAAA-3'
Antisense	5'-TTTTTTTTTTTTTAAATCGAGTTTAGCACTGTGTATAGGGGTTGACCACATACTTTAGTGAGGGT-TAATCAGTTTTTTTT-3'
Oligonucleotide tag 3	
Sense	5'-AAAAAAAACTGATTAACCCTCACTAAAGTAGTTCGCACATGCCTATACACAGTGCTAAACTCGATT-TAAAAAAAAAAAAA-3'
Antisense	5'-TTTTTTTTTTTTTAAATCGAGTTTAGCACTGTGTATAGGCATGTGCGAACTACTTTAGTGAGGGT-TAATCAGTTTTTTTT-3'
Oligonucleotide tag 4	
Sense	5'-AAAAAAAACTGATTAACCCTCACTAAAGTAGTCCAGACTGACCTATACACAGTGCTAAACTCGATT-TAAAAAAAAAAAAA-3'
Antisense	5'-TTTTTTTTTTTTTAAATCGAGTTTAGCACTGTGTATAGGTCAGTCTGGACTACTTTAGTGAGGGT-TAATCAGTTTTTTTT-3'

(continued)

Table 1
(continued)

Oligonucleotide tag 5	
Sense	5'-AAAAAAAACTGATTAACCCTCACTAAAGTACTGAAAGTGCCCTATACACAGTGCTAAACTCGATT-TAAAAAAAAAAA-3'
Antisense	5'-TTTTTTTTTTTTAAATCGAGTTTAGCACTGTGTATAGGGCCACTTTCAGTACTTTAGTGAGGGT-TAATCAGTTTTTTTT-3'
PCR primers	
Sense	5'-CCCCAAGTACAGCCTTCGCAGCCTGATTAACCCTCACTAAAG-3'
Antisense	5'-CTGGCTTCTTGATCGAGTTTAGCACTGTGTATAG-3'
Capture oligonucleotide 1	
	5'-CGCTAGATGGGACTATAC-3'
Capture oligonucleotide 2	
	5'-TGTGGTCAACCCCTATAC-3'
Capture oligonucleotide 3	
	5'-GTTCGGCACATGCCTATAC-3'
Capture oligonucleotide 4	
	5'-GTCCAGACTGACCTATAC-3'
Capture oligonucleotide 5	
	5'-CTGAAAGTGGCCCTATAC-3'

Oligonucleotides used for the generation of double-stranded oligonucleotide tags, for PCR amplification and as capture oligonucleotides (18-mer) for the sequence-specific quantitation of the different oligonucleotide tags by PCR-ELISA. Oligonucleotide tags and capture oligonucleotides are labeled by 5'-biotin. The double-stranded oligonucleotides tags (80-mer) are composed of an SH2 domain specific, internal recognition sequence of 18 bp with primer binding sites of 20 bp (5'-primer) and 23 bp (3'-primer) followed by terminal A/T-spacers of 8 bp and 15 bp, respectively. PCR primers carry a 5'-overhang of 22 bp (5'-primer) and 11 bp (3'-primer), respectively, for improved purification and analysis of PCR products by gel electrophoresis. The antisense primer is labeled by 5'-digoxigenin

domain. Weak or no binding indicates protein degradation or aggregate formation, possibly caused by improper bacterial expression and purification. Only use those SH2 domains for phosphotyrosine profiling that bind in a concentration- and phosphotyrosine-dependent manner.

8. Alternatively, far-Western blotting can be performed with nonbiotinylated GST-SH2 domains. For detection (step 5), 1 μg of SH2 domain is precomplexed (instead of streptavidin-HRP) with 0.2 μg of glutathione-HRP conjugate or anti GST-HRP (Sigma) for 30 min at room temperature. Proceed with **step 6**.

9. When cutting out the band of interest from the gel, to reduce the amount of extraction buffer needed do not cut out an unnecessary large piece of acrylamide, as the total volume to be loaded to the NAP-5 column is 0.5 ml. Weigh the gel slice and add a corresponding volume of gel extraction buffer. Take into account an additional volume of 100 μl needed for extra washing of the crushed gel.

10. Thorough crushing of the gel slice is important as this will significantly increase the yield of the extracted amount of DNA. Typically three cycles of repeated freezing and crushing are sufficient.

11. To prevent any contamination or carryover of oligonucleotide tags, use aerosol resistant filter tips suitable for pipetting of PCR reactions. For complex formation, use 1 μg of each SH2 domain and incubate the SH2 domain with 0.4 μg (7 pmol) of streptavidin (4 μl of a 1:100 dilution of ImmunoPure streptavidin (Pierce) in PBS). Add PBS to a final volume of 50 μl. After incubation, add 14 pmol of the corresponding double-stranded oligonucleotide tag.

12. When five SH2 domains are used in a multiplex reaction, adjust the volume by 5 ml of TBST (final volume of the reaction 10 ml). To this point, OTM was tested with up to ten SH2 domains in a multiplex reaction but for the establishment of the method, experiments should be started with a lower number of domains e.g., five SH2 domains in a multiplex reaction. To obtain the binding information for more than ten SH2 domains, separate binding reactions with ten domains each can be performed in parallel. Before multiplexing is started, test the performance of the assay using single oligonucleotide-tagged SH2 domains at a concentration of 0.2 μg/ml.

13. Circular pieces of membrane are used to reduce the binding surface in order to obtain an optimal signal-to-noise ratio. Binding capacity of PVDF is estimated for globular proteins to be 50–100 μg/cm². A circular piece of membrane with a diameter of 5 mm will therefore bind a maximum amount of

10–20 µg of lysate. For each step, membrane pieces have to be treated very carefully to prevent damage of the surface resulting in substantial increase of background. Do not use sharp forceps or other sharp tools and minimize the number of steps of transferring the membrane pieces and touching the surface. Using two forceps to catch the small membrane pieces is helpful for transfer. During each transfer, remove excess liquid on the membrane by carefully placing one edge of the membrane on a dry filter paper. Do not let the membrane dry.

14. To prevent cross-contamination and mix-up between the unlabelled filter pieces, the different membrane pieces should be kept and processed in separate dishes at all times.

15. Use a hanging drop on the pipette tip to apply and distribute the liquid without touching the membrane. Incubation should be performed with a two- to fourfold excess of protein amount over the binding capacity of the membrane to guarantee that ~10 µg of lysate are immobilized per filter piece. During incubation samples start drying but do not get completely dry.

16. Keep the membrane circles in TBST as they can be transferred more easily when floated in buffer.

17. To prevent cross-contamination, rinse forceps after each transfer. The elution of a large number of wells may be time consuming. Do not start the elution of all wells at one time.

18. For quantitation by PCR–ELISA, PCR amplification must be performed within the linear range of the PCR amplification curve, as quantitative differences cannot be properly determined at saturation. The cycle number determined by quantitative real-time PCR should be used for the amplification of eluates in a conventional PCR reaction. We find linear amplification is obtained at an optimum of 25–28 PCR cycles for all experiments. Phenyl phosphate may inhibit the PCR amplification when present in higher concentrations. If problems emerge, eluates can be purified using G-50 spin columns (GE Healthcare). In case of purification of eluates, up to 10 µl of eluates can be added to the PCR reaction.

19. As an alternative, the sequence-specific quantitation of oligonucleotide tags present in the eluates can be performed by quantitative real time PCR using hybridization probes complementary to the internal recognition sequences of the oligonucleotide tags. Quantitation can be performed using the LightCycler fast start DNA master Hyb probes PLUS kit and a LightCycler instrument (Roche). Using quantitative real-time PCR for quantitation of oligonucleotide tags increases reproducibility of the assay as the PCR amplification and the quantitative readout are performed in one reaction, and

no additional PCR–ELISA is needed. PCR–ELISA typically involves additional pipetting steps that reduce reproducibility.

20. The template image being generated by vector-based drawing software, e.g., Claris Draw (Claris), should be customized to each experiment: as many as 19 spots can be fit in a single well, and up to four membranes can be printed on a guide film. Note that a nitrocellulose membrane may slightly expand after wetted in a solution, which should be considered in designing an array.

21. In addition to the membranes for the 96-well format, we usually prepare small membranes containing four replicates (**Fig. 4a** top left).

22. The viscosity of sample solution, mainly influenced by the concentration of SDS and glycerol is an important factor for successful spotting. To assess the activity of probes, we always include control samples within an array; for GST-SH2 domains, pooled lysates of pervanadate-treated cells may serve as a positive control and tyrosine phosphatase-treated lysates as a negative control.

23. Depending on the experimenter's ability, the accuracy of manual spotting may vary slightly. Usually the raw result is suitable for semiquantitative analysis. A typical spotting volume with this protocol is about $0.1\,\mu l$ ($0.4\,\mu g$ per spot). Variation in sample spotting should be assessed by incubating test membranes with an antibody recognizing a ubiquitous protein. Likewise, the membrane used for the Rosette assay may be stripped and reprobed to normalize for variations in sample loading. For stripping, briefly rinse the dot blot twice in TBST, and wash in $100\,mM$ Glycine–HCl solution (pH 2.0) for $20\,min$ at RT with agitation occasionally replacing the buffer. Then further wash in TBST for $45\,min$ followed by blocking and reprobing with an appropriate antibody.

24. Dried membranes can be stored at room temperature at least 6 months.

25. Alternatively the membrane can be blocked at 4°C overnight.

26. Aliquots of SH2 domains are stored at −80°C. For labeling, $5\,\mu g$ of GST-SH2 domain stock is quickly thawed at room temperature and precomplexed with $1\,\mu g$ of GSH-HRP conjugate or $0.2\,\mu l$ of anti-GST-HRP ($15\,\mu g/\mu l$) conjugate for $1\,h$ at 4°C. Optimal concentration of probes should be determined beforehand. For GST-SH2 domains, concentrations from 0.5 to $5\,\mu g/ml$ are favorable for the Rosette assay.

27. The small membrane (**Note 21**) may be used for control immunoblotting, e.g., antiphosphotyrosine blotting, by incubating

it with an antibody in a six-well dish. Likewise, binding reactions for SH2 domain probes may be performed using multiple six-well dishes. Cut a spotted 96-well size membrane into $20 \times 20\,mm^2$ size pieces so one piece contains four replicate arrays, and then binding reaction is performed in separate wells.

28. To eliminate the chance of cross-well contamination rinse the tip of the aspiration tube with TBST after every aspiration.

29. Do not wash for more than 20 min. Excess washing may wash out the signal.

30. Select an appropriate chemiluminescence kit with sufficient sensitivity (*see* **Note 6**). Specific signal detected by exposed X-ray films or phosphorimager is quantified using densitometry, e.g., integrated density measurement by ImageJ software.

Acknowledgments

We are grateful to C. Thompson for technical assistance. This work was partially supported by grants from Breast Cancer Alliance and Connecticut Breast Health Initiative (to K.M.) and NIH grant CA107785 (to P.N. and B.J.M.).

References

1. Hunter T. (2000) Signaling–2000 and beyond. *Cell* 100(1), 113–27.

2. Schlessinger J. (2000) Cell signaling by receptor tyrosine kinases. *Cell* 103(2), 211–25.

3. Manning G, Whyte DB, Martinez R, Hunter T, Sudarsanam S. (2002) The protein kinase complement of the human genome. *Science* 298(5600), 1912–34.

4. Alonso A, Sasin J, Bottini N, et al. (2004) Protein tyrosine phosphatases in the human genome. *Cell* 117(6), 699–711.

5. Blume-Jensen P, Hunter T. (2001) Oncogenic kinase signalling. *Nature* 411(6835), 355–65.

6. Hanahan D, Weinberg RA. (2000) The hallmarks of cancer. *Cell* 100(1), 57–70.

7. Krause DS, Van Etten RA. (2005) Tyrosine kinases as targets for cancer therapy. *N. Engl. J. Med.* 353(2), 172–87.

8. Waksman G, Shoelson SE, Pant N, Cowburn D, Kuriyan J. (1993) Binding of a high affinity phosphotyrosyl peptide to the Src SH2 domain: crystal structures of the complexed and peptide-free forms. *Cell* 72(5), 779–90.

9. Pawson T. (2004) Specificity in signal transduction: from phosphotyrosine-SH2 domain interactions to complex cellular systems. *Cell* 116(2), 191–203.

10. Jones RB, Gordus A, Krall JA, MacBeath G. (2006) A quantitative protein interaction network for the ErbB receptors using protein microarrays. *Nature* 439(7073), 168–74.

11. Songyang Z, Shoelson SE, Chaudhuri M, et al. (1993) SH2 domains recognize specific phosphopeptide sequences. *Cell* 72(5), 767–78.

12. Huang H, Li L, Wu C, et al. (2007) Defining the specificity space of the human src-homology 2 domain. *Mol. Cell. Proteomics* 7, 768–84.

13. Machida K, Mayer BJ. (2005) The SH2 domain: versatile signaling module and pharmaceutical target. *Biochim. Biophys. Acta* 1747(1), 1–25.

14. Liu BA, Jablonowski K, Raina M, Arce M, Pawson T, Nash PD. (2006) The human and

mouse complement of SH2 domain proteins-establishing the boundaries of phosphotyrosine signaling. *Mol. Cell* 22(6), 851–68.

15. Machida K, Thompson CM, Dierck K, et al. (2007) High-throughput phosphotyrosine profiling using SH2 domains. *Mol. Cell* 26(6), 899–915.

16. Stagljar I. (2003) Finding partners: emerging protein interaction technologies applied to signaling networks. *Sci. STKE* 213, pe56.

17. Johnson SA, Hunter T. (2005) Kinomics: methods for deciphering the kinome. *Nat. Methods* 2(1), 17–25.

18. Machida K, Mayer BJ, Nollau P. (2003) Profiling the global tyrosine phosphorylation state. *Mol. Cell. Proteomics* 2(4), 215–33.

19. Nollau P, Mayer BJ. (2001) Profiling the global tyrosine phosphorylation state by Src homology 2 domain binding. *Proc. Natl. Acad. Sci. U S A* 98(24), 13531–6.

20. Dierck K, Machida K, Voigt A, et al. (2006) Quantitative multiplexed profiling of cellular signaling networks using phosphotyrosine-specific DNA-tagged SH2 domains. *Nat. Methods* 3(9), 737–44.

Part III

Phosphoprotein Analysis by Mass Spectrometry

Chapter 12

An Overview of the Qualitative Analysis of Phosphoproteins by Mass Spectrometry

Philip R. Gafken

Summary

Protein phosphorylation is a reversible and frequently occurring posttranslational modification regulating a large number of biological functions. Understanding the role phosphorylation events play in biochemical pathways requires the detection of phosphorylated proteins and their phosphorylated amino acids. Mass spectrometry has developed as the premier method for characterizing phosphoproteins as it is sensitive to detecting phosphosites within a single protein or a complex protein mixture (phosphoproteomics). Here an overview is provided of the sample separation and mass spectrometry techniques commonly used for qualitative phosphoprotein analysis.

Key words: Phosphorylation, Mass spectrometry, Phosphoprotein, Phosphopeptide, Qualitative analysis.

1. Introduction

Addition and removal of phosphate groups to proteins, carried out by kinase and phosphatase proteins, respectively, is ongoing throughout the life of a cell. In-depth analysis of the human genome has identified more than 500 kinase genes *(1)* and predicted more than 100 phosphatase genes *(2)*. It is well established that kinases and phosphatases often recognize multiple substrates resulting in the estimation that one-third of the human proteome is phosphorylated at any given time *(3)*. The classes of proteins and the biological functions affected by phosphorylation are widespread making protein phosphorylation one of the most common and dynamic forms of protein modification.

Marjo de Graauw (ed.), *Phospho-Proteomics, Methods and Protocols, vol. 527*
© 2009 Humana Press, a part of Springer Science + Business Media, New York, NY
Book DOI: 10.1007/978-1-60327-834-8_12

Prior to the use of mass spectrometry for analyzing bio-molecules, phosphoproteins were traditionally characterized by radiolabeling phosphoproteins with [γ-^{32}P]ATP, followed by proteolytic digestion, purification, and Edman-based protein sequencing. Although this route has been used successfully to identify phosphoproteins and locate phosphorylated amino acids, this process is not conducive to multiplexing, generally requiring a single phosphoprotein to be analyzed in a single experiment. Overall, the traditional means to characterize phosphoproteins was limited in the biological experiments it could address as it was unable to analyze large numbers of phosphoproteins in a single experiment, such as cataloging the phosphoproteins within a cell or identifying groups of phosphoproteins that change as a result of a biological perturbation. Mass spectrometry (MS) has emerged as a powerful technique for identifying phosphorylated proteins and locating the site(s) of phosphorylation within a phosphoprotein. Modern mass spectrometers have routine levels of detection down to low femtomole quantities, the ability to detect large numbers of phosphoproteins from complex mixtures (phospho-proteomics), and MS-based quantitative techniques exist allowing for qualitative and quantitative phosphoprotein analysis to be performed in a single experiment. Here, an overview is provided for the use of mass spectrometry for the qualitative analysis of phosphoproteins.

2. Experimental Workflow

Two levels of information are typically sought during the qualitative analysis of phosphoproteins. First is the identity of the phosphoprotein and second is the location of the phosphate group within the phosphoprotein. Reaching these two levels of information can be accomplished by either analyzing purified phosphoproteins or by analyzing mixtures of phosphoproteins, with the same general strategy being used in both situations. The strategy (**Fig. 1**) starts with treating the phosphoprotein or phosphoprotein mixture with a protease to create peptides. The peptide mixture is then analyzed by the mass spectrometer resulting in mass spectra that contain sequence information about the analyzed peptides. The mass spectral information is used to identify the peptides back to the proteins from which they originated and locate sites of phosphorylation. Under most circumstances the amount of data produced by the mass spectrometer can be enormous and very complex requiring computational assistance to analyze the data.

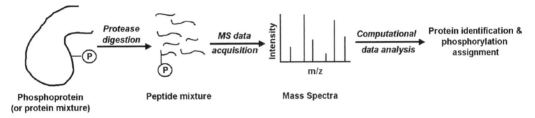

Fig. 1. General experimental workflow for qualitative analysis of phosphopeptides. A purified phosphoprotein (shown) or a protein mixture containing phosphoproteins (not shown) is treated with a protease to create a peptide mixture. The peptide mixture is then analyzed by mass spectrometry, or if desirable the peptide mixture can be enriched for phosphopeptides (not shown), and analyzed. The resulting mass spectrometry data are subjected to computational analysis to identify the protein(s) from which the peptides originated and to locate sites of phosphorylation.

It is worth noting that the workflow outlined above is referred to as "bottom–up" proteomics. This workflow is in contrast to "top-down" proteomics in which whole proteins, or phosphoproteins, are analyzed by the mass spectrometer to provide sequence information to identify a protein and locate sites of phosphorylation *(4)*. The vast majority of mass spectrometry-based proteomics experiments, including those related to phosphoproteins, have historically taken place via the bottom-up workflow. This has been due to three reasons. First, peptides behave similarly to one another and their chemical characteristics can be predicted making separations and enrichment of peptides straightforward. Second, the detection of peptides by mass spectrometry is more sensitive than the detection of proteins. Third, mass spectrometers easily provide the necessary spectral information for characterizing peptides in a high throughput manner but characterizing whole proteins in a high throughput manner does not yet exist. Although bottom-up workflows are currently the dominating approach, recent advances in mass spectrometry are making top–down workflows more prevalent (*see* **Subheading 6**). The remainder of this overview will focus on characterizing phosphoproteins via bottom-up proteomics.

3. Sample Preparation

Generating quality data from a mass spectrometer is directly linked to the quality of the sample used. In general, samples need to be prepared in such a fashion that materials incompatible with mass spectrometry are removed prior to introduction of the sample into the instrument. Often encountered reagents that are incompatible with mass spectrometry are high salts, detergents, phosphate buffers, and solvents with low volatility. It is important to remember that the context of a biological experiment will influence the preparation of a sample, and MS-incompatible reagents may be used during biochemical processing (e.g., use of detergents to isolate membrane proteins); however, it is important that the incompatible reagents are removed before mass spectrometry. It is highly recommended that a mass spectrometry lab be contacted prior to conducting biochemical processing to ensure compatibility.

The extent of phosphorylation of any given protein varies markedly within a cell and phosphorylation stoichiometry is often much less than one. Because of the nature of the bottom-up workflow for processing samples for phosphoprotein analysis, whether it is the processing of a single protein or a mixture of proteins, a smaller number of phosphorylated peptides are typically in the background of a larger number of nonphosphorylated peptides. This adds unwanted complexity to the mass spectrometry experiment, making it difficult to detect mass spectrometry signals from phosphopeptides within the large background of nonphosphorylated peptides. Therefore, robust separation and enrichment strategies for phosphoproteins and phosphopeptides are beneficial. Below are four separation and enrichment techniques that are useful for working with phosphoproteins and phosphopeptides.

3.1. SDS-PAGE

A single phosphorylation site on a protein will increase the mass of a protein by 80 Da, a mass increase not typically resolved on a gel. However, it has often been observed that a phosphoprotein will have an altered electrophoretic migration as compared with the nonphosphorylated form, and when treated with alkaline phosphatase the phosphoprotein band collapses to the nonphosphorylated form. The explanation of this observation is that uniform binding of SDS to the phosphoprotein is disrupted resulting in a decrease in the charge density of the phosphoprotein relative to the nonphosphorylated form. If the difference in charge densities between the two forms is large enough, a phosphorylated protein would have a retarded migration and appear at a higher molecular weight in the gel relative to the nonphosphorylated form. The widespread use of SDS-PAGE makes it a useful tool for separating phosphorylated from nonphosphorylated forms of the same protein. An added advantage of this technique is the ability to readily perform proteolytic digestions "in-gel" to convert phosphoproteins into phosphopeptides for subsequent analysis by mass spectrometry (5). The most common protease used for in-gel digestions is trypsin, as the digestion efficiency is high and the resulting peptides are of a size (700–3,000 Da) that are readily conducive to mass spectrometry. Other proteases that are less commonly used for in-gel digestions are endoproteinase Arg-C, endoproteinase Lys-C, and chymotrypsin, but in-gel digestion efficiency using these proteases is often times low and the proportion of peptides within the optimum size range for mass spectrometry may be low.

3.2. Purification Via Antibodies

An antibody raised against a protein can be used to immuno-precipitate the protein and search for phosphorylation sites. Immunoprecipitation can isolate a protein under a number of biological conditions (e.g., different stages of the cell cycle) to assess changes in phosphorylation on that protein. Antibodies can also be raised against a specific phosphosite on a protein and used for immunoprecipitation. Here, other phosphosites on a protein can be assessed when one phosphosite is known (the epitope of the antibody), and proteins that bind to the phosphoprotein can be determined for protein–phosphoprotein interaction studies. Also, antibodies with high specificity for phosphotyrosine are commercially available and able to immunoprecipitate phospho-tyrosine-containing proteins on a "phosphotyrosine-ome" scale. There have been reports where antiphosphotyrosine antibod-ies have been used after proteolytic digestion of cell lysates to immunoprecipitate phosphotyrosine-containing peptides prior to mass spectrometry (6). This enrichment of phosphotyrosine-containing peptides can make mass spectrometry analysis easier by markedly reducing the number of spectra associated with non-phosphotyrosine-containing peptides collected during the analy-sis. Unfortunately, antibodies that recognize phosphoserine and phosphothreonine with high specificity are not available making large-scale analyses of phosphoserine- and phosphotheronine-containing proteins much more involved.

3.3. SCX

A recently described technique for enriching phosphopeptides on a proteomics-scale utilized strong cation exchange (SCX) chro-matography, which separates peptides primarily based on charge. Based on a bioinformatic study of a human protein database, it was recognized that most peptides (68%) produced from complete trypsinization of the proteins in the database would have a net charge of +2 at pH 2.7 while a very small number of the peptides (3%) would have a net solution charge of +1 at pH 2.7 (7). It was surmised that a phosphate group at an acidic pH maintains a neg-ative charge, and thus most fully tryptic peptides that are phos-phorylated would have their net solution charge reduced from +2 to +1, resulting in most phosphopeptides containing single phosphate group eluting in the less complex +1 solution charge region during SCX separation. Highly enriched phosphopeptide fractions are produced with a large fraction of the nonphospho-rylated peptide background being removed. The enriched frac-tions are suitable for analysis by mass spectrometry or they can be further purified by other techniques, such as metal-assisted enrichment (see below). The power of combing SCX with mass spectrometry has been demonstrated with the identification of over 2,000 phosphorylation sites from a HeLa cell lysate *(7)* and over 500 phosphorylation sites in proteins from the developing mouse brain *(8)*.

3.4. Metal-Assisted Enrichment

The nature of bottom-up proteomics workflows is such that complex peptide mixtures are created when proteolytically digesting proteins and protein mixtures. With phosphoprotein analysis, phosphopeptides reside within a background of unphosphorylated peptides following proteolytic digestion of a protein or protein mixture. Simplifying the mixture, by removing unphosphorylated peptides from the mixture while enriching for phosphorylated peptides, enhances the detection of phosphorylated peptides by the mass spectrometer. An increasingly used technique for enriching phosphopeptides is immobilized metal affinity, or IMAC. This technique uses a metal, such as iron (III), copper (II), or gallium (III), chelated to a metal-binding ligand on a resin allowing for the positively-charged metal to attract the negative charge on the phosphate group of the phosphopeptide. This results in unphosphorylated peptides being washed away from bound phosphopeptides and the bound phosphopeptides being eluted from the resin. A similar technique to IMAC that uses titanium oxide based solid phase resin has recently been developed for phosphopeptide enrichment. Both techniques are extremely valuable in identifying phosphopeptides and they are described in detail in other chapters of this volume of Methods in Molecular Biology.

4. The Mass Spectrometer

Mass spectrometers are invaluable tools for characterizing proteins and peptides. These instruments measure the mass-to-charge ratio (*m/z*) of gas-phase ions that are created in the ionization source of the mass spectrometer. The two most common ionization sources for proteins and peptides are matrix-assisted laser desorption/ionization (MALDI) and electrospray ionization (ESI), as they are the most proficient techniques for ionizing macromolecules. Ions produced in the source are sorted in a mass analyzer and then detected. Both MALDI and ESI are coupled with numerous types of mass analyzers, including triple quadrupole, ion trap, time-of-flight, and Fourier-transform ion cyclotron resonance. Most modern instruments are capable of detecting nanogram quantities of proteins or peptides with mass errors less than ten parts per million (e.g., 0.01 Da error for a peptide of 1,000 Da). As it is difficult to discuss the strengths and limitations of each instrument within the broad field of qualitative phosphoprotein analysis, consultation with experienced biological mass spectrometry labs during the planning stages of an experiment is highly recommended to determine the most appropriate instrumentation to provide the desired results.

4.1. Multi-Stage Mass Analysis

Phosphopeptides that are interrogated by the mass spectrometer undergo a sequencing process performed in two stages called tandem mass spectrometry or MS/MS. In the first stage of analysis (**Fig. 2a**), peptides ionized in the ionization source are sorted by the mass analyzer and detected. This first stage measures the molecular mass of peptides and phosphopeptides present and it is termed the MS or MS1 analysis. In the second stage of analysis (**Fig. 2b**), an ion of interest from the MS analysis is isolated by the mass spectrometer and then subjected to collision induced dissociation (CID) in which the selected ion collides with inert gas molecules (such as helium or argon) to fragment the peptide ion, with fragmentation occurring mostly along the peptide backbone. The resulting fragment ions are then separated by a second mass analyzer and detected. This second stage is called the MS/MS or MS2 analysis and it provides mass information about the peptide fragments produced from CID that can be interpreted into amino acid sequence information. Ultimately, the sequence information generated by the mass spectrometer (**Fig. 2c**) provides sequence information to identify the protein from which the phosphopeptide originated and it provides information about the location of the phosphorylated amino acid.

The time required to perform a tandem mass spectrometry analysis of a single peptide ion, a MS1 analysis followed by a MS2 analysis, ranges from approximately 0.2 to 2 s depending on the type of mass spectrometer used. Virtually all tandem mass spectrometry experiments are computer controlled where software settings selected by the instrument operator determine which ions in a MS analysis are further analyzed by MS/MS analysis. This is called data dependent analysis and allows mixtures of peptides to be continuously analyzed by the mass spectrometer in an expedient manner without continuous operator intervention.

4.2. Neutral-loss Scanning

Successful identification of a phosphorylated peptide requires that fragmentation of the phosphopeptide during CID takes place along the peptide backbone in order to create mass sequence ladders in the mass spectrum that can be interpreted into amino acid sequence. Quite often with phosphopeptides, but not always, the major fragmentation event from CID is cleavage of the phosphate group or phosphoric acid (phosphate group and water) from the peptide (**Fig. 3a**), termed a neutral-loss as the phosphate group is not charged and not detected by the mass spectrometer, with minor fragmentation taking place along the

(a)

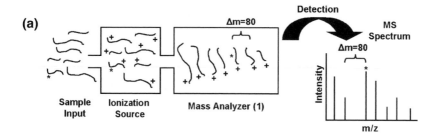

(b)

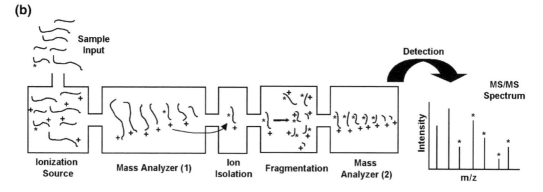

(c)

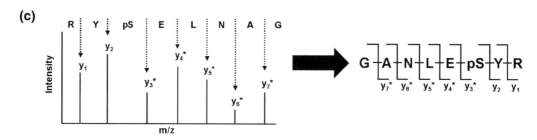

Fig. 2. Use of tandem mass spectrometry to identify phosphorylation sites within phosphopeptides. (**a**) The first stage of mass analysis, or MS, is performed where peptides created from the enzymatic digestion of a phosphoprotein (**Fig. 1**) are converted to positive-charged gas-phase ions (depicted as a *positive sign* attached to a *wavy line*) in the ionization source, separated in the mass analyzer, and detected. Tentative identification of a phosphopeptide (depicted as a *positive sign* and an *asterisk* attached to a *wavy line*) is made due to the presence of an 80 Da difference between two peaks in the mass spectrum, the mass difference between the phosphorylated and unphosphorylated forms of a peptide. (**b**) Determining the sequence of a phosphopeptide and locating the site of phosphorylation is performed by a second stage of mass analysis, or MS/MS, in which the phosphopeptide is selected by the first mass analyzer and subjected to fragmentation. The fragments of the phosphopeptide are then separated in a second mass analyzer and detected. (**c**) In its simplest form (MS/MS spectra are typically much more complex), the MS/MS spectrum shows an ion series in which the mass differences between adjacent peaks corresponds to the masses of amino acids (*Note*: fragments that occurs at the peptide bonds of a peptide are termed b-ions if the fragments contain the N-terminus or y-ions if the fragments contain the C-terminus; for simplicity, only y-ions are illustrated). The mass difference between the y_3^* and y_2 ions does not equal the mass of an amino acid, but equals the mass of an amino acid plus a phosphate group indicating the position of the phosphorylated amino acid within the peptide sequence.

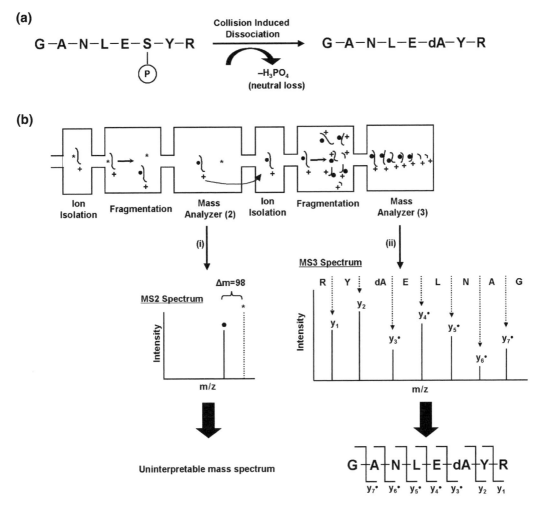

Fig. 3. Qualitative analysis of phosphopeptides by neutral-loss scanning. (**a**) The neutral loss of a phosphate group or a phosphate group and water (phosphoric acid) frequently occurs for phosphoserine-, phosphothreonine-, and phosphotyrosine-containing phosphopeptide during collision induced dissociation. In this example, during collision induced dissociation of a phosphopeptide, phosphoric acid is removed from serine as a neutral loss (phosphoric acid is not charged and cannot be detected by the mass spectrometer) to produce a peptide that contains dehydroalanine in place of phosphoserine. (**b**) In the neutral-loss mode of the mass spectrometer, a phosphopeptide (denoted with an *asterisk*) that is fragmented and has undergone a neutral loss can be detected by comparing the mass of the phosphopeptide with fragments that are detected in the tandem mass spectrum. In this example, a fragment in the MS2 spectrum that has a mass corresponding to the loss of phosphoric acid (98 Da) from the phosphopeptide is present indicating a neutral loss has occurred. The resulting dehydroalanine containing peptide (denoted as a *wavy line* with a *dot*) can be isolated in the second round of mass analysis, subjected to fragmentation to produce fragments of the peptide, with fragments being separated with a third round of mass analysis and detected. The analysis of the resulting MS/MS/MS, or MS3, spectrum is identical to that of the MS2 spectrum in **Fig. 2**, except a mass difference between adjacent ions corresponding to the mass of dehydroalanine indicates the presence of phosphoserine at that position in the phosphopeptide.

peptide backbone. The resulting fragmentation spectra provide little sequence information making it difficult or impossible to identify the protein from which the phosphopeptide originated and difficult or impossible to locate the site of phosphorylation within the peptide. Typical neutral-loss for phosphoserine and phosphotheronine is phosphoric acid to form dehydroalanine and β-methyldehydroalanine, respectively, while the typical neutral loss for phosphotyrosine is a phosphate group to form tyrosine *(9)*.

The mass spectrometer is able to detect the occurrence of a neutral loss during tandem mass spectrometry of the phosphopeptide in a technique called neutral loss scanning. Here, the mass of a precursor ion selected for tandem mass spectrometry is compared with fragment masses in the resulting MS2 spectrum. If a fragment occurs in the MS2 spectrum that is 80 Da or 98 Da less than the mass of the phosphopeptide measured in the MS1 spectrum, due to the loss of a phosphate group or phosphoric acid (phosphate group and water), respectively, then a neutral loss event has occurred (**Fig. 3bi**). An additional round of tandem mass spectrometry, or MS3, can be performed (often times with ion trap mass analyzers) on the resulting peptide. With the phosphate group removed from the phosphopeptide, the resulting fragmentation spectrum usually contain ions that were produced from fragmenting the peptide backbone, and thus provide sufficient sequence information to identify the peptide and locate the site of phosphorylation (**Fig. 3bii**). In the case of phosphoserine- and phosphothreonine-containing peptides, locating sites of dehydroalanine and beta-methyldehydroalanine, respectively, within the peptide sequence often indicate sites of phosphorylation.

5. Analytical Schemes for Identifying Sites of Phosphorylation

The qualitative analysis of phosphoproteins usually falls into one of two catagories: identifying sites of phosphorylation on a single protein or identifying sites of phosphorylation on proteins in a mixture. Although mass spectrometry can be used for both types of experiments, the analytical scheme for each type of experiment is different.

5.1. Peptide Mapping

The approach for identifying sites of phosphorylation on a single protein amounts to a thorough characterization of the protein's primary sequence. Here, an isolated phosphoprotein with a known amino acid sequence is digested with a protease and the resulting peptides are analyzed by mass spectrometry.

The resulting mass spectrometry data, though complex, is often amenable to manual analysis or it can be can be analyzed with the aid of automated protein database search algorithms (e.g., MASCOT, X!TANDEM, SEQUEST). Detected peptides are then "mapped" to the primary sequence of the protein (**Fig. 4**). If coverage of the map is low, due to sizes or chemical characteristics of peptides that are not suitable for detection by mass spectrometry, then a parallel proteolytic digestion using a different protease can be performed to produce additional sequence coverage. The more protein sequence coverage that is obtained, the more that is known about the phosphorylation status of the protein; conversely, the phosphorylation status of protein regions that are not mapped remain unknown assuming the unmapped regions contain serine, threonine, or tyrosine.

One example of a peptide mapping experiment is the determination of kinase specificity in which purified kinase with ATP phosphorylates a purified protein or peptide with subsequent identification of the site(s) of phosphorylation. Another example is identifying sites of phosphorylation on an immunoprecipitated protein. In both cases, a known protein sequence is being mapped so the selection of proteases that will provide optimum sequence

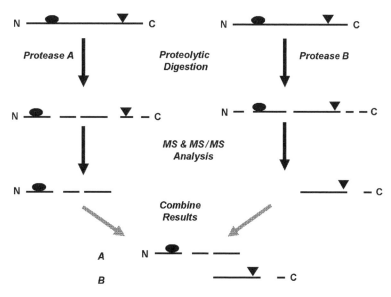

Fig. 4. Peptide mapping scheme. A phosphoprotein containing two sites of phosphorylation (denoted by an *oval* and a *triangle*) is digested with protease A to produce peptides that are identified by mass spectrometry. The resulting map of the peptide located the site of one phosphorylation site (*the oval*), but due to low sequence coverage the second phosphorylation site was not identified. A second digestion of the protein can be performed with protease B to produce a second map that identifies a second site of phosphorylation (*the triangle*). Combining the results from the two protease digestions yields larger sequence coverage to reveal more information about the phosphorylation status of the protein.

coverage can be determined prior to performing the experiment. The quantity of phosphoprotein needed for an experiment is dependent upon on factors such as solubility of the phospho-protein and the stoichiometry of phosphorylation. In general, it is recommended that 1–10 pmol of phosphoprotein are used to generate extensive peptide maps. This differs markedly to protein identification where only a few peptides sequenced back to a protein are necessary, and thus, as little as 100 femtomoles of a protein are required.

5.2. Shotgun Identification

In contrast to peptide mapping which analyzes a single protein, the shotgun identification scheme refers to the analysis of peptide mixtures produced from the proteolytic digestion of complex protein mixtures (e.g., yeast lysate). In phospho-proteomics, this scheme is often used to catalog or detect novel phosphorylation sites within a protein mixture. After protease digestion of the protein mixture to produce peptides, the peptide mixture can be enriched for phosphopeptides (e.g., IMAC, TiO_2) prior to analysis by mass spectrometry. Unlike peptide mapping in which data analysis may be conducted in a manual fashion, shotgun identification produces datasets with very large number (greater than 10,000) of spectra from unknown proteins requiring the use of automated protein database search algorithms for data analysis. Also, unlike peptide mapping where a single protein sequence is thoroughly analyzed, a complex mixture of unknown proteins is typically analyzed. This results in sequence coverage of identified proteins being quite low, oftentimes a single peptide identifying a protein; this means the phosphorylation status for large portions of protein sequences is unknown. Using this scheme, hundreds to thousands of phosphorylation sites can be identified in a single experiment (10).

6. Developing Technology

As stated earlier, the bottom-up approach for characterizing phosphoproteins and proteomics experiments in general is the dominant proteomics strategy. Even though mass spectrometry and separation technologies are most conducive to working with peptides, there are drawbacks to this approach. The primary drawback is the loss of information associated with the whole protein in the bottom-up approach due to the loss of peptides during sample preparation or peptides may not be an optimal size for mass spectrometric detection. Also, the nature of the bottom-up approach makes alternatively-spliced and post-translationally truncated proteins difficult to detect. Due to these drawbacks, serious research efforts are being made to make top-down proteomics more routine. First, mass spectrometers are being developed to facilitate top-down analyses. Second, methods to make large-scale or proteomics-scale protein separations that are compatible with mass spectrometry are in development. Third, methodologies to perform top-down experiments in a high throughput manner are being developed. As these developments progress, top-down strategies will be used more widely.

An important area of development for phosphopeptide and phosphoprotein analysis is peptide dissociation techniques for mass spectrometry. While CID of peptides has been extremely successful for sequencing peptides, during CID phosphopeptides are prone to partial or complete loss of the phosphate group while minimal fragmentation takes place along the peptide backbone, yielding little or no information about the peptide's sequence (**Fig. 3a**). Recently developed peptide fragmentation techniques called electron capture dissociation (ECD) and electron transfer dissociation (ETD) have been shown to fragment the peptide backbone while leaving the phosphorylated amino acid intact. Both ECD and ETD have been used to sequence phosphopeptides *(11, 12)* while ECD has been used to sequence phosphoproteins in top-down experiments *(11)*. Even though both techniques have shown their utility to fragmenting phosphopeptides, it is most likely that ETD will be more widely used in the future since this technology can be coupled to cost-efficient ion trap mass spectrometers that are commonly used in peptide and protein analysis and since ETD reactions take place on time scales that allow it to be coupled to LC-MS (liquid chromatography coupled in-line with mass spectrometry). Both ETD and ECD are now available on commercial mass spectrometers and their increased use will soon make an impact in characterizing phosphoproteins.

References

1. Manning, G., Whyte, D. B., Martinez, R., Hunter, T., and Sudarsanam, S. (2002) The protein kinase complement of the human genome. *Science* 298, 1912–1934.

2. Venter, J. C., et al. (2001) The sequence of the human genome. *Science* 291, 1304–1351.

3. Cohen, P. (2000) The regulation of protein function by multisite phosphorylation- a 25 year update. *Trends Biochem. Sci.* 25, 596–601.

4. Kelleher, N. (2004) Top-down proteomics. *Anal. Chem.* 76, 197A–203A.

5. Shevchenko, A., Wilm, M., Vorm, O., and Mann, M. (1996) Mass spectrometric sequencing of proteins silver-stained polyacrylamide gels. *Anal. Chem.* 68, 850–858.

6. Rush, J., Moritz, A., Lee, K. A., Guo, A., Goss, V. L., Spek, E. J., et al. (2005) Immunoaffinity profiling of tyrosine phosphorylation in cancer cells. *Nat. Biotechnol.* 23, 94–101.

7. Beausoleil,S.A.,Jedrychowski,M.,Schwartz,D., Elias, J. E., Villen, J., Li, J. et al. (2004) Large-scale characterization of hela cell nuclear phosphoproteins. *Proc. Natl. Acad. Sci. USA* 101, 12130–12135

8. Ballif, B. A., Villen, J., Beausoleil, S. A., Schwartz, D., and Gygi, S. P. (2004)

Phosphoproteomic analysis of the developing mouse brain. *Mol. Cell. Proteomics* 3, 1093–1101.

9. DeGnore, J. P. and Qin, J. (1998) Fragmentation of phosphopeptides in an ion trap mass spectrometer. *J. Am. Soc. Mass Spectrom.* 9, 1175–1188.

10. Ficarro, S. B., McCleland, M. L., Stukenberg, P. T., Burke, D. J., Ross, M. M., Shabanowitz, J., et al. (2002) Phosphoproteome analysis by mass spectrometry and its application to *Saccharomyces cerevisiae. Nat. Biotechnol.* 20, 301–305.

11. Shi, S. D., Hemling, M. E., Carr, S. A., Horn, D. M., Lindh, I., and McLafferty, F. W. (2001) Phosphopeptide/phosphoprotein mapping by electron capture dissociation mass spectrometry. *Anal. Chem.* 73, 19–22.

12. Syka, J. E., Coon, J. J., Schroeder, M. J., Shabanowitz, J. and Hunt, D. F. (2004) Peptide and protein sequence analysis by electron transfer dissociation mass spectrometry. *Proc. Natl. Acad. Sci. USA* 101, 9528–9533.

Chapter 13

The Analysis of Phosphoproteomes by Selective Labelling and Advanced Mass Spectrometric Techniques

Angela Amoresano, Claudia Cirulli, Gianluca Monti, Eric Quemeneur, and Gennaro Marino

Summary

This chapter focuses on the development of new proteomic approaches based on classical biochemical procedures coupled with new mass spectrometry methods to study the phosphorylation, the most important and abundant PTMs in modulating protein activity and propagating signals within cellular pathways and networks. These phosphoproteome studies aim at comprehensive analysis of protein phosphorylation by identification of the phosphoproteins, exact localization of phosphorylated residues, and preferably quantification of the phosphorylation. Because of low stoichiometry, heterogeneity, and low abundance, enrichment of phosphopeptides is an important step of this analysis. The first section is focused on the development of new enrichment methods coupled to mass spectrometry. Thus, improved approach, based on simple chemical manipulations and mass spectrometric procedures, for the selective analysis of phosphoserine and phosphothreonine in protein mixtures, following conversion of the peptide phosphate moiety into DTT derivatives, is described. However the major aim of this work is devoted to the use of isotopically labelled DTT, thus allowing a simple and direct quantitative MS analysis. The final part of the work is focused on the development of a strategy to study phosphorylation without preliminary enrichment but using the high performance of a novel hybrid mass spectrometer linear ion trap.

Key words: DTT labelling, Dansyl chloride labelling, Mass spectrometry, Protein phosphorylation, Precursor ion, MS[3].

1. Introduction

1.1. The Crucial Role of Phosphorylation

One of the descriptors of a protein, which is amenable to proteomics technology, is the delineation of post-translational modifications (PTMs). The delineation of a protein's function solely

Marjo de Graauw (ed.), *Phospho-Proteomics, Methods and Protocols, vol. 527*
© 2009 Humana Press, a part of Springer Science+Business Media, New York, NY
Book DOI: 10.1007/978-1-60327-834-8_13

from a change in its abundance provides a limited view, as numerous vital activities of proteins are modulated by PMTs that may not be reflected by the changes in a protein abundance.

Protein phosphorylation is one of the most important and abundant PTMs used to modulate protein activity and propagate signals within cellular pathways and networks (1). A primary role of phosphorylation is to act as a switch, to turn "on" or "off" a protein activity or a cellular pathway in a reversible manner. This modification adjusts folding and function of proteins, e.g., enzymatic activities or substrate specificities, and regulating protein localization, complex formation and degradation. Cellular processes ranging from cell cycle progression, differentiation, development, peptide hormone response, metabolic maintenance, and adaptation are all regulated by protein phosphorylation. The variety of functions in which phosphoproteins are involved necessitates a huge diversity of phosphorylations. Several amino acids can be phosphorylated by the four known types of phosphorylation. Serine, threonine, and tyrosine residues and also unusual amino acids such as hydroxy-proline can be O-phosphorylated (2). Moreover, protein phosphorylation is generally present at substoichiometric level in the cell. In fact, some residues are always quantitatively phosphorylated, whereas others may only be transiently modified (3).

Protein phosphorylation is tightly regulated by a complex set of kinases and phosphatases, the enzymes responsible for protein phosphorylation and dephosphorylayion (4). It has been estimated that as many as one-third of the proteins within a given eukaryotic proteome undergoes reversible phosphorylation and there are more than 100,000 estimated phosphorylation sites in the human proteome (5). The cooperation of kinases and phosphatases is highly dynamic, intensely regulated, and phosphorylation cycle may take place on a very shot timescale. In addition to the problems concerning regulation, the analysis is complicated by the complexity of phosphorylation patterns: In many cases, the effects of phosphorylation are combinatorial and multiple sites are phosphorylated.

Thus, the highlighting of phosphoproteome is a huge and challenging task with regard to the dynamics and different kinds of phosphorylation generating a variety of phosphoproteins that are not accessible to a single analytical method.

1.2. Phosphoproteome Tools

One classic approach for characterizing protein phosphorylation relies on (^{32}P)-labelling, followed by two-dimensional gel electrophoresis (6). Typically gel spots of interest containing phosphorylated proteins are excised, digested, and analyzed by Edman sequencing or mass spectrometry. The signals can be absolutely quantified. Thus, the quantitative problem is approached by comparing difference in relative abundance of phosphoproteins,

^{32}P-labelled, visualized by autoradiography measuring relative intensities of the spot on the gel *(7)*. Nevertheless, the ^{32}P labelling is not the method of choice for high-throughput proteome-wide analysis because of issues with handling radioactive compounds and the associated contamination of analytical instrumentation.

As an alternative approach, commercially available phospho-stains are used. They are less sensitive than radioactive methods but the handling of these "inactive" reagents is more convenient so far. Antibodies can be used to discriminate among serine, threonine, and tyrosine phosphorylation. The specificity and sensitivity of immunostaining are strongly dependent on the respective antibodies. Various phosphotyrosine antibodies of good specificity are available and only little cross-reactivity is observed *(8)*. Over the years, several strategies have been developed that increase the sensitivity of phosphoproteome analysis and eliminate the need for radioactive and antibody labelling. MS-based methods have been developed that provide more effective tools to identify, and potentially quantify, specific sites of phosphorylation.

1.3. Mass Spectrometry and Phosphorylation on the Rise

Recently, mass spectrometry based methods have emerged as powerful and preferred tools for the analysis of post-translational modifications including phosphorylation due to higher sensitivity, selectively, and speed than most biochemical technique *(9)*. Once a protein has been isolated, several techniques can be used to identify and localize the modified amino acids. In some case, the precise measurement by mass spectrometry of the molecular weight of the intact protein can address the average number of the modified residues just by looking at the increment of the protein mass compared with the unmodified one. However, the exact location and nature of the modifying group cannot be achieved just by mass spectrometric measurement of the intact protein.

Phosphopeptides may be identified simply by examination of the list of observed peptide masses for mass increases of 80 Da compared with the list of expected peptide masses. Any ambiguities can be resolved by sequencing the peptide using tandem mass spectrometry (MS/MS) *(10)*.

Although this method is relatively straightforward, it also misses many phosphorylated peptides because peptide maps are frequently incomplete, even for non-phosphorylated proteins (some trypsin peptides are poorly ionized or poorly recovered *(11)*; the increased acidity of the phosphate group generally results in decreased ionization efficient of peptide, and competition for ionization peptides in a mixture results in suppression of signal for some peptide *(12)*). A mass spectrometry based approach to phosphopeptide analysis is especially powerful when used in conjunction with electrospray ionization (ES) *(13)*. Combined with on-line liquid chromatography, ESI-MS has formed the basis for several novel techniques for identifying phosphopeptides. Moreover, phosphopeptides have characteristic fragmentation patterns. In fact, PSD is a process where specific ions, called metastable, decompose in the flight tube because they are not sufficiently stable. In the case of serine and threonine phosphorylated peptides, the most common fragmentations are due to the loss of H_3PO_4 (more abundant) and HPO_3. These fragments do not appear in the spectrum at the exact masses because they are not properly focused by mirror.

However, despite the advances in mass spectrometry for the phosphorylation analysis, some difficulties still remain. First, the generally low phosphorylation stoichiometry of most of the proteins such that phosphopeptides are essentially present in low amounts in the generated complex peptide mixtures. Second, the increased hydrophilicity and hence reduced retention of phosphopeptides on reversed-phase materials. Finally, the selective suppression of their ionization/detection efficiencies in the presence of large amounts of unphosphorylated peptides during mass spectrometry analysis in positive mode.

1.4. Phosphopeptides Enrichment

Separation technologies such as affinity, liquid reverse-phase, ion-exchange chromatographies, and capillary electrophoresis prior to MS analysis have been gradually used in proteomics to enrich phosphoproteins that may otherwise be lost in detection. The enrichment step combined with the high sensitivity of MS technologies provides great potential for phosphoproteome characterization.

Enrichment at level of phosphoproteins and/or phosphopeptide becomes increasingly important when dealing with complex mixture. Specific antibodies can be used to enrich phosphoproteins by immunoprecipitation *(14)* from complex lysates. Recently, the use of miniaturized immobilized metal affinity chromatography (IMAC), in which phosphopeptides are bound noncovalently to resins that chelate Fe (III) or other trivalent metals followed by base elution, has proved to be a potentially valuable method in phosphopeptide enrichment. With further refinement, this technique may offer the best performance for large-scale phosphorylation analysis *(15)*.

Further several methods for selective enrichment of phosphoproteins and phosphopeptides use chemical modification of the phosphate group. This approach does not distinguish between O-glycosilated and phosphorylated analytes, therefore, requiring additional experiments to confirm phosphorylation. Zhou et al. *(16)* established a multi-step derivatization method that is capable of enriching not only Ser/Thr-phosphorylated but all types of phosphorylated peptides.

New methods that involve modifying phosphoproteins with affinity tags in combination with stable isotope incorporation *(17, 18)* have been developed for the specific enrichment and quantitation of phosphopeptides.

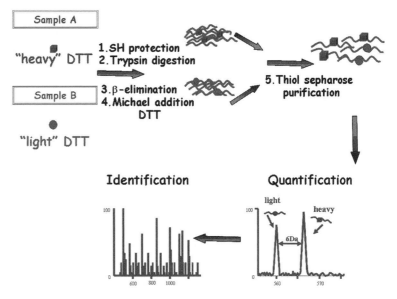

Scheme 1. Strategy to map phosphorylation sites by DTT labelling.

Here, an improved approach, based on simple chemical manipulations and mass spectrometric procedures (*see* **Scheme 1**), for the selective analysis of phosphoserine and phosphothreonine in protein mixtures following conversion of the peptide phosphate moiety into DTT derivatives *(19)* is described in the first part. The new residues in the sequence were shown to be stable and easily identifiable under general conditions for tandem mass spectrometric sequencing applicable to the fine localization of the exact site of phosphorylation. The suitability of the method with the principle of stable isotope coding is also demonstrated. However, the major aim of the present work is devoted to the use of isotopically labelled DTT thus allowing a simple and direct quantitative MS analysis.

The second part of this chapter relies on the selective detection and labelling of phosphopeptides *(20)*. In fact, several groups have reported the analytical potential of chemical derivatization in conjunction with mass spectrometry for the analysis of synthetic and biological compounds. Some derivatizing agent improves the ionization and the sensitivity of analyses by MS. Since 1968 *(21)*, the group discovered some interesting properties of 1-dimethylaminoaphthalene-5-sulphonyl- (DANS) amino acid derivatives. Since DANS derivatives give rise to a very intense fragment at *m/z* 170, the "metastable refocusing" (a primordial precursor ion scanning) was first applied for the analysis of amine mixtures *(22)*. Recently, revisiting these previous findings by using soft ionization procedures (ESI and MALDI), we have developed new strategies, introducing the novel acronym RIGhT

(Reporter Ion Generating Tag) based on dansyl labelling of the targeted residues aimed at the selective isolation and identification of the phosphorylation sites.

2. Materials

2.1. Protein Digestion

1. α-Casein water solution (1 μg/μl). Store aliquots at –20°C.
2. Buffer solution: 0.1 M ammonium bicarbonate (AMBIC), pH 8.5. Store at room temperature.
3. 10 mM dithiothreitol solution in AMBIC buffer.
4. 5 mM iodoacetamide solution in AMBIC buffer, freshly prepared in the dark.
5. Trypsin solution (TPCK-treated trypsin proteomic grade Fluka-Sigma-Aldrich): 1.0 ng/μl in 50 mM AMBIC, pH 8.5. Freshly prepared.

2.2. Phosphate Group Modification by Using DTT

1. Barium hydroxide Ba(OH)$_2$ solution: 55 M in water.
2. Solid carbon dioxide.
3. Hepes buffer solution: 10 mM in water, pH 7.5.
4. DTT solutions: In the light and heavy (CEA; Saclay, France) form, 30% w/v in Hepes buffer.

2.2.1. Isolation and Enrichment of Tagged Peptides

1. 1 ml thiol sepharose resin (Pierce Biotechnology; Rockford, IL) (aliquoted).
2. Binding buffer: 0.1 M Tris–HCl, pH 7.5.
3. Elution buffer: 20 mM DTT in 10 mM Tris–HCl, pH 7.5.

2.2.2. Mass Spectrometry

1. MALDI Applied Biosystem voyager DE-PRO instrument operating in reflector mode.
2. MALDI matrix solution: α-cyano-hydroxycinnamic acid (10 mg/ml) (Fluka Sigma-Aldrich) in 70% acetonitrile (ACN) (Baker Phillipsburg, NJ), 0.1% trifluoroacetic acid.
3. Peptide standard mixture (Applied Biosystems).
4. Single quadrupole ZQ electrospray (Waters Micromass) coupled to an HPLC (2690 Alliance purchased by Waters).
5. Reverse-phase HPLC column: Phenomenex 250 × 2.1 mm^2, i.d., 5 μm.
6. Solvent A: 0.05% TFA, 5% formic acid (Baker Phillipsburg) in water.
7. Solvent B: 95% ACN, 0.05% TFA 5% formic acid solution.

2.3. Phosphate Group Modification by DNS-Cl

1. Barium hydroxide $Ba(OH)_2$ solution: 55 M in water.

2. Solid carbonic dioxide.

3. DANSS reagent prepared by reaction of Dansyl chloride (0.1 mg/ml dissolved in ACN) with cystamine (molar ratio 3:1) (SIGMA; BioChemika >98.0%).

4. Agilent Zorbax C8 column ($150 \times 4.6\,mm^2$ i.d.) (Palo Alto, CA).

5. Solvent A: 0.1% formic acid, 2% ACN in water.

6. Solvent B: 0.1% formic acid, 2% water in ACN.

7. 10 mM Tris/HCl (Fluka-Sigma-Aldrich) buffer pH 8.5.

8. 20 mM tributylphosphine solution by dilution in water of the stock solution (200 mM). Freshly prepared

9. DANSH solution: 0.1 mg/ml, dissolved in ACN/water 3:1. Fresly prepared.

10. 1 mM tributylphosphine solution in Tris 2 M solution, pH 10.8.

2.3.1. Mass Spectrometry

1. MALDI-TOF voyager DE-PRO mass spectrometer (Applied Biosystem, Framingham, MA).

2. ZipTip pipette from Millipore (Billerica, MA) using the recommended purification procedure:

3. Wetting solution: 50% ACN in water.

4. Equilibration and washing solutions: 0.1% TFA.

5. Elution solution: 50% ACN, 0.1% TFA in water.

6. Matrix solution: α-cyano-hydroxycinnamic acid (10 mg/ml in 70% ACN and 0.1% TFA in water).

7. Peptide standard mixture from Applied Biosystem, containing des-Arg1-Bradykinin, Angiotensin I, Glu1-Fibrinopeptide B, ACTH (1–17), ACTH (18–39), and Insulin (bovine).

8. 4000Q-Trap (Applied Biosystems) mass spectrometer equipped with a linear ion trap coupled to 1100 nano HPLC system (Agilent Technologies).

9. Agilent reverse-phase pre-column cartridge (Zorbax 300 SB-C18, $5 \times 0.3\,mm^2$, 5 μm).

10. Agilent reverse-phase column (Zorbax 300 SB-C18, 150 mm × 75 μm, 3.5 μm).

11. Solvent A: 0.1% formic acid, 2% ACN in water.

12. Solvent B: 0.1% formic acid, 2% water in ACN.

13. Uncoated silica tip from NewObjectives (Ringoes, NJ) (O.D. 150 μm, i.d. 20 μm, tip diameter 10 μm).

3. Methods

3.1. DTT Labelling

The first part of this chapter reports a novel approach to phosphoprotein mapping based on site-specific modification of phosphoseryl/phosphotheonyl residues. The feasibility of this strategy for the identification of the phosphorylation sites in peptides was tested by MALDI-TOF/TOF mass spectrometer. Here, it is shown that an alternative conversion of the P-Ser and P-Thr residues into compounds, which are stable during collision induced dissociation, provided easily interpretable product ion spectra.

To test this procedure under realistic but controlled conditions, the reactions are carried out by using phosphorylated bovine α-casein, a normal protein utilised in these studies *(18)*. To address the problem of quantification *(23)*, having created a thiol group at the phosphoserine/phosphothreonine site, it is possible to basically follow the strategy outlined by Gygi et al. *(18)*. Because of the fact that we had the availability of fully deuterated DTT, we exploited a much simpler isotope tagging strategy **(Scheme 1)**.

3.1.1. Protein Digestion

1. Alkylate one aliquot of α-casein (*see* **Note 1**) by incubation in 5 mM iodoacetamide for 30 min at room temperature in the dark.

2. Carry out enzymatic digestion with 20 µl of trypsin solution in 50 mM ammonium bicarbonate pH 8.5 at 37°C for 18 h.

3. Directly analyse the peptide mixtures by MALDI-MS in reflector mode. Apply a mixture of 1 µl of analyte solution (1 pmol) and 1 µl of matrix solution to the metallic sample plate and dry at room temperature. Acquire spectra monitoring positive ions. On our spectrometer, we used accelerating voltage of 72 KV, grid 65% and gridwire 0.01% in respect of accelerating voltage; delay time 80 ns; mass range 400–4,000 *m/z*. Perform mass calibration using external peptide standards by Applied Biosystems. Analyse raw data using the computer software provided by the manufacturer and reported as monoisotopic masses. Identify the phosphopeptides by their 80 Da mass difference compared with the peptides expected from the sequence for the presence of the phosphate moiety. As an example, the signal at *m/z* 1660.6 was assigned to the peptide 106–119 within α-casein sequence carrying a phosphate group.

3.1.2. β-Elimination Reaction

1. Remove the phosphate moieties via barium hydroxide ion-mediated β-elimination from pSer and pThr (*see* **Note 2**). Carry out β-elimination reactions by incubating the peptide

mixture in 55 M Ba(OH)$_2$, at 37°C, for 90 min under nitrogen (*see* **Notes 3** and **4**).

2. Add Solid carbonic dioxide at room temperature to get read of the removal reagent excess. Remove the precipitated barium carbonate by centrifugation at 13,000 × *g*/min for 5 min (*see* **Notes 5–8**).

3. Directly analyse the resulting peptide mixture by MALDI-MS as described earlier. The mass spectral analysis carried out on the supernatant showed the occurrence of a series of signals corresponding to the β-eliminated peptides. As an example, the signal at *m/z* 1854.3 was assigned to the peptide 104–119 within α-casein occurring 98 Da lower than that expected on the basis of the amino acid sequence, thus indicating that the pSer115 was converted in dehydroalanyl after β-elimination reaction (*see* **Notes 8** and **9**).

3.1.3. DTT as a Bifunctional Reagent

1. Submit the β-eliminated peptide mixture to a Michael-type addition by using dithiothreitol (DTT) by following the procedure described *(19)*. Divide the mixture in different aliquots. Add DTT, in the light and heavy form, 30% w/v in Hepes buffer 10 mM, pH 7.5, to the lyophilised β-eliminated peptide mixtures (*see* **Note 3**).

2. Carry out the addition reaction for 3 h, at 50°C under nitrogen. The addition reaction results in the creation of a free thiol group in place of what was formerly a phosphate moiety.

3. Monitor the extent of reaction via MALDI-MS. Yield of the addition reaction thus revealed that the amount of DTT-modified peptide is about 70% (**Fig. 1a**), as expected for a typical Michael reaction addition *(19)*.

3.1.4. Affinity Capture of Thiolated Peptides

1. Achieve specific enrichment of thiolated peptides by using a sepharose thiopyridil resin treated after addition reaction the peptide mixtures with 1 ml of activated thiol sepharose resin directly in the eppendorf tube (*see* **Note 12**).

2. Wash the resin twice by using 5 ml of water and then twice with 5 ml of binding buffer.

3. Add the peptide mixture to the resin and wash with the binding buffer (5 ml).

4. Extract the modified peptides by using 5 ml of elution buffer and lyophilise.

5. Dissolve the enriched peptide mixture in 20 μl of 0.1% TFA and use 1 μl for the MALDI-MS analysis that reveals now major mass signals at *m/z* 1716.9, 2008.3, 2039.1, and 3001.2. These values were assigned to the β-eliminated

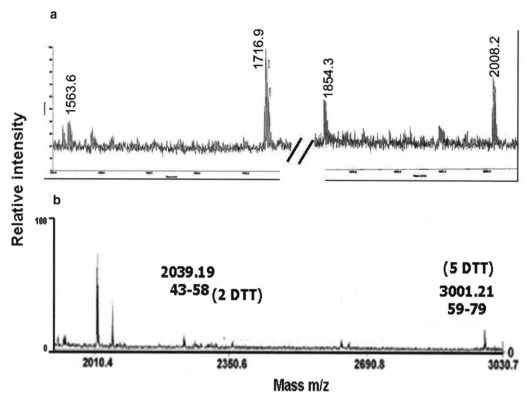

Fig. 1. MALDI-MS analysis. (a) Partial MALDI-MS spectrum of trypsin mixture from α-casein after DTT labelling reaction, showing the β-eliminated and DTT labelled peptides. (b) Partial MALDI-MS spectrum of the enriched peptide mixture. Multiphoshorylated peptides labelled with two and five DTT moieties are indicated.

phosphorylated fragments 106–119, 104–119, 43–58, and 59–79 modified by 1, 1, 2, and 5 DTT adducts respectively as indicated in **Fig. 1b**.

3.1.5. Differential Isotope Coded Analysis

1. Use DTT and DTT-D$_6$ (**Fig. 2**) to label samples containing stoichiometric concentrations of α-casein in ratios of 1:1, 2:1, and 4:1.

2. Analyse 1 µl of the newly generated peptide mixtures by MALDI-MS operating in reflector mode. Acquire the spectra related to the three different mixtures using about 2,000 shot/spectrum in order to produce a "stable" peak.

3. Quantify the peptides by measuring, in the mixture, the relative signal intensities for pairs of peptide ions of identical sequence differentially labelled with light or heavy DTT. The mass difference within the heavy DTT reagent (6 Da) reliably allows a good separation between the molecular ions.

4. Analyse aliquots (10 µl) of the same peptide mixtures (1:1, 2:1, and 4:1 DTT:DTT-D$_6$-labelled α-casein tryptic

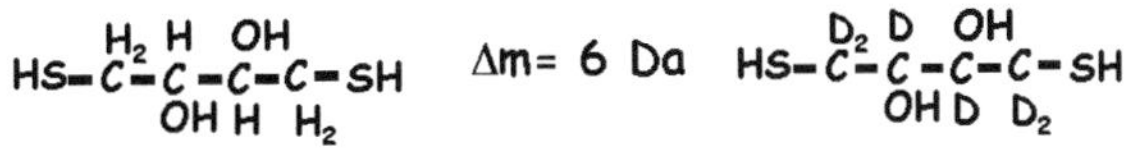

$$HS-\overset{H_2}{\underset{OH}{C}}-\overset{H}{\underset{H}{C}}-\overset{OH}{\underset{H_2}{C}}-C-SH \qquad \Delta m = 6\ Da \qquad HS-\overset{D_2}{\underset{OH}{C}}-\overset{D}{\underset{D}{C}}-\overset{OH}{\underset{D_2}{C}}-C-SH$$

Fig. 2. Light and heavy DTT.

digestions) by LC-MS on a single quadrupole ZQ electrospray coupled on a Phenomenex reverse-phase C18 column.

5. Elute peptide mixture at a flow rate of 0.2 ml/min with a 5–65% ACN-water gradient in 60 min. Spectra are acquired in positive mode using capillary voltage of 3.5 kV, cone voltage of 38 V, source temperature 150°C, and desolvatation gas temperature of 80°C. The flow of desolvatation gas was 300 l/h and cone gas 200 l/h. Mass range was 700–1,400 m/z.

6. Acquire and process data by profiling isotopic distribution using the Mass-Lynx program (Micromass).

7. Perform quantitative analysis by integrating the signals occurring in the LC-ESMS spectra of the doubly charged ions of the modified vs. unmodified peptides obtaining a linear progression.

3.2. Dansyl Labelling

As for the second labelling strategy illustrated in this chapter, the RIGhT strategy was introduced to indicate the general derivatization method based on dansyl labelling of the targeted residues. In the case of phosphopeptide (20) the procedure consists in the β-elimination of phosphor-Ser/-Thr residues and the Michael addiction of the resulting α,β-unsatured residues with DANSH. The dansyl derivatization introduces: (a) a basic secondary nitrogen into the molecule that enhances the efficiency of signal ionization; (b) a dansyl moiety that fragments according to previous data (21, 22). Using the great capabilities of a new hybrid mass spectrometer equipped with a linear ion trap analyzer, one can take advantage of the distinctive m/z 170 and 234 fragments in MS^2 and the diagnostic 234→170 fragmentation in MS^3 mode (**Scheme 2**).

3.2.1. α-Casein Digestion

1. Digest 5 pmol of α-casein with trypsin in 50 mM ammonium bicarbonate, pH 8.5, at 37°C overnight using an enzyme to substrate ratio of 1:50 w/w.

2. Dry trypsin digest under vacuum and store at 4°C.

3.2.2. Phosphate Group Modification

1. Remove the phosphate moieties from a tryptic α-casein digest from pSer and pThr via barium hydroxide ion-mediated β-elimination (19, 20).

2. Analyse the peptide mixture by MALDI-MS.

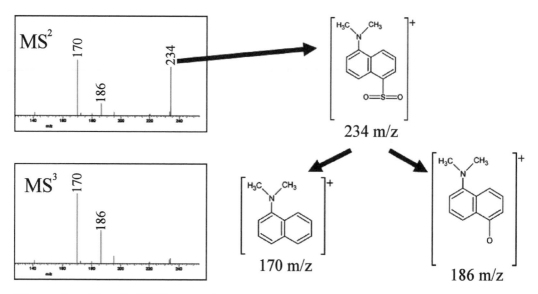

Scheme 2. Dansyl derivative ESMS fragmentation.

3. Purify the peptide mixtures using ZipTip pipette, under the recommended purification procedure. Elute the peptides using 20 µl of 50% ACN, 0.1% formic acid in water.

4. Apply a mixture of 1 µl of the eluted peptide solution and 1 µl of matrix solution to the metallic sample plate and dry at room temperature.

5. Perform mass calibration using a mixture of peptides containing des-Arg1-Bradykinin, Angiotensin I, Glu1-Fibrinopeptide B, ACTH (1–17), ACTH (18–39), and Insulin (bovine) as external standards. Analyse raw data and report as monoisotopic masses.

6. Modify the β-eliminated peptide mixture with DANSH via Michael-type addition.

7. Obtain DANSS by reaction of Dansyl chloride (0.1 mg/ml dissolved in ACN) with cystamine dissolved in 20 mM Na_2CO_3 pH 8 (molar ratio 3:1). Carry out reaction at 60°C for 2 h (*see* **Notes 17** and **18**).

8. Purify the product by RP-HPLC. We used an Agilent Zorbax C8 column using a 5–65% linear gradient in 30 min from water to ACN. Verify by ESI-MS analysis. Acquired spectra in positive mode using capillary voltage of 3.8 kV, cone voltage of 38 V, source temperature 150°C, and desolvatation gas temperature of 80°C. The flow of desolvatation gas was 150 l/h and cone gas 80 l/h. The acquisition mass range was 100–600 *m/z*.

9. Dissolve dried DANSS in 10 mM Tris/HCl Buffer and reduce in presence of 20 mM tributylphosphine. Carry out reaction for 30 min at room temperature under nitrogen (*see* **Note 19**).

10. Purify product by RP-HPLC in the same condition previously described and verify by ESI/MS analysis.

11. Add 100 μl of DANSH solution (0.1 mg/ml dissolved in ACN/water 3:1) to an equal volume of the β-eliminated peptide mixture solution (1 nmol/ml in deionized water) in presence of 1 mM tributylphosphine (*see* **Note 20**).

12. Add 2 M Tris solution to the sample at a final concentration of 100 mM (pH ~ 11). The reaction proceeds under nitrogen for 18 h at 50°C

13. Monitor the yield of addition reaction via MALDI-MS analysis carrying out as described earlier *(19)*.

3.2.3. Selective Isolation of Tagged Peptides

1. To investigate the feasibility of applying the method to proteomic analysis, prepare a peptide mixture from ten standard proteins (50 μg of a mixture in equimolar amount of myoglobin, bovine serum albumin, ovalbumin, carbonic anhydrase, RNase A, Lysozime, gluthatione S-trasferase, insulin, enolase, and trasferrin) by adding a trypsin solution (1 μg/ml) in 50 mM ammonium bicarbonate, pH 8.5, at 37°C overnight using an enzyme to substrate ratio of 1:50 w/w.

2. Spike the trypsin digest with 5 μg of α-casein trypsin mixture modified with dansyl-cysteamine as described earlier.

3. Submit 1 pmol of the peptide mixture to LC-MS analysis. We used a hybrid ion trap, 4000Q-Trap coupled to a 1100 nano HPLC system (*see* **Notes 21–22**).

4. Load the mixture on a reverse-phase pre-column cartridge at 10 μl/min (A solvent 0.1% formic acid, 2% ACN in water loading time 7 min).

5. Separate peptides on a reverse-phase column at a flow rate of 0.2 μl/min with a 5–65% linear gradient in 60 min (A solvent 0.1% formic acid, 2% ACN in water; B solvent 0.1% formic acid, 2% water in ACN).

6. We used a micro ionspray source at 2.4 kV with liquid coupling, with a declustering potential of 50 V, using an uncoated silica tip.

7. Acquire spectra by a survey Precursor Ion Scan for the ion *m/z* 170. The Q1 quadrupole was scanned from *m/z* 500 to 1,000 in 2 s, and ions were fragmented in q2 using a linear gradient of collision potential from 30 to 70 V (*see* **Note 23**).

8. Settle Q3 to transmit only ions at *m/z* 170. Follow this scan mode by an enhanced resolution experiment (ER) for the

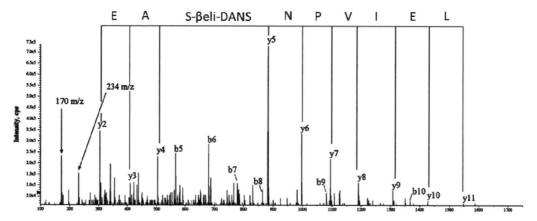

Fig. 3. MS/MS spectrum of the α-casein modified peptide 104–119. The fragment ions belonging to the b and y series are indicated.

ions of interest and then by MS³ and MS² acquisitions of the two most intense ions.

9. Acquire MS² spectra using the best collision energy calculated on the bases of m/z values and charge state (rolling collision energy).

10. Acquire MS³ spectra using Q0 Trapping, with a trapping time of 150 ms and an activation time of 100 ms, scanning from m/z 160 to 240 (*see* **Note 24**).

11. Acquire and process data. We used Analyst software from Applied Biosystems. The reconstructed ion chromatogram for the selective dansyl transition 234→170 in MS³ mode shows the presence of only two ions corresponding MS² spectra led to the reconstruction of the entire sequences of the β-eliminated phosphopeptides 104–119 and 106–119 carrying a DANSH moiety. As an example, the MS/MS spectrum of the modified peptide 104–119 is reported in **Fig. 3**. The modified ion is stable during collision induced dissociation provided easily interpretable product ion spectra. In fact, the y and b product ions still retains the modifying group linked to the β-eliminated Ser residue thus allowing the exact localization of the phosphorylation site (*see* **Notes 13–16, 25,** and **26**).

4. Notes

1. The analysis were carried out 1 μg of α-casein

2. Under strong alkaline conditions the phosphate moiety on Ser-P and Thr-P undergoes β-elimination to form

dehydroalanine (ΔSer) or dehydroalanine-2-butyric acid (ΔThr) respectively. The α,β-unsatured residues are potent Michael acceptor, which can readily react with a nucleophile *(24)*.

3. Reaction conditions were optimised for both the β-elimination and DTT addition reactions, taking in account the different amounts of protein.

4. The barium hydroxide was used in place of the previously reported NaOH *(25)* taking in account the higher purity of this reagent and the higher reactivity towards the pThr residues.

5. To get read of the removal reagent excess we simply made use of solid carbon dioxide. The use of this reagent let to remove the barium carbonate as a pellet on the bottom of a vial after centrifugation and it avoids the additional purification step by chromatography.

6. A gentle bubbling is observed at this stage due to barium carbonate precipitation.

7. Carbonic dioxide should be added until bubbling is produced, allowing the complete precipitation of barium salt. This is a very tricky step.

8. It is worth noting that when the pellet of barium carbonate was dissolved and analysed by MALDI-MS, no peptide was detected in the mass spectrum thus indicating the absence of a sort of non specific precipitation.

9. The spectra were obtained by acquiring about 2,000 shot/ spectrum in order to produce a "stable" peak. In our experience this is the average number of shots to have an even distribution of shots on the entire well surface, thus allowing a confident averaging independently of the peptide content of the single crystals.

10. As for DTT addition, the rational behind the choice of this reagent is due to its higher solubility and reactivity with respect to the already proposed EDT *(25)*. In fact EDT has limited solubility in aqueous medium, thereby requiring sample cleanup prior LC-MS analysis since the conversion is carried out in the presence of substantial amount of organic solvent *(26)*.

11. The direct mass spectral analysis of a complete peptide mixture suffered of suppression phenomena leading to the unappearance of signals due to most of the phosphorylated peptides.

12. Specific enrichment of thiolated peptides was carried out using an excess of resin considering the quantity of DTT presents in the peptide mixture.

13. The described procedure resulted in the identification of all the phosphorylated tryptic peptides, including the penta-phosphorylated one that has escaped in different attempts used *(18)*.

14. As for quantitative analysis, pairs of peptides tagged with the light and heavy DTT reagents are chemically identical and therefore serve as ideal mutual internal standards for quantitation.

15. The ratio between the intensities of the lower and upper mass components of these peaks provide an accurate measurement of the relative abundance of each peptide and hence of the related proteins in the original cell pool because the MS relative intensity of a given peptide is independent of the isotopic composition of the defined isotopically tagged reagent.

16. This integrated simple methodology, which involves chemical replacement of the phosphate moieties by affinity tags to locate and quantitate phosphorylated residues may result of interest in labelling of the O-glycosylation sites. In fact, the *O*-glycopeptides retained on a lectin affinity column can be eluted and submitted to the procedure described by Wells et al. *(27)*. The quantitative analysis can be performed by using light and heavy forms of DTT to selective label the modified sites thus leading to a quantitative description of O-glycosylation related to different cell growth conditions.

17. As for dansyl labelling, dansyl-cysteamine synthesis is particularly delicate. In fact cystamine chloride is soluble in water instead dansyl-chloride in organic solvent. The correct compromise was a solution 3:1, ACN–Na_2CO_3.

18. It is important for the successful synthesis to check the pH value at 8.

19. For the reduction of dansyl cystamine to dansyl-cysteamine tributilphosphine solution should be freshly prepared under nitrogen to avoid explosion to contact with ossigen of air.

20. In the addition reaction, the pH value has to be carefully checked (pH 11.0).

21. Ion trap tandem mass spectrometry experiments offer high sensitivity because of the ability to accumulate precursor ions.

22. The novel introduction of a linear ion trap with greatly increased capture efficiency and storage capacity resulted in new inputs in the proteomic field.

23. As concerning the use of dansyl chloride as marker, (a) DANS moiety introduces a strong basic group in peptides thus leading to an enhanced ionization; (b) DANS peptide

can be easily detected in precursor ion scan mode; (c) DANS peptide can be selectively detected in MS3 mode (via transition $234 \rightarrow 170$) in even complex peptide mixture.

24. The Q0 Trapping tool is able to increase the sensitivity of the scan in the different ion trap scan types used.

25. *O*-glycosylated residues also can be targeted with the dansyl reagent thus widening the range of effectiveness of the method.

26. It is worth considering that α-casein also contains a diphosphorylated and a pentaphosphorylated tryptic peptide, which escaped to be detected using this approach probably due to the chemical and physical characteristics of the modified fragments. In fact the introduction of a high number of dansyl moieties increase the hydrofobicity of the peptides; moreover the low efficiency of Michael addition results markedly increased when more then one site is present in the same peptide. However, the ionization of multiphosphorylated peptides is a difficult challenge and among the proposed derivatization procedures only the modification strategy based on the use of dithiothreitol (previous described, **Subheading 3.1**) led to the detection of intense ions of multiphosphorylated peptides in mass spectrometric analyses.

Acknowledgments

This work was supported by grants from the Ministero dell'Universita' e della Ricerca Scientifica (Progetti di Rilevante Interesse Nazionale 2002, 2003, 2005, 2006; FIRB 2001). Support from the National Center of Excellence in Molecular Medicine (MIUR – Rome) and from the Regional Center of Competence (CRdC ATIBB, Regione Campania – Naples) is gratefully acknowledged.

References

1. Ben-Levy R., Leighton I.A., Doza Y.N., Attwood P., Morrice N., Marshall C.J., Cohen P. (1995) Identification of novel phosphorylation sites required for activation of MAPKAP kinase-2. *EMBO J.* 14, 5920–30.

2. Meyer H.E., Eisermann B., Heber M., Hoffmann-Posorske E., Korte H., Weigt C., Wegner A., Hutton T., Donella-Deana A., Perich J.W. (1993) Strategies for nonradioactive methods in the localization of phosphorylated amino acids in proteins. *FASEB J.* 7, 776–82.

3. Schlessinger J. (1993–1994) Cellular signaling by receptor tyrosine kinases. *Harvey Lect.* 89, 105–23.

4. Lander E.S., et. al. (2001) Initial sequencing and analysis of the human genome. *Nature* 409, 860–921. Erratum in: Nature (2001) 412(6846):565; Nature (2001) 411(6838):720.

5. Kalume D.E., Molina H., Pandey A. (2003) Tackling the phosphoproteome: tools and strategies. *Curr. Opin. Chem. Biol.* 7, 64–9.

6. Bendt A.K., Burkovski A., Schaffer S., Bott M., Farwick M., Hermann T. (2003) Towards a phosphoproteome map of Corynebacterium glutamicum. *Proteomics* 3, 1637–46.

7. Cohen P. (2000) The regulation of protein function by multisite phosphorylation–a 25 year update. *Trends Biochem. Sci.* 25, 596–601.

8. McLachlin D.T., Chait B.T. (2001) Analysis of phosphorylated proteins and peptides by mass spectrometry. *Curr. Opin. Chem. Biol.* 5, 591–602.

9. Pandey A., Andersen J.S., Mann M. (2000) Use of Mass Spectrometry to Study Signaling Pathways. *Sci. STKE.* 37, PL1.

10. Annan R.S., Huddleston M.J, Verma R., Deshaies R.J., Carr S.A. (2001) A multidimensional electrospray MS-based approach to phosphopeptide mapping. *Anal. Chem.* 73, 393–404.

11. Biemann K. (1990) Sequencing of peptides by tandem mass spectrometry and high-energy collision-induced dissociation. *Methods Enzymol.* 193, 455–79.

12. Janek K., Wenschuh H., Bienert M., Krause E. (2001) Phosphopeptide analysis by positive and negative ion matrix-assisted laser desorption/ionization mass spectrometry. *Rapid Commun. Mass Spectrom.* 15, 1593–9.

13. Ballif B.A., Villen J., Beausoleil S.A., Schwartz D., Gygi S.P. (2004) Phosphoproteomic analysis of the developing mouse brain. *Mol. Cell Proteomics* 3, 1093–101.

14. Marcus K., Immler D., Sternberger J., Meyer H.E. (2000) Identification of platelet proteins separated by two-dimensional gel electrophoresis and analyzed by matrix assisted laser desorption ionization-time of flight-mass spectrometry and detection of tyrosine-phosphorylated proteins. *Electrophoresis* 21, 2622–36.

15. Gaberc-Porekar V., Menart V.R. (2001) Perspectives of immobilized-metal affinity chromatography. *J. Biochem. Biophys. Methods.* 49, 335–60.

16. Zhou H., Watts J.D., Aebersold R. (2001) A systematic approach to the analysis of protein phosphorylation. *Nat. Biotechnol.* 19, 375–8.

17. Smolka M.B., Zhou H., Purkayastha S. and Aebersold R. (2001) Optimization of the isotope-coded affinity tag-labelling procedure for quantitative proteome analysis. *Anal. Biochem.* 297, 25–31.

18. Gygi S.P., Rist B., Gerber S.A., Turecek F., Gelb M.H., Aebersold R. (1999) Quantitative analysis of complex protein mixtures using isotope-coded affinity tags. *Nat. Biotechnol.* 17, 994–9.

19. Amoresano A., Marino G., Cirulli C., Quemeneur E. (2004) Mapping phosphorylation sites: a new strategy based on the use of isotopically labelled DTT and mass spectrometry. *Eur. J. Mass Spectrom.* (Chichester, Eng) 10, 401–12.

20. Amoresano A., Monti G., Cirulli C., Marino G. (2006) Selective detection and identification of phosphopeptides by dansyl MS/MS/MS fragmentation. *Rapid Commun. Mass Spectrom.* 20, 1400–4.

21. Marino G., Buonocore V. (1968) Mass-spectrometric identification of 1-dimethyl-aminoaphthalene-5-sulphonyl-amino acids. *Biochem. J.*; 110, 603–4.

22. Addeo F., Malorni A., Marino G. (1975) Analysis of dansyl-1-amides in mixtures by mass spectrometry, using metastable defocusing. *Anal. Biochem.* 64, 98–101.

23. Mann M. (1999) Quantitative proteomics? *Nat. Biotechnol.* 17, 954–63.

24. Goering H.L., Relyea D.I., Larsen D.W. (1956) Preparation of thiols: formation from alkenes. *J. Am. Chem. Soc.* 78, 348–353.

25. Li W., Boykins R.A., Backlund P.S., Wang G., Chen H.C. (2002) Identification of phosphoserine and phosphothreonine as cysteic acid and beta-methylcysteic acid residues in peptides by tandem mass spectrometric sequencing. *Anal. Chem.* 15, 5701.

26. Jaffe H, Veeranna H, Pant C. (1998) Characterization of serine and threonine phosphorylation sites in beta-elimination/ethanethiol addition-modified proteins by electrospray tandem mass spectrometry and database searching. *Biochemistry.* 17, 16211–24.

27. Wells L., Vosseller K., Cole R.N., Cronshaw J.M., Matunis M.J. and Hart G.W. (2002) Mapping sites of O-GlcNAc modifications using affinity tags for serine and threonine post-translational modifications. *Mol. Cell. Proteomics.* 1, 791–804.

Chapter 14

On-Line Liquid Chromatography Electron Capture Dissociation for the Characterization of Phosphorylation Sites in Proteins

Steve M.M. Sweet and Helen J. Cooper

Summary

Electron capture dissociation (ECD) allows fragmentation of the phosphopeptide backbone while keeping the labile phospho-amino acid intact. This feature of ECD fragmentation, coupled with the acquisition of mass spectra at high mass accuracy, makes ECD well-suited to phosphorylation mapping. The following methods are designed to focus ECD events on phosphopeptides within a complex peptide sample, either by using phosphoric acid neutral loss peaks as a trigger or by targeted analysis of predetermined precursor masses.

Key words: ECD, Neutral loss, Phosphorylation, FT-ICR, LTQ-FT, Xcalibur, Site localization, AGC, Targeted, OMSSA.

1. Introduction

Phosphorylation is a widespread and biologically significant post-translational modification (1). Although mass spectrometry is well suited for the identification of sites of phosphorylation, problems stemming from phosphopeptide abundance and fragmentation behavior must be overcome. The low abundance of phosphopeptides relative to unmodified peptides can result in phosphopeptides not being selected for fragmentation in data-dependent analyses. This problem can be addressed by phosphopeptide enrichment (2). A further difficulty is that collision-induced dissociation (CID) fragmentation of serine and threonine phosphopeptides results in the neutral loss of phosphoric acid,

Marjo de Graauw (ed.), *Phospho-Proteomics, Methods and Protocols, vol. 527*
© 2009 Humana Press, a part of Springer Science + Business Media, New York, NY
Book DOI: 10.1007/978-1-60327-834-8_14

H_3PO_4, and correspondingly fewer informative backbone cleavages, impairing phosphopeptide identification and site localization *(2)*. Neutral loss of H_3PO_4 is not observed upon radical-driven fragmentation, such as ECD or electron transfer dissociation (ETD) *(3, 4)*. The retention of the phosphate group on the backbone chain, combined with high resolution, high mass accuracy, FT-ICR measurement of backbone fragments, makes ECD well-suited for phosphorylation site assignment *(5)*. However, ECD is currently significantly less sensitive than CID, requiring larger precursor ion populations which take longer to accumulate. Given this disadvantage, it makes sense to focus ECD events on putative phosphopeptides. The following two methods, neutral loss dependent ECD (NL ECD) and targeted ECD, are designed to achieve this. In NL ECD, the ECD events are restricted to peptides showing neutral losses of 98 Da (H_3PO_4) following CID *(6)*, i.e., peptides containing phosphoserine or phosphothreonine. In targeted ECD, the ECD events are confined to particular precursor masses, usually previously identified phosphopeptides where the exact site of phosphorylation is ambiguous *(7)*. The ECD parameters are adjusted to maximize the probability of site localization.

2. Materials

2.1. Mass Spectrometer and Software

1. Thermo Scientific LTQ-FT with ECD cathode.
2. Xcalibur 2.0.5 software (Thermo Fisher Scientific).
3. Bioworks 3.3 (Thermo Fisher Scientific).

2.2. ECD Calibration

1. Dissolve 1 mg of substance P peptide in 715 µl 50:50 methanol:water with 0.1% formic acid, to create a stock solution of 1 nmol/µl. Solvents should be HPLC grade. This solution can be stored at –20°C.
2. Add 20 µl of 1 nmol/µl stock to 10 ml of the same buffer to give a calibration solution of 2 pmol/µl.

2.3. Liquid Chromatography (LC) Buffers

1. All buffers are as for standard CID LC MS/MS analyses, e.g., H_2O with 0.1% formic acid and acetonitrile with 0.1% formic acid.

2.4. ECD Test Sample

1. α-Casein tryptic digest, at a concentration of 100 fmol/µl in 0.1% formic acid.

2.5. Database Search Software

1. Download OMSSA browser from http://pubchem.ncbi.nlm.nih.gov/omssa/.

3. Methods

3.1. Neutral Loss Dependent ECD for Phosphopeptide Discovery

3.1.1. ECD Calibration

1. Place ECD calibration solution into a syringe and infuse into the instrument.
2. Open LTQ Tune and click on the calibration icon.
3. Click on the FT ECD tab and select Calibrate: Potentials and Timing. Click Start.
4. Check that the ECD efficiency is above 15% (*see* **Note 1**).

3.1.2. Tune File Creation

1. Open LTQ Tune.
2. Open a Tune file used for standard LC MS/MS analysis.
3. Choose File, "Save As" and save with a new name.
4. Click on the cartoon of the FT-ICR cell ("FT Inject Ctrl").
5. Click on the FT tab. Change the AGC Target Settings as follows: "Full MS" to 1e + 06; "ECD" to 1e + 6.
6. Tick "Enable Full Scan Injection Waveforms" box.
7. Click on the Ion Trap tab.
8. Change the MSn AGC Target Setting to 5e + 04.
9. Tick "Enable Full Scan Injection Waveforms" box.
10. Click on ScanMode. Click on "Scan Time Settings…". Click on the FT tab. Change the ECD settings to 4 microscans, with 1,000 ms maximum inject time (*see* **Note 2**).

3.1.3. Mass Spectrometric Method Creation

1. Open Xcalibur. Choose instrument setup. Choose Data dependent MS/MS. Choose "yes" to "Initialize method with LQ FT support". The instrument setup window should now open.
2. Choose "Save As" and save with an appropriate name.
3. Change the Tune method to the Tune file created in the previous section.
4. Change scan events from 2 to 3 (*see* **Note 3**).

NL ECD Event Creation
1. Click on box for scan event 3.
2. Change the analyzer from ion trap to FTMS.
3. Change the Resolution from 100,000 to 25,000.
4. Change the Data type from Centroid to Profile.
5. Tick the "Dependent scan" box. Click on "Settings" next to this box.
6. In the Global section:

    a. Click on Dynamic exclusion. Tick "Enabled" box. Change exclusion list size to 500.

    b. Change Exclusion duration to 60 (*see* **Note 4**).

    c. Change Exclusion mass width to "Relative to reference mass (ppm)" and set the low and high values to 7.

    d. Click on Neutral Loss. Set the Neutral Loss mass width low and high values to 0.7 and 0.5, respectively (*see* **Note 5**).

7. In the Segment section:

    a. Click on Charge States. Tick "Enable charge state screening" box. Tick "Enable monoisotopic precursor selection" box.

    b. Tick "Enable charge-state dependent ECD time" box.

    c. In the Charge State Rejection section, tick the "Enabled" box and the "Reject charge states: 1" box.

    d. Click on neutral loss. Insert *m/z* values 49 and 32.67 (*see* **Note 6**).

    e. Change "Within top N:" to 5.

8. In the Scan Event section:

    a. Click on activation. Change the isolation width to 6 *m/z* (*see* **Note 7** and **Fig. 1**).

    b. Tick "ECD active" box.

    c. Click on FT ECD/IRMPD. Change the ECD Duration (ms) from 70 to 60.

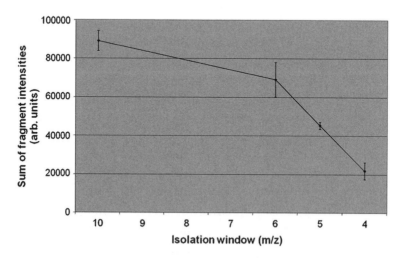

Fig. 1. The effect of varying the precursor ion isolation window on ECD fragment intensities. [M + 2H]$^{2+}$ ions of the DADEY'LIPQQG phosphopeptide were selected for ECD, with various isolation window sizes. All fragments, excluding precursor ion and charge-reduced precursor ion, were summed for each spectrum. Error bars show plus and minus one standard deviation ($n = 4$).

  d. Click on Current Scan Event. Change the minimum signal threshold (counts) from 500 to 1,000 (*see* **Note 8**).

  e. Choose "From neutral loss list."

  f. Tick "Repeat previous scan event with ECD" box.

9. Click "Ok" to close the window.

CID Event Creation

1. Click on the bar for Scan Event 2.

2. Click on "Settings" next to the Dependent scan box.

3. In the Scan Event section:

  a. Click on Current Scan Event. Change the Minimum signal threshold (counts) from 500 to 40,000 (*see* **Note 2**).

  b. Click on Activation. Change the Isolation width (*m/z*) from 2 to 6.

Configuring the Remaining Settings

1. Configure the remaining settings (Acquire time, Divert Valve, Contact Closure, etc.) as for a normal LC MS/MS run.

3.1.4. Testing the Method

1. Run the NL ECD method using 500 fmol α-casein tryptic digest as a test sample.

2. Check the resulting .raw file:

  a. Check that ECD events have been triggered (*see* **Note 9**).

  b. Check that in each ECD spectrum peaks corresponding to the precursor ion, reduced precursor ion and fragment ions are present. Fragment ions are usually at 1–15% of the intensity of the precursor ion (*see* **Note 10** and **Fig. 2**).

3. Search the resulting ECD mass spectra using OMSSA *(8)* (*see* **Note 11**):

  a. To create ECD.dta files, open the .raw file in Bioworks.

  b. Choose Sequest Search. In the dialog box select "Perform DTA Generation" and untick the boxes "Perform Search" and "Use a unified search file (.SRF)".

  c. Combine the resulting .dta files into a single file using OMSSA.

  d. Search ECD files with the following settings: c, z, and y fragments; higher mass accuracy (0.02 Da); remove precursor and charge-reduced precursor.

3.2. Targeted ECD

3.2.1. Identification of Putative Phosphopeptides

1. Identify the masses of phosphopeptides in your sample using a standard CID LC MS/MS analysis (*see* **Note 12**).

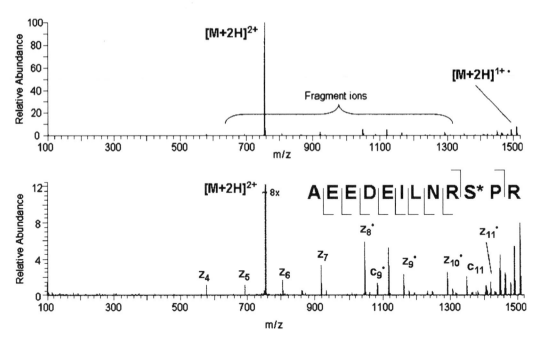

Fig. 2. On-line nanoLC ECD of [M + 2 H]$^{2+}$ ions of phosphorylated AEEDEILNRS*PR. (**a**) Full-scale ECD spectrum (**b**) Enlarged view of fragment ions. ECD mass spectrum comprises four co-added microscans.

3.2.2. Tune File Creation

1. Open LTQ Tune.
2. Open the Tune file created in **Subheading 3.1.1**.
3. Choose File, "Save As…" and save with a new name.
4. Click on ScanMode.
   a. Click on "Scan Time Settings…". Click on the FT tab. Change the ECD settings to: 8 microscans, with 1,000 ms maximum inject time (*see* **Note 2**).

3.2.3. Mass Spectrometric Method Creation

1. Open the method created in **Subheading 3.1.2** and choose "Save as" and save with a different name.
2. Click on the bar for Scan Event 2. Click on "Settings" (next to Dependent scan).
3. Within the "Segment" section:
   a. Click on Mass Lists. Enter precursor *m/z* values for all target phosphopeptides.
4. Within the "Scan Event" section:
   a. Click on Current Scan Event. Change the minimum signal threshold (counts) to 20,000.
   b. Choose "nth most intense from list":1.
5. Click on the bar for Scan Event 3. Click on "Settings" (next to Dependent scan).

6. Within the "Scan Event" section:

   a. Click on Current Scan Event. Change the minimum signal threshold (counts) to 20,000.

   b. Change "Mass determined from scan event:" from 2 to 1.

   c. Choose "nth most intense from list":1.

3.2.4. Testing the Method   1. Test the method as mentioned in **Subheading 3.1.4**.

4. Notes

1. The ECD efficiency should be ~20% (15–23% range). If the calibration fails or the efficiency is low, first check that an appropriate ECD duration (e.g., 70 ms) is set in the Define Scan dialog box. If this setting is correct, use the cathode activation procedure. Cathode activation takes approximately 6 h, therefore can be conveniently carried out overnight.

2. The minimum signal threshold and the number of microscans should be adjusted such that triggering an ECD event is likely to give a high quality ECD spectrum, resulting in a phosphopeptide identification. To a certain extent the minimum signal threshold and the number of microscans are interdependent, i.e. a lower signal threshold can be used with a larger number of microscans, while a high signal threshold will require fewer microscans. The chromatographic peak width of a peptide will give an upper limit for the number of microscans. With 1,000 ms inject time: four microscans can take up to 6.1 s, depending upon the time taken to reach the AGC target.

3. The minimum number of scan events is three: a survey scan, a CID event and a NL ECD event. Due to the longer time required for an ECD event, the optimum configuration should result in no more than one ECD event in most cycles. The number of events per cycle to achieve this will depend upon the sample type: a higher number of events is appropriate for a sample containing mainly unmodified peptides. If phosphopeptide enrichment has been performed, a method with three events will be optimal.

4. The appropriate duration of dynamic exclusion depends upon the chromatographic peak width of the eluting peptides: relatively broad peaks require longer exclusion time. Choosing a repeat count of 1 and using an exclusion duration slightly greater than the chromatographic peak width will maximize the number of peptides of distinct mass that can be identified.

It may be preferable to use a larger repeat count in order to maximize the likelihood of peptide identification. Relaxing the dynamic exclusion settings can allow identification of approximately co-eluting differentially phosphorylated forms of the same peptide.

5. Empirical observations in our laboratory show that recorded neutral losses for higher charge-state (and therefore in general mass) precursors tend to be slightly smaller than $98/z$, an extreme example being an apparent loss of 32.01 for a 3+ precursor, rather than 32.67. Consequently, the neutral loss m/z window is skewed towards the low m/z end.

6. 49 and 32.67 correspond to the neutral loss of H_3PO_4 from 2+ and 3+ precursors, respectively. Depending upon the choice of protease and sample type, it may be appropriate to include additional neutral losses, e.g., 24.5 (4+ precursor). Neutral loss of $H_2O + H_3PO_4$ may also be observed. Increasing the number of neutral loss masses is also likely to increase the number of false-positives, leading to ECD fragmentation of non-phosphorylated peptides.

7. For ECD events a larger isolation window (up to 10 m/z) gives better transfer of precursor ions into the FT-ICR cell (*see* **Fig. 1**). However, increasing the isolation window also increases the possibility of unwanted ions being selected, alongside the precursor of interest. The optimum size will depend on the sample complexity.

8. A threshold of 1,000 for the neutral loss peak ensures that most phosphopeptides will be selected for ECD. Choosing a higher threshold, in combination with requiring the NL peak to be one of the top two peaks, would limit ECD events to those phosphopeptides showing a particularly strong CID neutral loss.

9. If no ECD events have been triggered, check that ions corresponding to phosphopeptides are detected above the chosen threshold (40,000) and that these ions show a neutral loss upon CID fragmentation. If this is the case, check for errors in the method. If no α-casein peptides are detected, check that the LC system is operating correctly, using a standard peptide mixture.

10. If the ECD mass spectrum does not contain precursor and reduced precursor peaks, check whether the precursor isolation window is 6 m/z and check the percent AGC achieved. To do this, right click on ECD spectrum and choose View > Scan Header: scroll down to check MS2 Isolation Width; if the AGC target has not been achieved there will be a line stating, for example, FT Analyzer Message: Unfill = 0.14 (i.e., 14% of target was achieved).

11. Bioworks 3.3 (Thermo Fisher Scientific) and Mascot (Matrix Sciences) search engines can also accept ECD data, however in our experience OMSSA provides the best results.

12. To write a targeted ECD method, it is necessary to first identify the precursor masses of phosphopeptides in your sample. This can be via a standard LC CID MS/MS experiment or via a NL ECD experiment. In theory this could be on the basis of previously reported phosphopeptides, although it would be advisable to at least check that a peak corresponding to the expected phosphopeptide *m/z* value is present in your sample (LC MS survey scan).

Acknowledgments

We are grateful to Lewis Geer for assistance with the OMSSA algorithm for ECD. This work was funded by the EU (FP6 Endotrack; www.endotrack.org) (SMMS) and the Wellcome Trust (074131) (HJC).

References

1. Pawson, T., and Scott, J. D. (2005) Protein phosphorylation in signaling – 50 years and counting. *Trends Biochem. Sci.* 30, 286–90.

2. Ficarro, S. B., McCleland, M. L., Stukenberg, P. T., Burke, D. J., Ross, M. M., Shabanowitz, J., Hunt, D. F., and White, F. M. (2002) Phosphoproteome analysis by mass spectrometry and its application to *Saccharomyces cerevisiae*. *Nat. Biotechnol.* 20, 301–05.

3. Stensballe, A., Jensen, O. N., Olsen, J. V., Haselmann, K. F., and Zubarev, R. A. (2000) Electron capture dissociation of singly and multiply phosphorylated peptides. *Rapid Comm. Mass Spectrom.* 14, 1793–800.

4. Syka, J. E., Coon, J. J., Schroeder, M. J., Shabanowitz, J., and Hunt, D. F. (2004) Peptide and protein sequence analysis by electron transfer dissociation mass spectrometry. *Proc. Natl. Acad. Sci. USA* 101, 9528–33.

5. Sweet, S. M. M., and Cooper, H. J. (2007) Electron capture dissociation in the analysis of protein phosphorylation. *Exp. Rev. Proteomics* 4, 149–59.

6. Sweet, S. M. M., Creese, A. J., and Cooper, H. J. (2006) Strategy for the identification of sites of phosphorylation in proteins: Neutral loss triggered electron capture dissociation. *Anal. Chem.* 78, 7563–69.

7. Sweet, S. M. M., Mardakheh, F. K., Ryan, K. J. P., Langton, A. J., Heath, J. K., and Cooper, H. J. (2008) Targeted on-line liquid chromatography electron capture dissociation mass spectrometry for the localization of sites of in vivo phosphorylation in human Sprouty2. *Anal. Chem.* 80, 6650–57.

8. Geer, L. Y., Markey, S. P., Kowalak, J. A., Wagner, L., Xu, M., Maynard, D. M., Yang, X., Shi, W., and Bryant, S. H. (2004) Open mass spectrometry search algorithm. *J. Proteome Res.* 3, 958–64.

Chapter 15

Quantification of Protein Phosphorylation by μLC-ICP-MS

Ralf Krüger, Nico Zinn, and Wolf D. Lehmann

Summary

Determination of the protein amount and of the extent of protein phosphorylation is crucial for a variety of research fields, but is not always straightforward. We describe the application of capillary LC-ICP-MS (liquid chromatography–inductively coupled plasma–mass spectrometry) for quantification of phospho-proteins and their phosphorylation degree. Element mass spectrometry is ideally suited for monitoring and quantification of compounds with heteroelements such as phosphorus and sulphur, particularly because the ICP-MS response is virtually independent from the chemical form of the element. Determination of the phosphorylation stoichiometry, i.e. the relative abundance of the phosphorylated isoforms, can be assessed by the relative abundance of phosphorus compared with sulphur as a marker for the protein amount. Moreover, isotope dilution analysis by post-column addition of a ^{34}S-Spike provides absolute protein quantification with exceptionally high accuracy. Phosphoprotein analysis by capillary LC-ICP-MS may be applied to isolated proteins or protein digests and may include separation of impurities by 1D-SDS-PAGE followed by enzymatic digestion. Alternatively, digestion of complex protein mixtures such as cellular protein extracts allows determination of global, tissue-specific phosphorylation degrees.

Key words: μLC, ICP-MS, Protein phosphorylation, Protein quantification, Isotope dilution, Element analysis, Mass spectrometry.

1. Introduction

Depending on sample complexity, accurate and precise quantification of proteins is often still a challenge, since a number of matrix effects are encountered with established methods such as photometry, nephelometry, and turbidimetry. Antibody tests such as ELISA are highly sensitive, but sometimes specificity is limited. This is the main reason why mass spectrometry (MS) has gained much attraction for protein identification, and a variety of protocols for protein quantification by MS has been published,

Marjo de Graauw (ed.), *Phospho-Proteomics, Methods and Protocols, vol. 527*
© 2009 Humana Press, a part of Springer Science + Business Media, New York, NY
Book DOI: 10.1007/978-1-60327-834-8_15

most of them utilizing liquid chromatography coupled to tandem MS (e.g. LC-ESI-MS/MS) *(1–4)*. One major limitation is that the majority of MS-based approaches relies on internal standards, because the ionization efficiency – and thus the signal intensity – is strongly influenced by solvent composition, ionization source conditions, and accompanying substances causing suppression effects. Isotope labels are usually used to compensate for these effects *(5–7)*, but labelling of analytes or synthesizing of labelled standards is both laborious and costly.

Quantification of modified proteins, such as phosphoproteins, is even more challenging, no matter whether absolute quantification of modified proteins or quantification of the modification degree is required. Both parameters are highly relevant in a number of application fields, since relative or absolute abundance of the modified protein variant is usually closely correlated to protein function. Consequently, a variety of highly sophisticated methods for protein phosphorylation analysis by tandem MS has been developed *(1, 8–11)*.

In recent years, several examples have been published employing element MS as an alternative or supplementary approach for peptide and protein analysis *(12–30)*. In most cases, inductively coupled plasma mass spectrometry (ICP-MS) is used as element selective detector for liquid chromatography (LC) separations *(31)*. In contrast to the soft ion sources ESI or MALDI, ICP-MS is a destructive technique where atomic ions of heteroelements (metals, semi-metals, and some non-metals such as S and P) are generated and detected. However, the nature of ICP-MS implicates that protein identification or pinpointing of modification sites has to be conducted in parallel by other methods. The main advantage of ICP-MS is that usually no labelling is necessary and that the signal response is virtually compound independent and does only reflect the number of incorporated heteroatoms. Thus, ICP-MS is ideal for quantification purposes, provided the influence of varying solvent composition is considered.

We have presented a number of striking examples on how capillary LC-ICP-MS (liquid chromatography–inductively coupled plasma–mass spectrometry) can be used successfully for the analysis of protein phosphorylation *(12, 13, 19–21, 28, 29)*. The technique can be very helpful for spotting of phosphorylated peptides in a peptide mixture (e.g. protein digests) as shown in **Fig. 1**, for differentiation of cell states by quantification of phosphorylated proteins, and for determination of the protein phosphorylation degree. The latter case is particularly interesting, and knowledge of the relative amount of phosphorylated isoforms is often sufficient to answer scientific questions. Moreover, analysis of the phosphorylation degree is highly robust, since it relies on simultaneous monitoring of phosphorus (for the modified form) and sulphur (for the total protein amount), thus eliminating

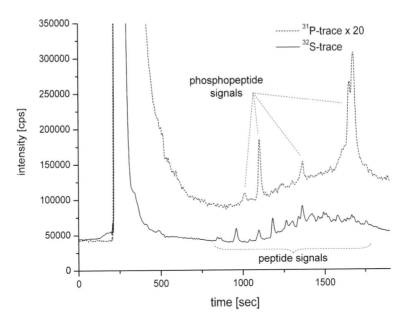

Fig. 1. μLC-ICP-MS chromatogram of ubiquitin-conjugating enzyme (CDC34) after tryptic digestion. The sample was purified by 1D-SDS-PAGE, and digested with modified trypsin. Subsequently, peptides were extracted from the gel, desalted by RP-C$_{18}$ micropipet tips, and injected into the capillary LC coupled online to a sector field ICP-MS.

fluctuations during analysis and incomplete recovery during sample preparation. The approach can be applied to isolated proteins and protein digests (13, 28, 29). Usually the protein or protein mixture of interest has to be purified from contaminants that contain the elements of interest (P, S), unless comparably pure proteins are analysed. Protein mixtures may be analyzed as well, although information about a certain protein requires complete chromatographic separation. Alternatively, analysis of crude protein mixtures (e.g. cellular protein extracts) allows determination of a "global" phosphorylation degree typical for the present condition and cell type.

Alternatively, if absolute instead of relative quantification is desired, ICP-MS offers the excellent possibility of monitoring several isotopes of the same element for isotope dilution analysis (IDA) (32–34). IDA is performed by adding a known amount of isotopically enriched standard and subsequent determination of the isotope ratio in the mixture. This approach has been applied successfully to peptide and protein analysis employing hyphenation of LC or CE to ICP-MS (18, 23–26). Since the ICP-MS response is independent from the chemical form in which the element is added, IDA of proteins can be conducted by post-column addition of a ^{34}S-enriched spike, without the need of ^{34}S-labelled protein standards (35). IDA requires no calibration curve and usually delivers highly precise and accurate values; therefore, we have included this method into our protocol. However, the absolute quantity of phosphorylation can only be assessed indirectly via the ratio of phosphorus to sulphur, due to the monoisotopic nature of phosphorus.

2. Materials

2.1. Sample Purification by 1D-SDS-PAGE (see Note 1)

1. Precast 1-dimensional polyacrylamide gradient gels (e.g. NuPage Novex 4–12% Bis-Tris gel, 12 cm length, Invitrogen).

2. Sample buffer: 8% lithium dodecyl sulphate (LDS), 40% glycerol, 564 mM Tris-base, 424 mM Tris–HCl, 2,041 mM EDTA, 0.88 mM SERVA Blue G250, 0.7 mM phenol red, pH 8.5 (e.g. NuPage LDS sample buffer (4×), Invitrogen).

3. Running buffer: 50 mM 3-(N-morpholino)propanesulphonic acid, 0.1% sodium dodecyl sulphate (SDS), 50 mM Tris-base, 1 mM EDTA, pH 7.7 (1× composition) (e.g. NuPage MOPS SDS running buffer (20×), Invitrogen). Dilute 50 ml with 950 ml of deionized water before use; the diluted solution may be re-used.

4. Prestained molecular weight markers (e.g. Rainbow, GE healthcare).

5. Coomassie staining solution: 0.01% Coomassie brilliant blue G250, 4.7% ethanol, 8.5% phosphoric acid in deionized water (e.g. Simply Blue, Invitrogen).

2.2. Protein Digestion and Peptide Desalting

1. Destaining solution I: 30% acetonitrile, 70% 0.1 M NH_4HCO_3.

2. Destaining solution II: 50% acetonitrile, 0.1% trifluoroacetic acid.

3. Reducing solution: 0.01 M dithiothreitol (DTT), 0.1 M NH_4HCO_3.

4. Neutralisation buffer: 0.1 M NH_4HCO_3.

5. Alkylation solution: 0.055 M iodoacetamide, 0.1 M NH_4HCO_3.

6. Trypsin solution: add 250 µl of 1 mM HCl to 25 µg modified trypsin (Roche Diagnostics). Store aliquots of 10 µl at −20°C.

7. RP-C_{18} micropipet tips for desalting (e.g. ZipTips, Millipore).

8. Extraction solution: 5% formic acid.

9. Elution solution: 50% acetonitrile, 2% formic acid.

10. Conditioning solution: 2% formic acid.

2.3. Protein Analysis by μLC-ICP-MS

1. ICP-MS suitable for analysis of elements that are interfered by polyatomic species such as NO$^+$, NOH$^+$, and O$_2^+$ in case of phosphorus and sulphur (e.g. high-resolution sector field ICP-MS Element II, Thermo Electron, *see* **Note 2**).

2. Syringe pump and microlitre syringe (100–250 μl) with capillary adapter. (Do not use a sharp needle like they are used for punctuation of septa!)

3. Low-flow interface including spray chamber and capillary-based nebuliser (e.g. DS5/CEI-100, CETAC, developed by Schaumlöffel et al. *(36)*, *see* **Note 3**).

4. Calibration solution: 1 μg/l multielement standard including Ba, B, Co, Fe, Ga, In, K, Li, Lu, Na, Rh, Sc, Tl, U, and Y.

5. Tuning solution (sulphur and phosphorus): 250 μM cysteine, 50 μM bis(4-nitrophenyl)phosphate (BNPP), 0.1% trifluoroacetic acid, 50% acetonitrile.

6. Capillary LC-system capable of flow rates below 10 μl/min.

7. Second LC-pump and y-piece for fused silica capillaries (e.g. NanoTight Y connector, Upchurch Scientific) for post-column spike addition (only necessary for IDA).

8. Capillary LC-column (suitable dimensions are: inner diameter (ID) 200 μm, length 15 cm) filled with RP-C$_{18}$ material, particle size 5 μm (e.g. from Grohm/Grace Vydac, *see* **Note 4**).

9. Eluent A: 2% acetonitrile, 0.065% trifluoroacetic acid (*see* **Note 5**).

10. Eluent B: 80% acetonitrile, 0.1% trifluoroacetic acid (*see* **Note 5**).

2.4. Characterisation of Quantification Parameters and Data Evaluation

1. Element standard solution for determination of relative sensitivities of sulphur and phosphorus: 250 μM cysteine, 50 μM phoshoserine.

2. Software for data evaluation (e.g. Origin or Excel).

3. Spike solution: ^{34}S elemental sulphur (IRE Diagnostic, Düsseldorf, Germany) digested with concentrated nitric acid (Acros, Geel, Belgium) in a microwave system and subsequent dilution with deionised water. The final concentration of the spike should be close to the expected concentration of the sample (only necessary for IDA).

3. Methods

Application of μLC-ICP-MS to analysis of (phospho)proteins typically comprises several steps. First of all, the protein or protein

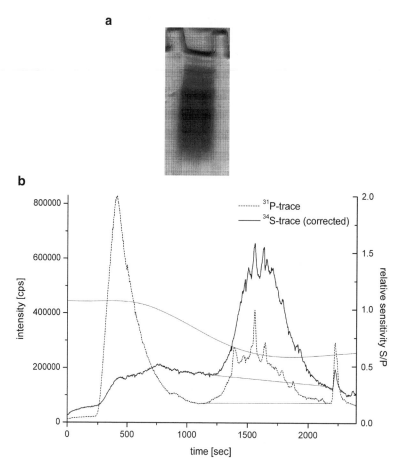

Fig. 2. Analysis of a cytosolic protein extract from mouse epidermis for determination of a global protein phosphoryla-tion stoichiometry. (**a**) Approximately 50 μg of the protein mixture was applied to 1D-SDS-PAGE. After a short electro-phoresis of 10 min (2 cm), which is usually long enough to efficiently remove small sulphur- or phosphorus-containing compounds, the gel was stained weakly. (**b**) The complete gel lane of approximately 1 × 2 cm² was cut, destained, and digested with trypsin. Peptides were extracted, desalted, and analysed with μLC-ICP-MS using a 30-min acetonitrile gradient (2–80%). Peptides are separated from small compounds such as phosphate, but elute together at medium organic content of the eluent. To correct for gradient-dependent sensitivity changes, the sulphur trace was divided by a sensitivity correction function, which was calculated from injection of element standards at different solvent composi-tion. The combined intensity (area) in the ³⁴S-trace gives a measure for protein amount, whereas the combined intensity in the ³¹P-trace is a measure for the abundance of phosphorylation. The late-eluting peak in the ³⁴S-trace may be residual Coomassie, whereas the late-eluting impurity in the ³¹P-trace is of unknown origin.

mixture of interest (*see* **Note 6**) has to be purified from con-taminants that may contain the elements to be monitored (P, S), unless the sample is only moderately contaminated. This is usu-ally achieved by a short 1D-SDS-PAGE and subsequent protein digestion and peptide extraction. This approach can be used also for crude protein extracts, e.g. protein precipitates from whole cells (*28, 29*). Such an example is shown in **Fig. 2** for the case of mouse epidermis cells. Before analysis of the purified samples, the ICP-MS system has to be optimized and calibrated by infusion of standards via a syringe pump. For relative quantification of the phosphorylation degree, the gradient-dependent response factors for sulphur and phosphorus have to be determined *(13, 28)*. Alternatively, absolute quantification by IDA requires determination of the isotope spike flow (this variant is described in **Subheading 4, Note 14**). Finally, the obtained data have to be evaluated based on the corrected element traces and the number of sulphur and phosphorus atoms incorporated in the respective protein.

3.1. Sample Purification by 1D-SDS-PAGE (see Note 7)

1. Prepare running buffer by diluting 50 ml of the 20× running buffer with 950 ml of water in a measuring cylinder and mix.

2. Unpack the gel, remove the comb, and insert it into the gel chamber. Fill the chamber with running buffer until the gel is covered and both electrodes have contact.

3. Dilute samples and molecular weight marker four times with sample buffer. Apply 2–20 μl of sample solution into each gel pocket. The final protein amount per lane should not exceed 50–100 μg.

4. Close the cover of the gel chamber and run the gel for 10–15 min at 200 V until the protein markers have migrated 2–3 cm. This is usually enough to remove the majority of small molecular weight contaminants, which may contain either sulphur or phosphorus. However, SDS-PAGE might be developed completely if isolation of a specific target protein from a mixture is desired.

5. Switch off the voltage and open the gel chamber. Take out the gel cartridge, remove the gel from the box and place it into a suitable tank for staining.

6. Wash the gel 5 min with deionised water. Apply Coomassie solution until the gel is covered and let it shake for about 1 h. Ensure that the bands are only moderately stained, since residual Coomassie appears as late-eluting peak in the ICP-MS sulphur trace (*see* **Note 8**). Remove the staining solution and wash the gel twice with deionised water. Shake overnight in deionised water.

3.2. Protein Digestion and Peptide Desalting (see Note 7)

1. Prepare a surface for gel slicing and a scalpel by washing with ethanol and water. Slice either the gel band of interest or the complete gel lane of 2–3 cm into pieces of approximately 1 × 1 mm and put them into an Eppendorf tube (100 μl or 250 μl depending on the gel volume).

2. Destain and wash the gel pieces twice with destaining solution I and twice with destaining solution II. If necessary, repeat with destaining solution II until the majority of Coomassie is removed (a faint blue is not critical).

3. Remove the supernatant and dry in vacuum using moderate heating assistance (30–40°C). A standard laboratory pump and a water bath may also be used instead of a speedvac (*see* **Note 9**). The gel pieces should be shrinked completely.

4. Add reducing solution until the gel pieces are covered completely. For single protein bands, 25 μl is usually sufficient, but more solution may be needed for whole gel lanes. Leave to react at 55°C for 45 min, e.g. in a thermomixer. Remove supernatant (if there is any) and dry in vacuum as described in **step 3**.

5. Add at least 25 μl of alkylation solution until the gel pieces are covered completely. Leave to react at room temperature for 30 min. Remove supernatant (if there is any) and dry in vacuum as described in **step 3**.

6. Wash two times with neutralisation buffer. Remove supernatant and dry in vacuum as described in **step 3**.

7. Prepare buffered trypsin solution by adding 10 μl of neutralisation buffer (pH 8) to each 10 μl of frozen trypsin aliquots to be used (*see* **Note 10**). Add 4–6 μl of this solution to each sample. Wait until the trypsin solution is completely sucked up by the gel pieces. This ensures that the majority of trypsin gets into the gel and reaches the proteins. After 10 min, add 10 μl of neutralisation buffer and wait for another 10 min. After rehydration, add 30 μl of neutralisation buffer or more until the gel pieces are covered. Close the tube and incubate at 37°C overnight.

8. Transfer the supernatant into a fresh tube. Add at least 20 μl of extraction solution (gel pieces should be covered). Shake for about 30 min and transfer combined supernatants. Repeat at least once to ensure complete peptide extraction (*see* **Note 11**). Dry combined supernatants under a gentle stream of nitrogen or with a speedvac until all solvent is removed.

9. Dissolve peptides in 10–20 μl of conditioning solution for desalting. Wash ZipTip by slowly aspirating 10 μl of elution solution. Repeat with 10 μl of conditioning solution. Load samples by repeatedly aspirating the sample solution. Wash with 10 μl of washing solution. Elute peptides with 2–10 μl of elution solution. The procedure may be repeated using the same ZipTip to enhance recovery. Desalted samples may be used directly for μLC-ICP-MS analysis and/or nanoESI-MS analysis. Alternatively, they can be dried under nitrogen or in a speedvac and stored at ≤20°C for later use.

3.3. Protein Analysis by μLC-ICP-MS

1. Prepare the LC eluents and start the LC pumps (*see* **Note 12**). Adapt the flow rate to your column size and properties. 2–5 μl/min is the optimum for the applied capillary columns (200 μm ID, 5 μm particle size). Equilibrate for at least 30 min.

2. Ignite the plasma of the ICP-MS, start scanning and let the instrument (source and magnet) warm up for at least 30 min.

3. Fill the syringe with calibration solution and connect it via an appropriate adapter to the DS5 interface. Infuse with a syringe pump at a flow rate matching your LC parameters (2–10 µl/min in case of 200 µm ID columns). Tune the instrument to get maximum sensitivity by monitoring elements covering the whole mass range (e.g. Li, In, and U). In case of a sector field instrument such as the Element II, repeat in medium resolution (4,000) and perform a mass calibration if necessary (*see* **Note 13**).

4. Fill the syringe with tuning solution. The percentage of organic in the tuning solution should match the conditions at the time when the majority of analytes is eluting from the LC column (e.g. 50% acetonitrile). The use of small polar organic compounds as element standard for tuning of the ICP-MS avoids memory effects at the fused silica wall. Tune for maximum sensitivity on the target elements sulphur and phosphorus using medium resolution. Typical values are sample gas flow, 0.9–1.3 l/min; auxiliary gas flow, 0.6–0.8 l/min; cool gas flow, 16 l/min; and RF power, 1,250–1,350 W (*see* **Note 13**). The spray efficiency of the DS5 nebulizer may be tuned also from time to time by adjusting the tip position with a screwdriver.

5. Disconnect the syringe pump from the interface and screw in the outlet of the LC column via a fused silica capillary. To minimize peak broadening, be sure to use capillaries with low ID (≤50 µm) matching the rest of your analytical system.

6. Monitor the sulphur and phosphorus traces (chromatogram view) using fast scan cycles (200–400 ms per element). The following details refer to magnetic sector field MS only (Element II, Thermo Electron): Use electric scan mode and 20% peak window for monitoring of the peak top only. This will ensure a high duty cycle during data acquisition. If you are monitoring elements with very similar masses, e.g. sulphur and phosphorus, reduce all magnet settling times in the method to the smallest possible value (0.001). Longer magnet settling times are only necessary if the masses differ by more than 30%. Be careful; settling times are calculated automatically any time the method is changed!

7. Choose the appropriate LC gradient depending on the complexity of your sample. Peptides may be separated with a gradient from 2% acetonitrile to 80% acetonitrile; for protein separation the gradient may start at higher organic content (e.g. 20% acetonitrile). Adapt the number of scans in your ICP-MS method to the length of your gradient. Data acquisition may be started by a trigger signal from the LC system.

8. Test the LC system by injection of a blank and an appropriate phosphopeptide or phosphoprotein standard (*see* **Fig. 3**).

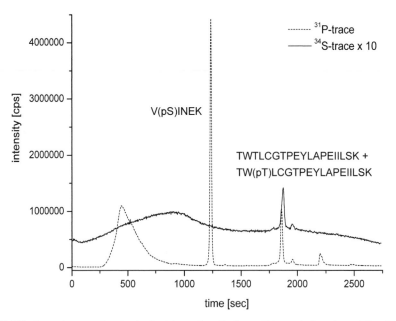

Fig. 3. µLC-ICP-MS chromatogram of a standard mixture of synthetic peptides and phosphopeptides. Phosphorylated and unphosphorylated isoforms of the peptide TWTLCGTPEYLAPEIILSK are not separated in the ³⁴S-trace, but the slightly different retention times in the ³¹P-trace and the ³⁴S-trace point to the fact that the phosphopeptide elutes somewhat earlier. The peptide V(pS)INEK gives a signal in the ³¹P-trace, but no signal in the ³⁴S-trace, since it has no sulphur. The peptide VSINEK, which is also present, gives no signal in either trace.

Inject between 2 µl and 10 µl, depending on your LC-system (column, flow rate). Take into account sufficient equilibration time before injection of the next sample. This refers not only to LC column regeneration, but also to the stability of the ICP-MS signal. The main reason is that introduction of organic solvents leads to carbon deposition on the sample cones. Equilibration with aqueous solution removes deposits by oxidation, but the time varies depending on the amount of organic that has been introduced before (*see* **Note 3**).

9. Write a sequence list and inject your samples.

3.4. Characterisation of Quantification Parameters and Data Evaluation

1. Export all element traces (chromatogram data in *x, y* format) into an appropriate software tool for data evaluation, e.g. Origin or Excel.

2. The following description assumes the case where the ratio of P to S is used for determination of the phosphorylation stoichiometry. (Data evaluation in case of IDA is described in **Subheading 4**, *see* **Note 14**.) If you are using a solvent gradient, relative sensitivities are changing with retention time. Therefore, sensitivities have to be determined by infusion of an element standard at varying solvent compositions. Choose an eluent composition and inject the element standard directly into your LC-system (*see* **Note 15**). Repeat using different eluent compositions until you have enough values covering your

gradient, one value per 10% change of organic content is usually sufficient. You may use the obtained sensitivity values for calculating a correction function that describes your gradient-dependent sensitivity ratio continuously, e.g. by a polynomial fit.

3. Divide the sulphur intensities by the relative sensitivity of sulphur to phosphorus, like it is shown in **Fig. 4**. You may smooth the element traces afterwards.

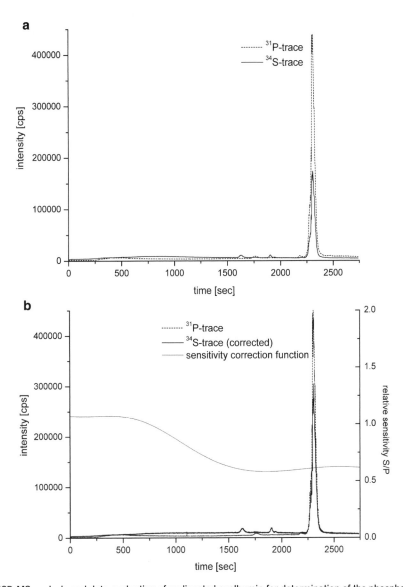

Fig. 4. µLC-ICP-MS analysis and data evaluation of undigested ovalbumin for determination of the phosphorylation stoichiometry. (**a**) Raw data. (**b**) Processed data after division of the ³⁴S-trace by the correction function (relative sensitivity S/P). The peaks in both traces were integrated and the ³⁴S-intensity was divided by the number of sulphur atoms (22) to get the relative protein abundance. Division of the ³¹P-intensity by the protein abundance results in a phosphorylation stoichiometry of 1.45 mol phosphorus per mole protein.

4. Integrate the peak of interest in both element traces. Divide the phosphorus area by the sulphur area to get the relative amounts of phosphorus and sulphur. In case of protein digests, you may combine the area of all peaks in each element trace in order to calculate the ratio P to S of the entire protein or protein mixture.

5. For determination of the stoichiometry of phosphorus to protein, i.e. the phosphorylation degree, the abundance of sulphur in your peptide or protein of interest has to be considered (*see* **Note 16**). Divide the intensity of the sulphur trace by the number of sulphur atoms of your given protein to achieve the intensity representing your protein amount. Then divide the phosphorus intensity by this protein intensity, which will give the phosphorylation extent of your protein or protein mixture of interest.

4. Notes

1. Commercially available precast gels and buffers were used in our studies (NuPage, Invitrogen), but 1D-SDS-PAGE may be performed using homemade gels or equipment from any other supplier, although this was not explicitly tested. However, if problems with high sulphur or phosphorus background arise, a possible contamination of gel ingredients or solvents should be considered.

2. Alternatively, phosphorus and sulphur detection is also possible with quadrupole instruments that are equipped with a collision or reaction cell *(37, 38)*. These instruments can be operated in two different modes. In the collision mode, polyatomic interferences are eliminated by collisions with an inert gas, usually a mixture of helium and hydrogen. In the reaction mode, addition of oxygen leads to formation of PO$^+$ (*m/z* 47) and SO$^+$ (*m/z* 48). At these masses, interference is much less pronounced.

3. You can avoid carbon deposition by the introduction of an additional oxygen stream into the spray chamber (15–25 ml/min). Be sure to use Pt-Cones when introducing oxygen, since Ni-Cones are subject to damage. Oxygen addition is mandatory if you are using high organic content and comparably high flow rates (>5 µl/min); for instance, for analysis of intact proteins by IDA, where an additional spike flow is necessary. We used a homemade spray chamber with an extra oxygen inlet for this purpose *(35)*.

4. Depending on the application, utilization of other column types may be reasonable. For instance, RP-phenyl or RP-C$_4$ materials are in most cases better suited for separation of intact

proteins than RP-C$_{18}$. Of course, columns from other suppliers may also be used. In contrast, protein separation by ion chromatography or size exclusion chromatography usually requires higher flow rates. This rules out both the use of a low-flow interface and the use of a high content of organic solvents, unless membrane desolvation systems are used.

5. Eluent composition and gradient may be changed according to your application, for instance higher content of organic solvent in case of protein separations.

6. Since the choice of the protocol depends mainly on the sample, e.g. type of cells, this topic is outside the scope of the present method protocol. In principle, a variety of protein extraction methods should be compatible with the approach described here. The only precaution for phosphoprotein analysis is that phosphatase and protease inhibitors have to be added already at the first lysis or precipitation step in order to avoid protein dephosphorylation or degradation. As a precaution, DNAse might also be added, although it was not tested whether this is really essential.

7. This step may be omitted depending on protein source and purity. On the other hand, the protocol is also suited for sequence analysis with nanoESI-MS or LC-ESI-MS.

8. Staining usually does not interfere with peptide analysis by LC-ICP-MS, as long as the staining intensity is not too high (not more than a faint blue). Staining may also be omitted completely, if the whole gel piece is to be used for digestion and extraction and if staining is not necessary for protein localisation in the gel.

9. It is useful to prick a small whole with a needle into the lid of the Eppendorf cup and close the lid in order to prevent loss of analyte during evaporation, especially if no centrifuge is used.

10. Other enzymes may be used for certain purposes *(39)*. We found a more uniform digestion (sequence coverage) with enzyme mixtures, e.g. using trypsin and AspN simultaneously. A multiprotease approach using the combined extracts of four different enzymatic digests was found to yield higher recovery of phosphopeptides in nanoLC-ESI-MS *(40)*.

11. Other protocols use acetonitrile-containing solutions for peptide extraction. This might work here as well, but it was not explicitly tested. However, interfering compounds exhibiting sulphur or phosphorus may be extracted from the gel by using organic extraction solvents.

12. An additional, isocratic LC pump is necessary for post-column IDA. This should be connected via a y-piece with low dead volume and good mixing efficiency (*see* **Subheading 2.3, item** 7), since complete mixing of eluate and spike is mandatory for IDA.

13. The exact procedure of calibration and the tuning conditions is instrument dependent. For instance, the Element II software allows for automated correction of mass drift – be sure to activate this auto-lockmass feature.

14. For absolute quantification by post-column IDA, the following procedure applies. A graphical illustration of the process is given in **Fig. 5**. The theoretical background has been

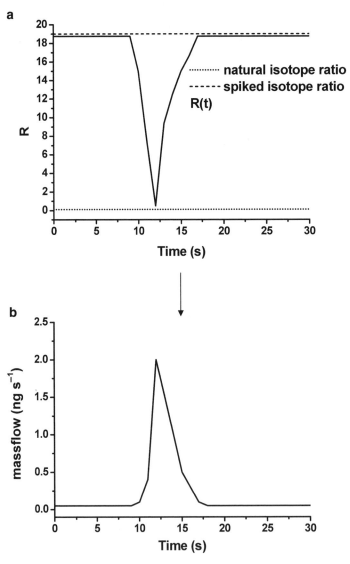

Fig. 5. Graphical illustration of the process for absolute quantification by post-column IDA (species-unspecific isotope dilution analysis). (**a**) Plot of the isotope ratios used for calculation of the sample amount. Depicted are the isotope ratios of the sample, of the spike and the experimental, time-dependent isotope ratio $R(t)$. (**b**) Mass flow chromatogram after conversion of the isotope ratio $R(t)$ to the massflow using the IDA formula $Mf_s(t) = Mf_{Sp} \times \dfrac{M_S}{M_{Sp}} \times \dfrac{h_{Sp}{}^{34s} - R(t) \times h_{Sp}{}^{32s}}{R(t) \times h_S{}^{32s} - h_S{}^{34s}}$. The integral over the signal gives the absolute amount of element in the sample.

described by Heumann and in a tutorial review by Sanz-Medel and co-workers *(32–34)*.

(a) Two isotopes of the respective element have to be monitored simultaneously, e.g. ^{32}S and ^{34}S in case of protein quantification. Determine the relative isotope abundances $h_{Sp}^{32_S}$ and $h_{Sp}^{34_S}$ in the spike solution, which is introduced via the second LC pump and y-Connector (*see* **Subheadings 2.3, item 7** and **2.4, item 3**). This can be done by injection of the spike using the following equation:

$$h_{Sp}^{32_S} = \frac{\text{Intensity}(^{32}S)}{\text{Intensity}(^{32}S + {}^{34}S)}.$$

(b) When running your IDA analysis, calculate the experimental isotope ratio $R(t) = \dfrac{\text{Intensity}(^{34}S)}{\text{Intensity}(^{32}S)}$ as a function of the retention time, as shown in **Fig. 5a**. $R(t)$ can then be converted into the time-dependent massflow $Mf_s(t)$ using

the equation $Mf_s(t) = Mf_{Sp} \times \dfrac{M_S}{M_{Sp}} \times \dfrac{h_{Sp}^{34_S} - R(t) \times h_{Sp}^{32_S}}{R(t) \times h_S^{32_S} - h_S^{34_S}}$,

where Mf_{Sp} is the massflow of the spike, and M_S and M_{Sp} are the average molecular weight of the element in sample and spike, respectively (*see* **Fig. 5b**). The integral over a distinct signal gives directly the absolute amount of element in it, corresponding to the equation

$$m_s = Mf_{Sp} \times \frac{M_S}{M_{Sp}} \times \int \frac{h_{Sp}^{34_S} - R(t) \times h_{Sp}^{32_S}}{R(t) \times h_S^{32_S} - h_S^{34_S}}.$$

For calculation of the protein amount, the mass has to be converted into the molarity and divided by the number of sulphur atoms in the respective protein.

(c) The massflow of the spike Mf_{Sp} represents the amount of spike reaching the detector in a given timeframe. It can be obtained by injecting a well-known amount of a standard. With knowledge of the amount of standard m_S injected and the signal area it produces (which corresponds to the integral), Mf_{Sp} can now be calculated easily by transformation of the IDA equation to

$$Mf_{Sp} = m_S \times \frac{M_{Sp}}{M_S} \times \int \frac{R(t) \times h_S^{32_S} - h_S^{34_S}}{h_{Sp}^{34_S} - R(t) \times h_{Sp}^{32_S}}.$$

Complete recovery of the standard has to be guaranteed.

15. Alternatively, you may spike the elements of interest directly into the LC eluents and monitor a blank gradient for sensitivity determination.

16. The procedure may also be applied to complete cellular protein precipitates in order to obtain a global average phosphorylation degree typical for the present cell type and state. In this case, the average sulphur content of the proteins has to be estimated, e.g. based on proteome or DNA data.

Acknowledgement

The authors thank the following researchers from the German Cancer Research Center (DKFZ) in Heidelberg: T. Barz and W. Pyerin for providing the sample of in vitro phosphorylated CDC34, and K. Müller-Decker and D. Kübler for providing the protein extract of mouse epidermis. We thank J. Heilmann for his helpful discussion. Financial support of the German research ministry BMBF (Bundeministerium für Bildung und Forschung) within the Proteomics Program is also gratefully acknowledged.

References

1. Gerber, S. A., Rush, J., Stemman, O., Kirschner, M. W., Gygi, S. P. (2003) Absolute quantification of proteins and phosphoproteins from cell lysates by tandem MS. *Proc. Natl Acad. Sci. U. S. A.* 100, 6940–6945.

2. Ong, S. E., Mann, M. (2005) Mass spectrometry-based proteomics turns quantitative. *Nat. Chem. Biol.* 1, 252–262.

3. Anderson, L., Hunter, C. L. (2006) Quantitative mass spectrometric multiple reaction monitoring assays for major plasma proteins. *Mol. Cell. Proteomics* 5, 573–588.

4. Bantscheff, M., Schirle, M., Sweetman, G., Rick, J., Kuster, B. (2007) Quantitative mass spectrometry in proteomics: a critical review. *Anal. Bioanal. Chem.* 389, 1017–1031.

5. Zhang, H., Yan, W., Aebersold, R. (2004) Chemical probes and tandem mass spectrometry: a strategy for the quantitative analysis of proteomes and subproteomes. *Curr. Opin. Chem. Biol.* 8, 66–75.

6. Leitner, A., Lindner, W. (2004) Current chemical tagging strategies for proteome analysis by mass spectrometry. *J. Chromatogr. B* 813, 1–26.

7. Beynon, R. J., Pratt, J. M. (2005) Metabolic labeling of proteins for proteomics. *Mol. Cell. Proteomics* 4, 857–872.

8. Mann, M., Ong, S. Grønborg, M., Steen, H., Jensen, O. N., Pandey, A. (2002) Analysis of protein phosphorylation using mass spectrometry: deciphering the phosphoproteome. *Trends Biotechnol.* 20, 261–268.

9. Bonenfant, D., Schmelzle, T., Jacinto, E., Crespo, J. L., Mini, T., Hall, M. N., Jenoe, P. (2003) Quantitation of changes in protein phosphorylation: a simple method based on stable isotope labelling and mass spectrometry. *Proc. Natl Acad. Sci. U. S. A.* 100, 880–885.

10. Steen, H., Jebanathirajah, J. A., Springer, M., Kirschner, M. W. (2005) Stable isotope-free relative and absolute quantitation of protein phosphorylation stoichiometry by MS. *Proc. Natl Acad. Sci. U. S. A.* 102, 3948–3953.

11. Reinders, J., Sickmann, A. (2005) State-of-the-art in phosphoproteomics. *Proteomics* 5, 4052–4061.

12. Wind, M., Edler, M., Jakubowski, N., Linscheid, M., Wesch, H., Lehmann, W. D. (2001) Analysis of protein phosphorylation by capillary liquid chromatography coupled to element plasma mass spectrometry with 31P detection and to electrospray mass spectrometry. *Anal. Chem.* 73, 29–35.

13. Wind, M., Wesch, H., Lehmann, W. D. (2001) Protein phosphorylation degree: determination

by capillary liquid chromatography and inductively coupled plasma mass spectrometry. *Anal. Chem.* 73, 3006–3020.

14. Axelsson, B.-O., Jörnten-Karlsson, M., Michelsen, P., Abou-Shakra, F. (2001) The potential of inductively coupled plasma mass spectrometry detection for high-performance liquid chromatography combined with accurate mass measurements of organic pharmaceutical compounds. *Rapid. Commun. Mass Spectrom.* 15, 375–385.

15. Wang, J., Dreessen, D., Wiederin, D. R., Houk, R. S. (2001) Measurement of trace elements in proteins extracted from liver by size exclusion chromatography-inductively coupled plasma-mass spectrometry with a magnetic sector mass spectrometer. *Anal. Biochem.* 288, 89–96.

16. Zhang, C., Wu, F., Zhang, Y., Wang, X., Zhang, X. (2001) A novel combination of immunoreaction and ICP-MS as a hyphenated technique for the determination of thyroid-stimulating hormone (TSH) in human serum. *J. Anal. At. Spectrom.* 16, 1393–1139.

17. Quinn, Z. A., Baranov, V. I., Tanner, S. D., Wrana, L. J. (2002) Simultaneous determination of proteins using an element-tagged immunoassay coupled with ICP-MS detection. *J. Anal. At. Spectrom.* 17, 892–896.

18. Schaumlöffel, D., Prange, A., Marx, G., Heumann, K. G., Bratter, P. (2002) Characterization and quantification of metallothionein isoforms by capillary electrophoresis-inductively coupled plasma-isotope-dilution mass spectrometry. *Anal. Bioanal. Chem.* 372, 155–163.

19. Wind, M., Kelm, O., Nigg, E. A., Lehmann, W. D. (2002) Identification of phosphorylation sites in the polo-like kinases Plx1 and Plk1 by a novel strategy based on element and electrospray high resolution mass spectrometry. *Proteomics* 2, 1516–1523.

20. Wind, M., Gosenca, D., Kübler, D., Lehmann, W. D. (2003) Stable isotope phospho-profiling of fibrinogen and fetuin subunits by element mass spectrometry coupled to capillary liquid chromatography. *Anal. Biochem.* 317, 26–33.

21. Wind, M., Wegener, A., Eisenmenger, A., Kellner, R., Lehmann, W. D. (2003) Sulfur as key element for quantitative protein analysis by capillary liquid chromatography coupled to element mass spectrometry. *Angew. Chem. Int. Ed.* 42, 3425–3427.

22. Encinar, J. R., Ouerdane, L., Buchmann, W., Tortajada, J., Lobinski, R., Szpunar, J. (2005) Identification of water-soluble selenium-containing proteins in selenized yeast by size-exclusion-reversed-phase HPLC/ICPMS followed by MALDI-TOF and electrospray Q-TOF mass spectrometry. *Anal. Chem.* 75, 3765–3774.

23. Harrington, C. F., Vidler, D. S., Watts, M. J., Hall, J. F. (2005) Potential for using isotopically altered metalloproteins in species-specific isotope dilution analysis of proteins by HPLC coupled to inductively coupled plasma mass spectrometry. *Anal. Chem.* 77, 4034–4041.

24. del Castillo Busto, M. E., Montes-Bayon, M., Sanz-Medel, A. (2006) Accurate determination of human serum transferrin isoforms: exploring metal-specific isotope dilution analysis as a quantitative proteomic tool. *Anal. Chem.* 78, 8218–8226.

25. Lobinski, R., Schaumlöffel, D., Szpunar, J. (2006) Mass spectrometry in bioinorganic analytical chemistry. *Mass. Spectrom. Rev.* 25, 255–289.

26. Schaumlöffel, D., Giusti, P., Preud'Homme, H., Szpunar, J., Lobinski, R. (2007) Precolumn isotope dilution analysis in nanoHPLC-ICPMS for absolute quantification of sulfur-containing peptides. *Anal. Chem.* 79, 2859–2868.

27. Lopez-Avila, V., Sharpe, O., Robinson, W. H. (2006) Determination of ceruloplasmin in human serum by SEC-ICPMS. *Anal. Bioanal. Chem.* 386, 180–187.

28. Krüger, R., Kübler, D., Pallissé, R., Burkovski, A., Lehmann, W. D. (2006) Protein and proteome phosphorylation stoichiometry analysis by element mass spectrometry. *Anal. Chem.* 78, 1987–1994.

29. Krüger, R., Wolschin, F., Weckwerth, W., Bettmer, J., Lehmann, W. D. (2007) Plant protein phosphorylation monitored by capillary liquid chromatography-element mass spectrometry. *Biochem. Biophys. Res. Commun.* 355, 89–96.

30. Ballihaut, G., Mounicou, S., Lobinski, R. (2007) Multitechnique mass-spectrometric approach for the detection of bovine glutathione peroxidise selenoprotein: focus on the selenopeptide. *Anal. Bioanal. Chem.* 388, 585–591.

31. Montes-Bayon, M., DeNicolab, K., Caruso, J. A. (2003) Liquid chromatography-inductively coupled plasma mass spectrometry. *J. Chromatogr. A* 1000, 457–476.

32. Heumann, K. G. (1992) Isotope dilution mass spectrometry (IDMS) of the elements. *Mass Spectrom. Rev.* 11, 41–67.

33. Heumann, K. G. (2004) Isotope-dilution ICP-MS for trace element determination and speciation: from a reference method to a routine method? *Anal. Bioanal. Chem.* 378, 318–329.

34. Rodriguez-Gonzalez, P., Marchante-Gayon, J. M., Alonso J. I. G, Sanz-Medel, A. (2005) Isotope dilution analysis for elemental speciation: a tutorial review. *Spectrochim. Acta Part B* 60, 151–207.

35. Zinn, N., Krüger, R., Leonhard, P., Bettmer, J. (2008) μLC coupled to ICP-SFMS with postcolumn isotope dilution analysis of sulphur for absolute protein quantification. *Anal. Bioanal. Chem.* 391, 537–543.

36. Schaumlöffel, D., Ruiz Encicar, J., Lobinski, R. (2003) Development of a sheathless interface between reversed-phase capillary HPLC and ICPMS via a microflow total consumption nebulizer for selenopeptide mapping. *Anal. Chem.* 75, 6837–6842.

37. Bandura, D. R., Baranov, V. I., Tanner, S. D. (2002) Detection of ultratrace phosphorus and sulfur by quadrupole ICPMS with dynamic reaction cell. *J. Anal. At. Spectrom.* 74, 1497–1502.

38. Koppenaal, D. W., Eiden, G. C., Baringa, C. J. (2004) Collision and reaction cells in atomic mass spectrometry: development, status, and applications. *J. Anal. At. Spectrom.* 19, 561–570.

39. Schlosser, A., Pipkorn, R., Bossemeyer, D., Lehmann, W. D. (2001) Analysis of protein phosphorylation by a combination of elastase digestion and neutral loss tandem mass spectrometry. *Anal. Chem.* 73, 170–176.

40. Schlosser, A., Vanselow, J. T., Kramer, A. (2005) Mapping of phosphorylation sites by a multi-protease approach with specific phosphopeptide enrichment and nanoLC-MS/MS analysis. *Anal. Chem.* 77, 5243–5250.

Chapter 16

Reverse-Phase Diagonal Chromatography for Phosphoproteome Research

Kris Gevaert and Joël Vandekerckhove

Summary

We present a gel-free proteomics procedure for the specific isolation of phosphorylated peptides from whole proteome digests. Central is the use of diagonal, reverse-phase chromatography which consists of two consecutive reverse-phase peptide separations with a modification step in between. The latter alters the column retention of affected peptides, thereby allowing their specific isolation from the bulk of nonaffected peptides. To isolate phosphopeptides from complex mixtures, this modification step is a dephosphorylation reaction using a cocktail of broad-spectrum phosphatases. Upon dephosphorylation, peptides undergo a hydrophobic shift and are thereby sorted from in vivo nonphosphorylated peptides. To increase the overall yield of phosphopeptides, a pre-enrichment step was found necessary and to further distinguish true ex-phosphorylated peptides from nonphosphorylated peptides sorted artificially, differential isotope labeling was introduced. The complete COFRADIC sorting procedure is described here.

Key words: COFRADIC, Gel-free proteomics, Diagonal chromatography, Phosphatases, RP-HPLC, Column retention.

1. Introduction

In view of the high number of protein kinases encoded by the human genome – over 500 genes are predicted to encode for the human kinome *(1)* – not surprisingly, protein phosphorylation is one of the most common post-translational protein modifications found in higher eukaryotes. Following 1D- or 2D-polyacrylamide gel electrophoresis of proteome preparations, phosphoproteins are typically detected by ^{32}P-autoradiography *(2)*, Western blotting (especially for tyrosine-phosphorylation *(3)*) or using fluorescent labels *(4)*. However, since such gel-based approaches

Marjo de Graauw (ed.), *Phospho-Proteomics, Methods and Protocols, vol. 527*
© 2009 Humana Press, a part of Springer Science + Business Media, New York, NY
Book DOI: 10.1007/978-1-60327-834-8_16

hold a number of potential disadvantages – low copy number proteins and highly hydrophobic proteins are typically missed – novel gel-free or peptide-centric proteomic techniques have been introduced over the past years (for a recent review, *see* **ref.** *5*). These techniques focus on mass spectrometry based identification of (often) massive numbers of peptides, and link these peptides to their precursor proteins. As peptides are generally more soluble than intact proteins, even highly hydrophobic proteins could be identified and characterized this way (*see* for e.g., **ref.** *6*).

Several peptide-centric techniques for phosphoproteome research have been introduced and are briefly listed below. One of the most used techniques relies on immobilized metal ion affinity chromatography (IMAC, e.g., using Fe^{3+} ions) to complex and thereby isolate phosphorylated peptides *(7)*. More recently, titanium dioxide (TiO_2) particles were found to specifically retain phosphorylated peptides (e.g., *(8)*). A different strategy exploits the negative charge of the phosphate group to enrich phosphopeptides in the peptide fraction not binding to a strong cation exchange (SCX) column run at pH 2.7 *(9)*. Finally, different chemical methods have been described to isolate phosphopeptides (e.g., **ref.** *10, 11*) but generally failed to reach the broad scientific community probably due to the large numbers and often inefficient chemical steps needed.

Over the past few years, our lab adapted the "old" diagonal electrophoresis/chromatography techniques *(12, 13)* so that they can be used for contemporary, peptide-centric proteomics research. We termed our technique COFRADIC (combined fractional diagonal chromatography *(14)*) and it essentially consists of one primary peptide separation, a chemical or an enzymatic reaction altering the side-chain of specific set of peptides (the COFRADIC sorting reaction) followed by a series of secondary peptide separations identical to the primary one. The COFRADIC sorting reaction changes the column retention of the targeted set of peptides such that these peptides elute differently during the secondary separation and are in this way specifically isolated. In contrast to other gel-free proteomics techniques, one of the main advantages of COFRADIC is the fact that it allows different sets of peptides to be isolated, thereby opening up different aspects of dynamic proteomes for detailed investigation. Examples include protein processing *(15)*, N-glycosylation *(16)*, and sialylation of *N*-glycans *(17)*, and other examples are elaborated upon in recent reviews (e.g., **ref.** *18, 19*).

We recently published a COFRADIC protocol for analyzing protein phosphorylation *(20)*. Here, the sorting reaction is a dephosphorylation event using a cocktail of broad-spectrum phosphatases. Compared to their phosphorylated counterparts present in the primary fractions, dephosphorylated peptides, or rather ex-phosphorylated peptides as we tend to call them, undergo a hydrophobic shift and are thereby isolated from the majority of non-phosphorylated peptides. Since such COF-RADIC sorted peptides lost their phosphate group, MS/MS-based peptide identification is straightforward however one potential disadvantage is that all information on the exact site of phosphorylation is lost.

We also introduced an enzymatic isotope labeling step developed in our lab *(21)* to label tryptic peptides with either two oxygen-16 or two oxygen-18 atoms at their C-terminal carboxyl group, and in combination with a dephosphorylation step of one pool of labeled peptides prior to the COFRADIC steps (**Fig. 1**), this allows us a direct distinction between true ex-phosphorylated peptides carrying only one isotope label and peptides that were sorted independent of the sorting reaction (e.g., by methionine oxidation *(14)*) since the latter will be present as isotope couples. The overall sorting strategy shown in **Fig. 1** also includes an

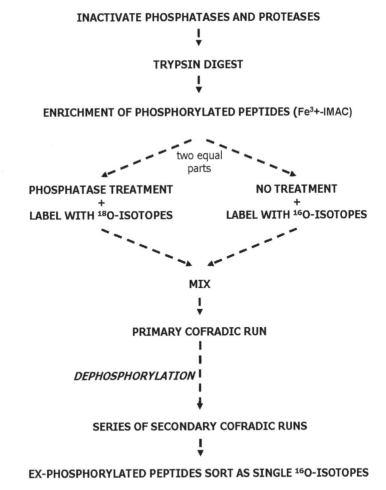

Fig. 1. COFRADIC isolation scheme of ex-phosphorylated peptides from proteome digests.

enrichment step for phosphopeptides simply to increase the overall likelihood of identifying ex-phosphorylated peptides.

The COFRADIC sorting protocol is detailed below and important notes and points of attention are indicated.

2. Materials

2.1. Materials for Proteome Preparation

1. Cell disruption buffer: 4 M urea, 0.625% (w/v) CHAPS, protease inhibitors (Complete EDTA-free protease inhibitor cocktail tablets, F. Hoffmann-La Roche Ltd.), phosphatase inhibitors (10 mM NaF, 200 μM sodium orthovanadate, 20 mM β-glycerophosphate, 5 μM phenylvalerate and 2 mM levamisole hydrochloride), 10 mM tris(2-carboxyethyl) phosphine (TCEP) and 100 mM iodoacetamide in 100 mM NaH_2PO_4 (pH 7.4) (*see* **Note 1**).

2. Protein digestion buffer: 2 M of fresh urea in 100 mM Tris–HCl (pH 8.7) and phosphatase inhibitors (see cell disruption buffer) (*see* **Note 2**).

3. PD-10 columns (Amersham Biosciences) for desalting.

4. Sequencing-grade modified trypsin (Promega).

5. IMAC-Select Affinity Gel (Sigma) loaded with Fe^{3+}-ions according to the supplier's protocol (*see* **Note 3**).

6. Phosphatases: calf intestinal alkaline phosphatase (New England Biolabs, Inc.), lambda protein phosphatase (Upstate Ltd.) and alkaline *Escherichia coli* phosphatase (Sigma).

7. $H_2{}^{18}O$ (93.7% (w/w) pure, ARC Laboratories).

2.2. COFRADIC RP-HPLC Column and Solvents

1. Analytical RP-HPLC column: 2.1 mm internal diameter × 150 mm (length) 300SB-C18 column, Zorbax® (Agilent).

2. RP-HPLC solvent A: 10 mM ammonium acetate (pH 5.5) in acetonitrile/water (2/98, v/v, both HPLC analyzed, Mallinckrodt Baker B.V.) (*see* **Note 4**).

3. RP-HPLC solvent B: 10 mM ammonium acetate (pH 5.5) in acetonitrile/water (70/30, v/v) (*see* **Note 4**).

3. Methods

The methods described below used cultured animal cell lines as the source for phosphorylated proteins. In fact, we here started with about 25 million cultured human hepatocytes (HegG2) that, upon cell lysis, yielded about 2.5 mg of protein material. Eventually, depending on the cell type/line, tissue or organism, and the degree of protein phosphorylation, numbers of cells should be adjusted so that similar amounts of protein starting material are reached.

3.1. Proteome and Peptide Preparation (see also Fig. 1)

1. Disrupt 25 million pelleted HegG2 cells in the dark in 5 mL of cell disruption buffer for 5 min at room temperature and transfer on ice for 30 min. To ensure adequate cell lysis, repeatedly vortex this solution.

2. Centrifuge for 10 min at $13,000 \times g$ at 4°C and discard the pellet containing insoluble material.

3. Desalt the protein solution on two PD-10 columns (2.5-mL protein solution per column) and collect each desalted protein fraction in 3.5 mL of protein digestion buffer.

4. Mix the two desalted protein solutions together.

5. Add 40 µg of trypsin and incubate overnight at 37°C.

6. Adjust the pH of the peptide solution to pH 3 by adding 1 M HCl.

7. Remove insolubilities by centrifugation for 10 min at $13,000 \times g$.

8. Load phosphopeptides on Fe^{3+}-loaded IMAC beads according to the manufacturer's protocol.

9. Elute phosphorylated peptides from the IMAC resin with 2 mL of 0.4 M ammonium hydroxide.

10. Split the peptide mixture in two equal parts (mixtures A and B) and dry to complete dryness in a centrifugal vacuum concentrator.

11. Re-dissolve mixture A in 50 µL of calf intestinal alkaline phosphatase (CIP) reaction buffer and add 10 units of CIP, 200 units of lambda protein phosphatase and 0.66 units of alkaline *Escherichia coli* phosphatase (*see* **Note 5**). Incubate for 1 h at 37°C and lower the pH to 5 by adding 80 µL 0.5 M KH_2PO_4. Add 4 µg sequencing-grade modified trypsin and dry to complete dryness. Redissolve mixture A in 150 µL $H_2^{18}O$ and incubate overnight at 37°C. Transfer this ^{18}O-labeled, dephosphorylated peptide mixture to an Eppendorf tube containing 1.5 µmol tris(2-carboxyethyl) phosphine (TCEP) and 15 µmol iodoacetamide both present as a dried pellet. Incubate for 1 h at 37°C in the dark and store at –20°C until further use.

12. Repeat **step 11** for mixture B but exclude phosphatases and replace $H_2^{18}O$ by natural water.

3.2. COFRADIC Isolation of Ex-Phosphorylated Peptides

1. Combine peptide mixtures A and B.

2. Fractionate the peptides by RP-HPLC using the gradient depicted in **Table 1** and collect primary fractions using the scheme shown in **Table 2** (*see* **Note 6**).

Table 1
Solvent gradient used for the primary and secondary
COFRADIC RP-HPLC separations

Time	Solvent A (%)	Solvent B (%)
0	100	0
10	100	0
110	0	100
120	0	100
140	100	0

Note that using the RP column indicated in the materials section, a stable flow rate of 80 µL/min is recommended and the gradient shown here is a linear one and includes column re-equilibration following the peptide separation window

3. Combine primary fractions separated by 16 min (*see* **Table 2**), dry and re-dissolve them in 50 µl of CIP reaction buffer containing 10 units of CIP, 200 units of lambda protein phosphatase and 0.66 units of alkaline *Escherichia coli* phosphatase (*see* **Note 5**).

4. Dephosphorylate for 1 h at 37°C and stop the reaction by adding 40 µL of HPLC solvent A and 10 µL of 50 mM acetic acid.

5. Refractionate each pool of dephosphorylated peptides individually on the same column and same solvent gradient (**Table 1**) as used for the primary RP-HPLC run and collect sorted, ex-phosphorylated peptides (hydrophobic shift) using the sorting scheme shown in **Table 2**.

6. Store sorted peptides at –20°C until further analysis.

4. Notes

1. Combining different broad-spectrum phosphatase inhibitors is an absolute requirement for preserving the phosphorylation status of proteins. Reduction and alkylation of cysteines was further found beneficial for disrupting the structures of different catabolic enzymes (e.g., for blocking trypsin activity, *see* **ref.** (**21**)) and is here included during cell lysis to additionally block enzymatic activity that would otherwise destroy the structural integrity of the extracted proteins.

Table 2
Fraction pooling scheme for COFRADIC isolation of ex-phosphorylated peptides

Pooling scheme of primary fractions

Primary fraction	Elution interval (min)	Primary fraction	Elution interval (min)	Primary fraction	Elution interval (min)
1	22–23	17	38–39	33	54–55
2	23–24	18	39–40	34	55–56
3	24–25	19	40–41	35	56–57
4	25–26	20	41–42	36	57–58
5	26–27	21	42–43	37	58–59
6	27–28	22	43–44	38	59–60
7	28–29	23	44–45	39	60–61
8	29–30	24	45–46	40	61–62
9	30–31	25	46–47	41	62–63
10	31–32	26	47–48	42	63–64
11	32–33	27	48–49	43	64–65
12	33–34	28	49–50	44	65–66
13	34–35	29	50–51	45	66–67
14	35–36	30	51–52	46	67–68
15	36–37	31	52–53	47	68–69
16	37–38	32	53–54	48	69–70
Example of the elution of secondary fractions					
Primary fraction 6	30–40	Primary fraction 22	46–56	Primary fraction 38	62–72

Using the RP-HPLC solvent gradient shown in **Table 1**, primary fractions separated by 16-min each can be pooled together prior to the dephosphorylation reaction (COFRADIC sorting reaction). One example of the elution windows of ex-phosphorylated peptides is given in the lower part of the table (i.e., sorting of ex-phosphorylated peptides from primary fractions 6, 22, and 38)

2. Urea solutions should be prepared fresh and never heated. Urea in solution always forms ammonium cyanate and its concentration increases with time and increasing temperature. Ammonium cyanate reacts with protein-bound free

primary amino groups, amongst other complicating peptide and protein identification by mass spectrometric techniques.

3. Several phosphopeptide enrichment kits are available and can all be used prior to the actual COFRADIC analyses. Alternatively, phosphopeptides can also be enriched by other means, for example using strong cation exchange chromatography (9).

4. For the ammonium acetate buffer system a 500 mM ammonium acetate stock solution was used that is prepared as follows: for 2 L of this stock solution, 57.19 mL acetic acid is titrated with NH_4OH to pH of 5.5 (about 60 mL of NH_4OH should be added).

5. A combination of different broad-spectrum phosphatases was found necessary to ensure near complete dephosphorylation of phosphopeptides.

6. Key to using diagonal chromatography for isolating specific subsets of peptides is the introduction of a sufficiently large chromatographic shift (22). This shift depends both on the nature of the affected peptide as well as on the chromatographic system and buffers used. For isolation of ex-phosphorylated peptides, a HPLC solvent system buffering at pH 5.5 was found to produce the largest shift.

Acknowledgements

The authors acknowledge support by research grants from the Fund for Scientific Research – Flanders (Belgium) (project numbers G.0156.05, G.0077.06, and G.0042.07), the Concerted Research Actions (project BOF07/GOA/012) from the Ghent University, the Inter University Attraction Poles (IUAP06), and the European Union Interaction Proteome (6th Framework Program).

References

1. Manning, G., Whyte, D. B., Martinez, R., Hunter, T., and Sudarsanam, S. (2002) The protein kinase complement of the human genome. *Science* 298, 1912–34.

2. Chu, G., Egnaczyk, G. F., Zhao, W., Jo, S. H., Fan, G. C., Maggio, J. E., Xiao, R. P., and Kranias, E. G. (2004) Phosphoproteome analysis of cardiomyocytes subjected to beta-adrenergic stimulation: identification and characterization of a cardiac heat shock protein p20. *Circ Res* 94, 184–93.

3. Maguire, P. B., Wynne, K. J., Harney, D. F., O'Donoghue, N. M., Stephens, G., and Fitzgerald, D. J. (2002) Identification of the phosphotyrosine proteome from thrombin activated platelets. *Proteomics* 2, 642–8.

4. Martin, K., Steinberg, T. H., Goodman, T., Schulenberg, B., Kilgore, J. A., Gee, K. R., Beechem, J. M., and Patton, W. F. (2003) Strategies and solid-phase formats for the analysis of protein and peptide phosphorylation employing a novel fluorescent phosphorylation sensor

dye. *Comb Chem High Throughput Screen* 6, 331–9.

5. Gevaert, K., Van Damme, P., Ghesquière, B., Impens, F., Martens, L., Helsens, K., and Vandekerckhove, J. (2007) A la carte proteomics with an emphasis on gel-free techniques. *Proteomics* 7, 2698–718.

6. Wu, C. C., MacCoss, M. J., Howell, K. E., and Yates, J. R., 3rd (2003) A method for the comprehensive proteomic analysis of membrane proteins. *Nat Biotechnol* 21, 532–8.

7. Ficarro, S. B., McCleland, M. L., Stukenberg, P. T., Burke, D. J., Ross, M. M., Shabanowitz, J., Hunt, D. F., and White, F. M. (2002) Phosphoproteome analysis by mass spectrometry and its application to *Saccharomyces cerevisiae*. *Nat Biotechnol* 20, 301–5.

8. Pinkse, M. W., Uitto, P. M., Hilhorst, M. J., Ooms, B., and Heck, A. J. (2004) Selective isolation at the femtomole level of phosphopeptides from proteolytic digests using 2D-NanoLC-ESI-MS/MS and titanium oxide precolumns. *Anal Chem* 76, 3935–43.

9. Beausoleil, S. A., Jedrychowski, M., Schwartz, D., Elias, J. E., Villen, J., Li, J., Cohn, M. A., Cantley, L. C., and Gygi, S. P. (2004) Large-scale characterization of HeLa cell nuclear phosphoproteins. *Proc Natl Acad Sci U S A* 101, 12130–5.

10. Oda, Y., Nagasu, T., and Chait, B. T. (2001) Enrichment analysis of phosphorylated proteins as a tool for probing the phosphoproteome. *Nat Biotechnol* 19, 379–82.

11. Zhou, H., Watts, J. D., and Aebersold, R. (2001) A systematic approach to the analysis of protein phosphorylation. *Nat Biotechnol* 19, 375–8.

12. Brown, J. R., and Hartley, B. S. (1966) Location of disulphide bridges by diagonal paper electrophoresis. The disulphide bridges of bovine chymotrypsinogen A. *Biochem J* 101, 214–28.

13. Cruickshank, W. H., Malchy, B. L., and Kaplan, H. (1974) Diagonal chromatography for the selective purification of tyrosyl peptides. *Can J Biochem* 52, 1013–7.

14. Gevaert, K., Van Damme, J., Goethals, M., Thomas, G. R., Hoorelbeke, B., Demol, H., Martens, L., Puype, M., Staes, A., and Vandekerckhove, J. (2002) Chromatographic isolation of methionine-containing peptides for gel-free proteome analysis: identification of more than 800 *Escherichia coli* proteins. *Mol Cell Proteomics* 1, 896–903.

15. Van Damme, P., Martens, L., Van Damme, J., Hugelier, K., Staes, A., Vandekerckhove, J., and Gevaert, K. (2005) Caspase-specific and nonspecific in vivo protein processing during Fas-induced apoptosis. *Nat Methods* 2, 771–7.

16. Ghesquière, B., Van Damme, J., Martens, L., Vandekerckhove, J., and Gevaert, K. (2006) Proteome-wide characterization of N-glycosylation events by diagonal chromatography. *J Proteome Res* 5, 2438–47.

17. Ghesquière, B., Buyl, L., Demol, H., Damme, J. V., Staes, A., Timmerman, E., Vandekerckhove, J., and Gevaert, K. (2007) A new approach for mapping sialylated N-glycosites in serum proteomes. *J Proteome Res* 6, 4304–12.

18. Gevaert, K., Van Damme, P., Ghesquière, B., and Vandekerckhove, J. (2006) Protein processing and other modifications analyzed by diagonal peptide chromatography. *Biochim Biophys Acta* 1764, 1801–10.

19. Gevaert, K., Impens, F., Van Damme, P., Ghesquiere, B., Hanoulle, X., and Vandekerckhove, J. (2007) Applications of diagonal chromatography for proteome-wide characterization of protein modifications and activity-based analyses. *Febs J* 274, 6277–89.

20. Gevaert, K., Staes, A., Van Damme, J., DeGroot, S., Hugelier, K., Demol, H., Martens, L., Goethals, M., and Vandekerckhove, J. (2005) Global phosphoproteome analysis on human HepG2 hepatocytes using reversed-phase diagonal LC. *Proteomics* 5, 3589–99.

21. Staes, A., Demol, H., Van Damme, J., Martens, L., Vandekerckhove, J., and Gevaert, K. (2004) Global differential non-gel proteomics by quantitative and stable labeling of tryptic peptides with oxygen-18. *J Proteome Res* 3, 786–91.

22. Liu, P., Feasley, C. L., and Regnier, F. E. (2004) Optimization of diagonal chromatography for recognizing post-translational modifications. *J Chromatogr A* 1047, 221–7.

Chapter 17

Chemical Tagging Strategies for Mass Spectrometry-Based Phospho-proteomics

Alexander Leitner and Wolfgang Lindner

Summary

The study of protein phosphorylation in combination with chemical methods may serve several purposes. The removal of the phosphate group from phosphoserine and -threonine residues by β-elimination has been employed to improve sensitivity for mass spectrometric detection and to attach affinity tags for phosphopeptide enrichment. More recently, phosphoramidate chemistry has been shown to be another promising tool for enriching phosphorylated peptides, and other phosphate-directed reactions may also be applicable to the study of the phosphoproteome in the future. In recent years, the combination of large-scale phospho-proteomics studies with stable isotope labeling for quantification purposes has become of growing importance, frequently involving the introduction of chemical tags such as iTRAQ. In this chapter, we will highlight several key strategies that involve chemical tagging reactions.

Key words: Mass spectrometry, Chemical tagging, Phosphorylation, Enrichment, Quantification, Peptide sequencing.

1. Introduction

Within the context of this volume, it is not necessary to further emphasize the importance of mass spectrometry (MS) to the field of phospho-proteomics. MS has allowed the localization of phosphorylation sites as well as their dynamics in a biological context, for example, as a response to external stimuli. Enormous progress has been made in recent years as a result of the advancements in instrumentation such as faster and more sensitive mass spectrometers and miniaturized chromatography. Together with

Marjo de Graauw (ed.), *Phospho-Proteomics, Methods and Protocols, vol. 527*
© 2009 Humana Press, a part of Springer Science + Business Media, New York, NY
Book DOI: 10.1007/978-1-60327-834-8_17

the development of dedicated software tools for dealing with the large amounts of experimental data that are generated in terms of data processing, interpretation, and validation, it is now possible to map hundreds and even thousands of phosphorylation sites in a complex sample and to obtain quantitative information from them.

However, one must not underestimate the importance of powerful sample preparation protocols that help to overcome the difficulties associated with the analysis of phosphoproteins and -peptides. These include the substoichiometric nature of phosphorylation at a given site and resulting sensitivity problems for MS detection and the challenging fragmentation behavior of many phosphopeptides under collision-induced dissociation conditions that interfere with the elucidation of peptide sequences and the assignment of phosphorylation sites. Several excellent reviews that have been published recently deal with the whole workflow of phosphoproteomic analysis *(1–5)*, and key steps are summarized in **Fig. 1**.

Briefly, a protein mixture may either be prefractionated by chromatographic or electrophoretic methods prior to an enzymatic digestion step. Phosphospecific enrichment is mostly performed by immunoprecipitation at the protein level, because this

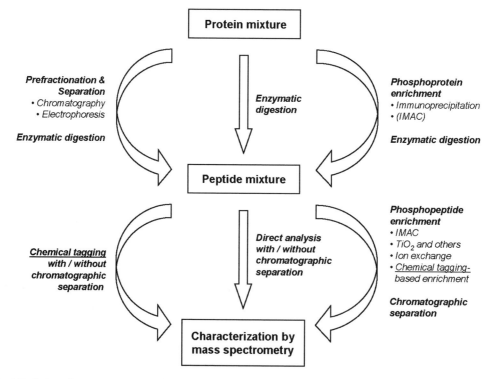

Fig. 1. Techniques in mass spectrometry-based phospho-proteomics. Overview of different methodologies to improve the identification and localization of phosphorylation sites in proteins.

modification is significantly less abundant than phosphorylation on serine or threonine residues. Following the digestion step, samples may be directly analyzed by MS if they are obtained following two-dimensional electrophoresis. At this stage, phosphospecific chemical tagging may be used to improve sensitivity. If more complex peptide mixtures need to be analyzed, especially when no prefractionation at the protein level has been performed, phosphopeptides almost always need to be enriched in order to allow their comprehensive characterization. For this purpose, a variety of techniques have been introduced: Immobilized metal affinity chromatography (IMAC), enrichment using metal oxide materials such as titanium dioxide or zirconium dioxide, ion exchange chromatography, and, finally, methods based on chemical tagging.

This chapter will focus on two main aspects of chemical tagging for phospho-proteomics, that is, those methods that involve chemical reactions in the workflow with the goal to facilitate the analysis of different aspects of protein phosphorylation: In **Subheadings 2** and **3**, chemical strategies that directly target phosphate groups are discussed. First, the possibilities of addressing the phosphate group by chemical means are discussed and the two most promising methods are presented in detail. This is followed by an overview of various recent applications involving the chemical modification of phosphate groups. In **Subheading 4**, methods that involve the chemical modification of other functional groups in proteins and peptides in connection with studying phosphorylation, particularly for quantitative phospho-proteomics, are highlighted.

2. Reaction Principles for Phosphate Group-Directed Tagging

2.1. Overview

As for any application of chemical reactions in proteomic applications (6–10), a number of prerequisites should also be fulfilled for phosphoproteomic tagging. These include, among others, a quantitative reaction yield, mild reaction conditions (so that the modification is preserved, if so desired), and sufficient specificity for the phosphate group. These and other factors have been described by us in a review on chemical tagging reactions in general (7). Some aspects are of particular relevance, for example, as phosphorylation is hardly ever quantitative on a given site, one always deals with unmodified peptides/proteins being present in a large excess. Side reactions that occur only to a small degree, say 1% or less, can reach levels where a discrimination between an unmodified and a modified site is no longer unambiguously possible.

Unfortunately, no tagging reaction that specifically modifies the phosphate group of a peptide or protein in the presence of other ubiquitous moieties, especially carboxyl groups has been demonstrated. Therefore, it is common to include a protection step into the tagging protocol that prevents the possible reaction of the reagent with other functional groups. One example is the (methyl) esterification of carboxylic acids with methanolic HCl to avoid modification of the C-terminus or the side chains of aspartic and glutamic acid (see below).

Figure 2 gives an overview of different chemical methods that have been described for phosphoproteomic applications. These include β-elimination/Michael addition, phosphoramidate chemistry, diazo chemistry, and oxidation–reduction condensation. The first two concepts are described in more detail below.

2.2. Concepts Based on β-Elimination

The most widely used chemoselective strategy that targets phosphorylated amino acids is the combination of β-elimination (BE) and Michael addition (MA). Under strongly basic conditions,

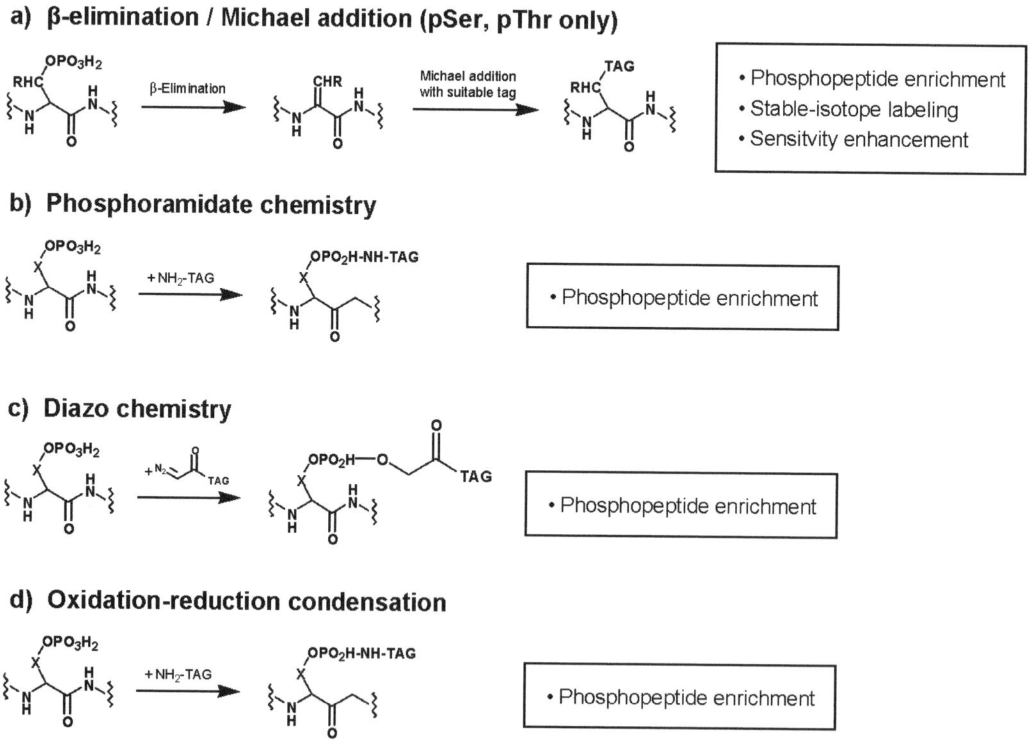

a) β-elimination / Michael addition (pSer, pThr only)

b) Phosphoramidate chemistry

c) Diazo chemistry

d) Oxidation-reduction condensation

Fig. 2. Reaction schemes and application fields for different phosphate group-directed tagging strategies. (**a**) β-elimination/ Michael addition, where R = H for pSer and CH$_3$ for pThr, (**b**) phosphoramidate chemistry, (**c**) diazo chemistry, (**d**) oxidation–reduction condensation. Methods (**b**) and (**d**) yield the same product, but differ in their reaction conditions (not shown); methods (**b**–**d**) are generally suitable for phosphorylated Ser, Thr, and Tyr residues. For details, see text.

pSer and pThr residues (but not pTyr) are dephosphorylated and converted into dehydroalanine and β-methyldehydroalanine, respectively (*see* **Fig. 2a**). The double bond of the reaction products serves as a Michael acceptor that may subsequently react with nucleophiles such as thiol reagents. Various protocols have been proposed for the elimination step, using different bases, but most commonly barium hydroxide serves as the reagent. It has been shown that phosphoserine residues tend to undergo elimination more rapidly than phosphothreonine residues, so that prolonged incubation under basic conditions may be required to ensure a complete conversion of both phosphoamino acids. This, however, may increase the risk of dehydratization of unphosphorylated serine and threonine residues, as observed by McLachlin and Chait *(11)*. This unwanted side-reaction is of particular relevance for the analysis of complex mixtures as it increases the risk of generating false-positives from unphosphorylated sites which may be present in a large excess. In addition, O-linked *N*-acetylglucosamine (O-GlcNAc) modifications are also susceptible to BE under typical reaction conditions (which has been exploited for O-GlcNAc profiling *(12–13)*), and differentiation of this modification and phosphorylation requires particular consideration. If O-GlcNAc profiling is of interest, phosphatases may be employed prior to the elimination step; however, a similar approach is not easily feasible for phosphorylation analysis, due to the lack of widely applicable deglycosylation methods for O-GlcNAc residues. Careful experimental design reportedly allows the discrimination of both modifications *(12, 14)*.

2.3. Concepts Based on Phosphoramidate Chemistry

Under certain conditions the phosphate group of pSer, pThr, and pTyr residues may react with an amide group to form a phosphoramidate (**Fig. 2b**). Typical coupling conditions include a water-soluble carbodiimide such as EDC [*N*-(3-dimethylaminopropyl)-*N'*-ethylcarbodiimide] and imidazole buffer, as originally described for the derivatization of nucleotides *(15)*. In principle, this method has the advantage over β-elimination in that it is also amenable to pTyr residues; however, the very low abundance of pTyr sites makes it difficult to investigate tyrosine phosphorylation without additional enrichment steps such as immunoprecipitation. Phosphoramidate chemistry also requires the protection of carboxyl groups which are usally converted to their methyl esters (see below).

3. Phosphate Group-Directed Applications of Chemical Tagging in Phospho-proteomics

3.1. β-Elimination-Based Methods for Phosphopeptide Enrichment

BE-based methods have been predominantly used for the enrichment of phosphopeptides, although there are other potential benefits such as improved detection sensitivity (these will be covered in the following section). Upon removal of the phosphate group, different affinity tags may be attached that enable the isolation of phosphopeptides, either by affinity chromatography or by reversible covalent bonding to a solid support (typically in bead format).

Examples for the first approach include different biotinylation methods that allow the enrichment of tagged phosphopeptides on avidin resins (12, 14, 16–20). Other proposed strategies have employed a poly-His-tag (21) for enrichment by IMAC, a fluorous affinity tag (22) or a fluorescent affinity tag that may be used in combination with immunoprecipitation of labeled peptides (23). Covalent capture is frequently performed by attaching a dithiol compound such as dithiothreitol or ethanedithiol that serves as a nucleophile for MA and provides an additional –SH group for thiol-based chemistry. This way, it is possible to isolate tagged peptides by binding to thiol resins (11, 13, 24–25) or capture them on functionalized solid-phases carrying maleimide (26) or iodoacetyl (14, 27) groups. Tseng et al. used a resin with 6-(2-mercapto-acetylamine)-hexanoic acid as a reactive group for the capture of β-eliminated peptides without prior modification with a different reagent (28).

Most of the methods listed above have only been applied to model protein mixtures or protein complexes, although their use for more complex samples seems feasible as has been demonstrated by Smith et al. who were able to identify 28 phosphopeptides from a human breast cancer cell line (27).

In a number of cases, the BE/MA protocol includes the use of differentially isotope-coded reagents, enabling the relative quantitation of different samples by comparing abundances of "light" and "heavy" forms of peptides (see below). Such an approach may also be used without a dedicated enrichment step (29) and can be considered complementary to isotope-coded tagging strategies that target other functional groups in peptides or proteins, which are discussed in detail in **Subheading 4**.

3.2. β-Elimination-Based Methods for Improved MS Analysis

BE chemistry not only provides a possibility of enriching phosphopeptides from peptide mixtures, but also helps to deal with challenges related to MS and MS/MS analysis. The dominant neutral loss of phosphoric acid from pSer and pThr residues (30) as observed on different MS platforms employing collision-induced dissociation frequently makes it impossible to obtain peptide sequence information because of the lack of informative fragment ions. Upon removal of the labile phosphate group, amide bond cleavages within the peptide backbone are facilitated and the quality of MS/MS spectra is usually improved without having to resort to electron-mediated dissociation techniques (electron capture and electron transfer dissociation) that are not yet widespread. In addition, as peptide sequencing is carried out in positive ionization mode, replacing the acidic phosphate group with a positively charged moiety (for example, quaternary ammonium or guanidine groups) may lead to improved sensitivity for detecting phosphopeptides.

To this end, various chemical tagging methods have been reported that use BE/MA for purposes other than phosphopeptide enrichment. Addition of alkanethiols was already shown to aid in phosphopeptide identification almost a decade ago *(29, 31–32)*. By using a mixture of alkanethiols for derivatization, phosphopeptides were identified in MALDI peptide mass fingerprints because of their characteristic pattern (e.g., separated by 14 Da when reagents differing by a –CH$_2$– group were used) and their increased sensitivity *(32)*. More recently, enhanced sensitivity for MS detection was demonstrated using different tagging reagents, such as 2-phenylethanethiol *(33)*, mercaptoethylpyridine *(34)*, guanidinoethanethiol *(35–36)* or 2-mercaptoethyltrimethylammonium chloride *(37)*. In these cases, enhanced MS response is commonly the result of the introduction of strongly basic or permanently positively charged moieties in place of the phosphate group. Similarly, such an effect was observed for His-tagged former phosphopeptides *(21)*. Sensitivity increases of up to tenfold and more were reported in these cases, especially when using MALDI as the ionization technique, while less data is available from ESI platforms.

3.3. Methods Based on Phosphoramidate Chemistry

Aebersold and co-workers have described several methods employing phosphoramidate chemistry (PAC) for the enrichment of phosphopeptides. In contrast to protocols based on BE, this strategy is generally applicable to the probing of tyrosine phosphorylation as well, given sufficient sensitivity and dynamic range of the downstream MS analysis or if pTyr-specific enrichment is employed.

Initial work demonstrated the feasibility of PAC-based enrichment for studying phosphorylation in yeast *(38)*. To avoid side reactions, both amino and carboxyl groups were protected prior to the reaction of phosphate groups with cysteamine. Enrichment was achieved by first capturing modified phosphopeptides on a solid phase in the form of iodoacetyl beads and cleavage of the phosphoramidate bond under controlled acidic conditions that did not result in cleavage of peptide (amide) bonds. Twenty-four phosphopeptides were identified at that time, covering mainly highly abundant yeast proteins. These results may be attributed both to the lengthy experimental protocol and the less sophisticated MS instrumentation available at the time.

In 2005, an improved strategy involving dendrimer conjugation was presented *(39)*. In this approach, phosphopeptides were coupled to an amine-functionalized polyamidoamine (PAMAM)-type dendrimer, which allowed the convenient enrichment of phosphopeptides from the digestion mixture by means of ultra-filtration. Recovery of the phosphopeptides was again achieved by acid hydrolysis, followed by LC-MS/MS. Dendrimer-based PAC enrichment was combined with phosphotyrosine-specific immunoprecipitation and stable isotope-coding using differential methyl esterification of carboxyl groups (*see* below) to study dynamic phosphorylation events in human T cells stimulated with pervanadate. NanoLC-MS/MS on a linear ion trap instrument allowed the identification of 97 Tyr-phosphorylated or associated proteins.

Recently, Bodenmiller et al. used a different variation of the PAC protocol to study phosphorylation in *Drosophila melanogaster* cell line *(40)*. The simplified solid-phase capture protocol involved protection/isotope-coding of carboxyl groups in tryptic peptides by methyl esterification, EDC-catalyzed attachment of cystamine to the phosphate groups, a reduction step to yield free thiol groups on the tagged phosphates and capture of the derivatized peptides on maleimide-functionalized glass beads. The enriched phosphopeptides were again recovered by acidic cleavage from the support and analyzed by nanoflow LC-MS/MS.

In a separate work, the group also compared the performance of the PAC approach with two other phosphopeptide enrichment methods, namely IMAC and titanium dioxide enrichment *(41)*. A result of this study was that, although there was some overlap in the phosphorylation sites that were discovered with the three approaches, a significant number of sites were only identified with one particular technique. This observation suggests that only the use of several, at least partially orthogonal methods may allow the probing of the whole phosphoproteome of a cell line or body fluid.

3.4. Other Methods

Concepts based on reactions other than BE and PAC have not yet advanced beyond the developmental stage. Lansdell and Tepe have demonstrated the enrichment of a model peptide by coupling the phosphate group to diazo-functionalized resin (**Fig. 2c**) and cleavage under alkaline conditions *(42)*. Pflum and co-workers have employed oxidation-reduction condensation (**Fig. 2d**) to isolate synthetic phosphopeptides and phosphopeptides from a β-casein digest and recombinant CREB (cAMP response element binding protein) *(43)*. Phosphopeptides were directly coupled to a glycine-resin via the amino group, and – following washing steps to eliminate nonspecifically bound compounds – recovered by a cleavage step using 95% TFA. It will be interesting to see whether these or other approaches will show to be viable for the use with complex biological samples to probe complete phospho-proteomes.

4. Tagging of Other Functional Groups for Phosphopro-teomic Applications

The application of differential stable isotope labeling (SIL) strategies (44–46) has further increased the importance of mass spectrometry in the field of phospho-proteomics. When the analyte of interest is present in two forms that differ only in their isotopic composition, they may be differentiated in the mass spectrometer based on their different molar masses but otherwise possess identical physicochemical properties (although incorporation of multiple deuterium atoms (^{2}H) may result in partial chromatographic separation from the isotopomer containing only ^{1}H). Thus, if differentially labeled forms of proteins or peptides can be generated, hundreds or thousands of peptides are amenable to relative quantitation and thus also their phosphorylation sites.

SIL may be achieved at different levels of the proteomic workflow. Samples generated in cell culture can be metabolically labeled, as e.g., in the SILAC (stable isotope labeling of amino acids in cell culture) strategy where isotope labeled amino acids are added to the growth medium (47, 48). Although such a method offers the most straightforward strategy to labeling, essentially as a "byproduct" of cell growth, it is obviously not applicable to biofluids and tissue samples, and the labeling protocol is not easily applicable to all cell lines as specific growth media are required. Therefore, alternative ways to introduce isotope labels are frequently used. ^{18}O isotopes may be introduced on the C-termini of peptides during an enzymatic digestion step by performing proteolysis in ^{18}O-labeled water (applied to phosphorylation analysis in (49)), but the achievable mass shift for the "heavy" peptide is limited to 4 Da and experimental conditions need to be carefully controlled to ensure quantitative labeling.

A more versatile method that is equally applicable to all sample types is chemical tagging for SIL purposes. A number of different labeling chemistries have been employed to introduce isotope labeled groups at different reactive sites in a peptide or protein. In the following, we will restrict the discussion to studies with focus on phosphorylation analysis, but a number of recently published reviews provide a more general overview on the topic (44–46). To quantify phosphopeptides, isotope labels are usually introduced on ubiquitous functional groups, i.e., amino or carboxyl groups, and phosphopeptides are enriched with the help of IMAC or metal oxide affinity materials. Alternatively, the isotope tag may be directly introduced at the phosphorylation site via BE/MA chemistry, as has already been mentioned earlier.

Isotope coding by differential esterification of carboxyl groups can be conveniently combined with enrichment techniques that rely on the protection of acidic functionalities. In particular, IMAC-based protocols have been shown to benefit from this reaction *(50–53)* as nonspecific binding of acidic peptides was found to be reduced. Different chemical tagging reactions discussed in **Subheadings 2** and **3** are dependent on the protection of COOH groups as well, and several applications of the PAC enrichment strategy include differential esterification (see above). Practically, labeling is achieved in a straightforward manner by dissolving the dried sample in methanolic hydrochloric acid prepared either from normal of deuterated methanol. However, especially for very small sample amounts the drying step may lead to some irreversible absorption on surfaces. Furthermore, complete labeling of multiple carboxyl groups may require extended incubation times that increase the risk of degradation of the sample.

Amino groups on the N-terminus of peptides and on lysine side chains can be tagged by a number of isotope-coded tagging reagents, such as acetic *(54)* or propionic *(55–57)* anhydrides, *N*-acetoxysuccinimide *(58)* or trimethylammonium butyrate *(58)*. Other examples include the reductive dimethylation with formaldehyde *(59–60)* and the N-isotag concept, an activated γ-aminobutyric acid derivative (tBoc-γ-aminobutyric acid *N*-hydroxysuccinimide *(61–63)*).

A particular kind of amine-specific tagging strategy involves iTRAQ chemistry. iTRAQ stands for isobaric tags for relative and absolute quantitation and uses MS/MS rather than MS data for quantitation. This is made possible by a dedicated amine-reactive reagent that upon reaction with a peptide creates isobaric derivatives for peptides originating from different samples, which then yield different characteristic fragment ions ("reporter ions") for each sample *(64)*. Quantitation on the tandem MS level has its advantages and drawbacks: At the MS level, signals from differentially labeled peptides all contribute to one signal, therefore increasing sensitivity, on the other hand, selection of the iTRAQ peptides for MS/MS is required to obtain the reporter signals. The commercially available kit (Applied Biosystems, Foster City, CA) allowed the labeling and simultaneous analysis of four different samples in its initial version *(64)*, and in 2007, an 8-plex reagent with a different structure has been introduced *(65–66)*. Studies thus far published in the literature *(67–74)*, however, made use of the 4-plex iTRAQ exclusively.

5. Summary and Outlook

Chemical enrichment methods for phosphorylation analysis are in competition with a number of different affinity techniques of which IMAC and metal oxide affinity enrichment are currently the most widely used. Although there is an increasing amount of data for the application of chemical tagging reactions to samples of limited complexity (such as single proteins or protein complexes), there have been a few applications for the comprehensive analysis of highly complex samples. However, it has been shown in several cases that orthogonal enrichment techniques are helpful to obtain a more complete picture of the phoshoproteome. Therefore, chemical tagging techniques offer a valuable addition to the proteomic "toolbox" and further developments in this field are important.

The analysis of the phosphoproteome typically requires high-end instrumentation that helps to overcome critical issues for MS analysis of phosphopeptides, such as low abundance and unfavorable fragmentation properties. In addition, particular care is required for data analysis and interpretation, because the inference of the originating protein(s) and the localization of the particular phosphorylation site is challenging. Frequently, appropriate facilities are not available to experts in the chemical methodology, which may explain the lack of data for some methods described above. In addition, the step from a less complex sample to whole biofluids or cell lysates may cause unexpected difficulties (which is of course not restricted to chemical methods). It will be interesting to see whether more data will become available from some of the promising techniques described herein, also because some methods not only allow enrichment of phosphopeptides from digests but also address some of the other limitations in mass spectrometry-based phospho-proteomics by improving sensitivity and peptide fragmentation.

In particular, methods that involve the solid-phase capture of tagged phosphopeptides appear most promising as they allow for stringent washing steps that reduce the amount of nonspecifically bound peptides. Examples for such strategies include both BE- and PAC-based protocols. Furthermore, the replacement of the phosphate group with tags that facilitate peptide ionization is a promising strategy for the profiling of individual phosphoproteins that are only available in limited amounts so that an enrichment strategy would not be useful.

Finally, it is evident that stable isotope labeling by chemical tagging is an accepted technique in the community that will also benefit from the further development of new and improved reagents allowing the comparison of multiple samples such as the 8-plex iTRAQ reagent or the similarly designed ExacTag (PerkinElmer, Waltham, MA).

References

1. Mukherji, M. (2005) Phosphoproteomics in analyzing signaling pathways. *Exp. Rev. Proteomics* 2, 117–128.

2. Reinders, J. and Sickmann, A. (2005) State-of-the-art in phosphoproteomics. *Proteomics* 5, 4052–4061.

3. Delom, F. and Chevet, E. (2006) Phospho-protein analysis: from proteins to proteomes. *Proteome Sci.* 4, 15 (article number).

4. Pinkse, M. W. H. and Heck, A. J. R. (2006) Essential enrichment strategies in phospho-proteomics. *Drug Discov. Today: Technol.* 3, 331–337.

5. Collins, M. O., Yu, L. and Choudhary, J. S. (2007) Analysis of protein phosphorylation on a proteome-scale. *Proteomics* 7, 2751–2768.

6. Julka, S. and Regnier, F. (2004) Quantification in proteomics through stable isotope coding: a review. *J. Proteome Res.* 3, 350–363.

7. Leitner, A. and Lindner, W. (2004) Current chemical tagging strategies for proteome analysis by mass spectrometry. *J. Chromatogr. B* 813, 1–26.

8. Mirzaei, H. and Regnier, F. (2005) Structure specific chromatographic selection in targeted proteomics. *J. Chromatogr. B* 817, 23–34.

9. Leitner, A. and Lindner, W. (2006) Chemistry meets proteomics: The use of chemical tagging reactions for MS-based proteomics. *Proteomics* 6, 5418–5434.

10. Gevaert, K., Damme, P. V., Ghesquière, B., Impens, F., Martens, L. et al. (2007) A la carte proteomics with an emphasis on gel-free techniques. *Proteomics* 7, 2698–2718.

11. McLachlin, D. T. and Chait, B. T. (2003) Improved beta-elimination-based affinity purification strategy for enrichment of phos-phopeptides. *Anal. Chem.* 75, 6826–6836.

12. Wells, L., Vosseller, K., Cole, R. N., Cronshaw, J. M., Matunis, M. J. et al. (2002) Mapping sites of O-GlcNAc modification using affinity tags for serine and threonine post-transla-tional modifications. *Mol. Cell. Proteomics* 1, 791–804.

13. Vosseller, K., Hansen, K. C., Chalkley, R. J., Trinidad, J. C., Wells, L. et al. (2005) Quan-titative analysis of both protein expression and serine/threonine post-translational modifi-cations through stable isotope labeling with dithiothreitol. *Proteomics* 5, 388–398.

14. Poot, A. J., Ruijter, E., Nuijens, T., Dirksen, E. H. C., Heck, A. J. R. et al. (2006) Selective enrichment of Ser-/Thr-phosphorylated pep-tides in the presence of Ser-/Thr-glycosylated peptides. *Proteomics* 6, 6394–6399.

15. Chu, B. C. F., Wahl, G. M. and Orgel, L. E. (1983) Derivatization of unprotected polynu-cleotides. *Nucleic Acids Res.* 11, 6513–6529.

16. Goshe, M. B., Conrads, T. P., Panisko, E. A., Angell, N. H., Veenstra, T. D. et al. (2001) Phosphoprotein Isotope-Coded Affinity Tag Approach for Isolating and Quantitating Phosphopeptides in Proteome-Wide Analyses. *Anal. Chem.* 73, 2578–2586.

17. Oda, Y., Nagasu, T. and Chait, B. T. (2001) Enrichment analysis of phosphorylated pro-teins as a tool for probing the phosphopro-teome. *Nat. Biotechnol.* 19, 379–382.

18. Adamczyk, M., Gebler, J. C. and Wu, J. (2001) Selective analysis of phosphopeptides within a protein mixture by chemical modifi-cation, reversible biotinylation and mass spec-trometry. *Rapid Commun. Mass Spectrom.* 15, 1481–1488.

19. Goshe, M. B., Veenstra, T. D., Panisko, E. A., Conrads, T. P., Angell, N. H. et al. (2002) Phosphoprotein isotope-coded affinity tags: application to the enrichment and identifi-cation of low-abundance phosphoproteins. *Anal. Chem.* 74, 607–616.

20. Veken, P. v. d., Dirksen, E. H. C., Ruijter, E., Elgersma, R. C., Heck, A. J. R. et al. (2005) Development of a novel chemical probe for the selective enrichment of phosphorylated serine- and threonine-containing peptides. *ChemBioChem* 6, 2271–2280.

21. Jalili, P. R., Sharma, D. and Ball, H. L. (2007) Enhancement of ionization efficiency and selective enrichment of phosphorylated pep-tides from complex protein mixtures using a reversible poly-histidine tag. *J. Am. Soc. Mass Spectrom.* 18, 1007–1017.

22. Go, E. P., Uritboonthai, W., Apon, J. V., Trauger, S. A., Nordstrom, A. et al. (2007) Selective metabolite and peptide capture/mass detection using fluorous affinity tags. *J. Proteome Res.* 6, 1492–1499.

23. Stevens, S. M., Jr, Chung, A. Y., Chow, M. C., McClung, S. H. et al. (2005) Enhancement of phosphoprotein analysis using a fluorescent affinity tag and mass spectrometry. *Rapid Commun. Mass Spectrom.* 19, 2157–2162.

24. Amoresano, A., Marino, G., Cirulli, C. and Quemeneur, E. (2004) Mapping phospho-rylation sites: a new strategy based on the use of isotopically-labelled dithiothreitol and mass spectrometry. *Eur. J. Mass Spectrom.* 10, 401–412.

25. Thaler, F., Valasina, B., Baldi, R., Xie, J., Stewart, A. et al. (2003) A new approach to

phosphoserine and phosphothreonine analysis in peptides and proteins: chemical modification, enrichment via solid-phase reversible binding, and analysis by mass spectrometry. *Anal. Bioanal. Chem.* 376, 366–373.

26. Chowdhury, S. M., Munske, G. R., Siems, W. F. and Bruce, J. E. (2005) A new maleimide-bound acid-cleavable solid-support reagent for profiling phosphorylation. *Rapid Commun. Mass Spectrom.* 19, 899–909.

27. Qian, W.-J., Goshe, M. B., II, Camp, D. G., Yu, L.-R., Tang, K. et al. (2003) Phosphoprotein isotope-coded solid-phase tag approach for enrichment and quantitative analysis of phosphopeptides from complex mixtures. *Anal. Chem.* 75, 5441–5450.

28. Tseng, H.-C., Ovaa, H., Wei, N. J. C., Ploegh, H. and Tsai, L.-H. (2005) Phosphoproteomic analysis with solid-phase capture-release-tag approach. *Chem. Biol.* 12, 769–777.

29. Weckwerth, W., Wilmitzer, L. and Fiehn, O. (2000) Comparative quantification and identification of phosphoproteins using stable isotope labeling and liquid chromatography/mass spectrometry. *Rapid Commun. Mass Spectrom.* 14, 1677–1681.

30. DeGnore, J. P. and Qin, J. (1998) Fragmentation of phosphopeptides in an ion trap mass spectrometer. *J. Am. Soc. Mass Spectrom.* 9, 1175–1188.

31. Jaffe, H., Veeranna and Pant, H. C. (1998) Characterization of serine and threonine phosphorylation sites in beta-elimination ethanethiol addition-modified proteins by electrospray tandem mass spectrometry and database searching. *Biochemistry* 37, 16211–16224.

32. Molloy, M. P. and Andrews, P. C. (2001) Phosphopeptide derivatization signatures to identify serine and threonine phosphorylated peptides by mass spectrometry. *Anal. Chem.* 73, 5387–5394.

33. Klemm, C., Schröder, S., Glückmann, M., Beyermann, M. and Krause, E. (2004) Derivatization of phosphorlyated peptides with S- and N-nucleophiles for enhanced ionization efficiency in matrix-assisted laser desorption/ionization mass spectrometry. *Rapid Commun. Mass Spectrom.* 18, 2697–2705.

34. Arrigoni, G., Resjö, S., Levander, F., Nilsson, R., Degerman, E. et al. (2006) Chemical derivatization of phosphoserine and phosphothreonine containing peptides to increase sensitivity for MALDI-based analysis and for selectivity of MS/MS analysis. *Proteomics* 6, 757–766.

35. Ahn, Y. H., Ji, E. S., Lee, J. Y., Cho, K. and Yoo, J. S. (2007) Arginine-mimic labeling with guanidinoethanethiol to increase mass sensitivity of lysine-terminated phosphopeptides by matrix-assisted laser desorption/ionization time-of-flight mass spectrometry. *Rapid Commun. Mass Spectrom.* 21, 2204–2210.

36. Ahn, Y. H., Ji, E. S., Kwon, K. H., Lee, J. Y., Cho, K. et al. (2007) Protein phosphorylation analysis by site-specific arginine-mimic labeling in gel electrophoresis and matrix-assisted laser desorption/ionization time-of-flight mass spectrometry. *Anal. Biochem.* 370, 77–86.

37. Li, H. and Sundararajan, N. (2007) Charge switch derivatization of phosphopeptides for enhanced surface-enhanced raman spectroscopy and mass spectrometry detection. *J. Proteome Res.* 6, 2973–2977.

38. Zhou, H., Watts, J. D. and Aebersold, R. (2001) A systematic approach to the analysis of protein phosphorylation. *Nat. Biotechnol.* 19, 375–378.

39. Tao, W. A., Wollscheid, B., O'Brien, R., Eng, J. K., Li, X.-j. et al. (2005) Quantitative phosphoproteome analysis using a dendrimer conjugation chemistry and tandem mass spectrometry. *Nat. Methods* 2, 591–598.

40. Bodenmiller, B., Mueller, L. N., Pedrioli, P. G. A., Pflieger, D., Jünger, M. A. et al. (2007) An integrated chemical, mass spectrometric and computational strategy for (quantitative) phosphoproteomics: application to *Drosophila melanogaster* Kc167 cells. *Mol. BioSyst.* 3, 275–286.

41. Bodenmiller, B., Mueller, L. N., Mueller, M., Domon, B. and Aebersold, R. (2007) Reproducible isolation of distinct, overlapping segments of the phosphoproteome. *Nat. Methods* 4, 231–237.

42. Lansdell, T. A. and Tepe, J. J. (2004) Isolation of phosphopeptides using solid phase enrichment. *Tetrahedron Lett.* 45, 91–93.

43. Warthaka, M., Karwowska-Desaulniers, P. and Pflum, M. K. H. (2006) Phosphopeptide modification and enrichment by oxidation-reduction condensation. *ACS Chem. Biol.* 1, 697–701.

44. Ong, S.-E. and Mann, M. (2005) Mass spectrometry-based proteomics turns quantitative. *Nat. Chem. Biol.* 1, 252–262.

45. Julka, S. and Regnier, F. E. (2005) Recent advancements in differential proteomics based on stable isotope coding. *Brief. Funct. Genom. Proteom.* 4, 158–177.

46. Chen, X., Sun, L., Yu, Y., Xue, Y. and Yang, P. (2007) Amino acid-coded tagging approaches in quantitative proteomics. *Exp. Rev. Proteomics* 4, 25–37.

47. Beynon, R. J. and Pratt, J. M. (2005) Metabolic labeling of proteins for proteomics. *Mol. Cell. Proteomics* 4, 857–872.

48. Mann, M. (2006) Functional and quantitative proteomics using SILAC. *Nat. Rev. Mol. Cell Biol.* 7, 952–958.

49. Bonenfant, D., Schmelzle, T., Jacinto, E., Crespo, J. L., Mini, T. et al. (2003) Quantitation of changes in protein phosphorylation: A simple method based on stable isotope labeling and mass spectrometry. *Proc. Natl. Acad. Sci. USA* 100, 880–885.

50. Ficarro, S. B., McCleland, M. L., Stukenberg, P. T., Burke, D. J., Ross, M. M. et al. (2002) Phosphoproteome analysis by mass spectrometry and its application to *Saccharomyces cerevisiae*. *Nat. Biotechnol.* 20, 301–305.

51. Ficarro, S., Chertihin, O., Westbrook, V. A., White, F., Jayes, F. et al. (2003) Phosphoproteome analysis of capacitated human sperm - evidence of tyrosine phosphorylation of a kinase-anchoring protein 3 and valosin-containing protein/p97 during capacitation. *J. Biol. Chem.* 278, 11579–11589.

52. Brill, L. M., Salomon, A. R., Ficarro, S. B., Mukherji, M., Stettler-Gill, M. et al. (2004) Robust phosphoproteomic profiling of tyrosine phosphorylation sites from human T Cells using immobilized metal affinity chromatography and tandem mass spectrometry. *Anal. Chem.* 76, 2763–2772.

53. Smith, J. C., Duchesne, M. A., Tozzi, P., Ethier, M. and Figeys, D. (2007) A differential phosphoproteomic analysis of retinoic acid-treated P19 Cells. *J. Proteome Res.* 6, 3174–3186.

54. Kang, J.-H., Katayama, Y., Han, A., Shigaki, S., Oishi, J. et al. (2007) Mass-tag technology responding to intracellular signals as a novel assay system for the diagnosis of tumor. *J. Am. Soc. Mass Spectrom.* 18, 106–112.

55. Zhang, X., Jin, Q. K., Carr, S. A. and Annan, R. S. (2002) N-Terminal peptide labeling strategy for incorporation of isotopic tags: a method for the determination of site-specific absolute phosphorylation stoichiometry. *Rapid Commun. Mass Spectrom.* 16, 2325–2332.

56. Jin, M., Bateup, H., Padovan, J. C., Greengard, P., Nairn, A. C. et al. (2005) Quantitative analysis of protein phosphorylation in mouse brain by hypothesis-driven multistage mass spectrometry. *Anal. Chem.* 77, 7845–7851.

57. Zappacosta, F., Collingwood, T. S., Huddleston, M. J. and Annan, R. S. (2006) A quantitative results-driven approach to analyzing multisite protein phosphorylation: the phosphate-dependent phosphorylation profile of the transcription factor Pho4. *Mol.Cell. Proteomics* 5, 2019–2030.

58. Riggs, L., Seeley, E. H. and Regnier, F. E. (2005) Quantification of phosphoproteins with global internal standard technology. *J. Chromatogr. B* 817, 89–96.

59. Huang, S.-Y., Tsai, M.-L., Wu, C.-J., Hsu, J.-L., Ho, S.-H. et al. (2006) Quantitation of protein phosphorylation in pregnant rat uteri using stable isotope dimethyl labeling coupled with IMAC. *Proteomics* 6, 1722–1734.

60. Huang, S.-Y., Tsai, M.-L., Chen, G.-Y., Wu, C.-J. and Chen, S.-H. (2007) A systematic MS-based approach for identifying in vitro substrates of PKA and PKG in rat uteri. *J. Proteome Res.* 6, 2674–2684.

61. Smolka, M. B., Albuquerque, C. P., Chen, S.-h., Schmidt, K. H., Wei, X. X. et al. (2005) Dynamic changes in protein-protein interaction and protein phosphorylation probed with amine-reactive isotope tag. *Mol. Cell. Proteomics* 4, 1358–1369.

62. Smolka, M. B., Chen, S.-H., Maddox, P. S., Enserink, J. M., Albuquerque, C. P. et al. (2006) An FHA domain-mediated protein interaction network of Rad53 reveals its role in polarized cell growth. *J. Cell. Biol.* 175, 743–753.

63. Smolka, M. B., Albuquerque, C. P., Chen, S.-h. and Zhou, H. (2007) Proteome-wide identification of in vivo targets of DNA damage checkpoint kinases. *Proc. Natl. Acad. Sci. USA* 104, 10364–10369.

64. Ross, P. L., Huang, Y. L. N., Marchese, J. N., Williamson, B., Parker K. et al. (2004) Multiplexed protein quantitation on *Saccharomyces cerevisiae* using amine-reactive isobaric tagging reagents. *Mol. Cell. Proteomics* 3, 1154–1169.

65. Choe, L., D'Ascenzo, M., Relkin, N. R., Pappin, D., Ross, P. et al. (2007) 8-Plex quantitation of changes in cerebrospinal fluid protein expression in subjects undergoing intravenous immunoglobulin treatment for Alzheimer's disease. *Proteomics* 7, 3651–3660.

66. Pierce, A., Unwin, R. D., Evans, C. A., Griffiths, S., Carney, L. et al. (2008) Eight-channel iTRAQ enables comparison of the activity of 6 leukaemogenic tyrosine kinases. *Mol. Cell. Proteomics* 7, 853–863.

67. Zhang, Y., Wolf-Yadlin, A., Ross, P. L., Pappin, D. J., Rush, J. et al. (2005) Time-resolved mass spectrometry of tyrosine phosphorylation sites in the epidermal growth factor receptor signaling network reveals dynamic modules. *Mol. Cell. Proteomics* 4, 1240–1250.

68. Sachon, E., Mohammed, S., Bache, N. and Jensen, O. N. (2006) Phosphopeptide quantitation using amine-reactive isobaric tagging reagents and tandem mass spectrometry: applications to proteins isolated by gel electrophoresis. *Rapid Commun. Mass Spectrom.* 20, 1127–1134.

69. Wolf-Yadlin, A., Kumar, N., Zhang, Y., Hautaniemi, S., Zaman, M. et al. (2006) Effects of HER2 overexpression on cell signaling networks governing proliferation and migration. *Mol. Syst. Biol.* 2, 54 (article number).

70. Williamson, B. L., Marchese, J. and Morrice, N. A. (2006) Automated identification and quantification of protein phosphorylation sites by LC/MS on a hybrid triple quadrupole linear ion trap mass spectrometer. *Mol. Cell. Proteomics* 5, 337–346.

71. Jones, A. M. E., Bennett, M. H., Mansfield, J. W. and Grant, M. (2006) Analysis of the defence phosphoproteome of *Arabidopsis thaliana* using differential mass tagging. *Proteomics* 6, 4155–4165.

72. Zhou, F., Galan, J., Geahlen, R. L. and Tao, W. A. (2007) A novel quantitative proteomics strategy to study phosphorylation-dependent peptide-protein interactions. *J. Proteome Res.* 6, 133–140.

73. Wolf-Yadlin, A., Hautaniemi, S., Lauffenburger, D. A. and White, F. M. (2007) Multiple reaction monitoring for robust quantitative proteomic analysis of cellular signaling networks. *Proc. Natl. Acad. Sci. USA* 104, 5860–5865.

74. Bantscheff, M., Eberhard, D., Abraham, Y., Bastuck, S., Boesche, M. et al. (2007) Quantitative chemical proteomics reveals mechanisms of action of clinical ABL kinase inhibitors. *Nat. Biotechnol.* 25, 1035–1044.

Part IV

Arrays to Study Protein-Phosphorylation

Chapter 18

Antibody Array Platform to Monitor Protein Tyrosine Phosphorylation in Mammalian Cells

Alicia S. Chung and Y. Eugene Chin

Summary

Protein tyrosine phosphorylation plays a central role in cell-signaling and is a focus of biomedical studies and cancer therapy. However, it is still challenging to identify or characterize the coordinated changes of many candidate proteins of one particular pathway or multiple pathways simultaneously. Antibody array is a recently developed approach applied for differential analysis of multiple protein posttranslational modification events in mammalian cells. It is based on the highly specific recognition between the immobilized antibodies on the array and their specific target proteins in a high-throughput screening format. Here we have described in detail two methods for differential analysis of protein tyrosine phosphorylation in cells by (1) using a single fluorescent protein capture format on membrane array and (2) a competitive protein capture method on glass surface array.

Key words: Antibody, Protein array, Posttranslational, Phosphorylation, Fluorescent antibody technique, Fluorescent dyes.

1. Introduction

Mammalian cells respond to environmental changes with protein tyrosine phosphorylation, which is a reversible process mediating cellular function involved with regulating development, proliferation, differentiation, and growth. Aberrant protein tyrosine phosphorylation can lead to disease states including neoplastic transformation of many cell types. The past few years have seen the development of a number of antibody microarray platforms for the analysis of complex mixtures of proteins with posttranslational modifications in mammalian cells. Antibody arrays have

Marjo de Graauw (ed.), *Phospho-Proteomics, Methods and Protocols, vol. 527*
© 2009 Humana Press, a part of Springer Science + Business Media, New York, NY
Book DOI: 10.1007/978-1-60327-834-8_18

been applied for detection of cytokines *(1)*, bacteria and bacterial toxins *(2)*, potential biomarkers in human serum *(3)*, and protein expression profiling *(2, 4)*. Array based methods are also employed in clinical settings in diagnoses from infection to heart disease *(4, 5)*. Thus, antibody arrays hold a potential in a variety of applications including proteomics research, drug discovery, and diagnostics for disease research and treatment. Here we have discussed two fluorescence-based methods for detection of tyrosine phospho-protein profiles using an antibody array approach *(6)*. Briefly, phospho-tyrosine modified proteins are isolated from treated or untreated cell lysates, followed by single or dual fluorescent dye labeling for incubation over antibodies arrayed and immobilized on membranes. The antibody array membranes are washed and used to detect and profile tyrosine-phosphorylated proteins. The image of signals is captured using a laser scanner such as the LI-COR BioCOR Odyssey Fluorescent Scanner and analyzed by fluorescence densitometry.

2. Materials

2.1. Cell Culture and Lysis

1. Dulbecco's modified Eagle's medium (Sigma, St. Louis, MO) supplemented with 10% fetal bovine serum (Invitrogen, Carlsbad, CA).

2. Penicillin (5,000 U/ml) and streptomycin (5,000 U/ml) solution were purchased from Invitrogen and stored at –20°C. Both were used at 500 U/ml.

3. Solutions of trypsin (0.25%) and ethylenediamine tetraacetic acid (EDTA) (1 mM) were purchased from Invitrogen and stored at –20°C.

4. Epidermal growth factor (EGF) was purchased from R & D Systems and stored in aliquots at –20°C.

5. Radioimmune precipitation assay (RIPA) buffer for cell lysis and array incubation: 50 mM Tris–HCl pH 7.4, 1% Nonidet P-40, 0/25% sodium deoxycholate, 150 mM NaCl, 1 mM EDTA, 1 mM PMSF, 1 μg/ml each of aprotinin (Sigma), leupeptin (Sigma), and pepstatin (Sigma), 1 mM Na_3VO_4, and 1 mM NaF (*see* **Note 1**).

6. Tris-buffered saline with Tween (TBS-T): Prepare 10× stock with 1.37 M NaCl, 27 mM KCl, 250 mM Tris–HCl, pH 7.4, 1% Tween 20. Dilute 100 ml with 900 ml deionized water for use.

7. Bio-Rad (Hercules, CA) protein assay reagent was used to quantify protein concentration and used according to manufacturer's protocol.

8. VWR (West Chester, PA) 3-D rotator waver with microcentrifuge tube adapters

9. Teflon cell scrapers (Fisher).

2.2. Antibodies and Antibody Array

1. All antibodies tested were purchased from Santa Cruz Biotechnology, Inc. (Santa Cruz, CA): Anti-His6 antibody (sc-803), EGFR (sc-03), JAK2 (sc-294), STAT1 (sc-346), STAT3 (sc-482), STAT5a (sc-1,081), AKT1/2 (sc-8,312), and SOCS3 (sc-9,023).

2. Protein G/A agarose beads were purchased from Santa Cruz Biotechnology, Inc.

3. Alexa680 (excitation/emission maxima of 679/702 nm) was obtained from Molecular Probes (Invitrogen). Cy2 (489/506 nm) and Cy5 (650/667 nm) were purchased from Amersham Biosciences (Piscataway, NJ) (*see* **Note 2**).

4. PVDF membrane was purchased from Bio-Rad. Poly-L-lysine coated glass slides were purchased from CEL & Associates (Pearland, TX) (*see* **Note 3**).

5. Microcaster™ Hand-Held microarray spotter system used for printing antibodies onto glass slides was purchased from Schleicher & Schuell (Keene, NH).

6. HybriSlip Hybridization cover (Electron Microscopy Services, Hatfield, PA).

7. Ni-NTA Ni²⁺ bead resin (Qiagen, Valencia, CA).

2.3. Antibody Array Signal Detection

1. Fluorescent detection of PVDF arrays done on LI-COR BioCOR Odyssey Fluorescent Scanner manufactured by LI-COR Biosciences (Lincoln, NB).

2. Immobilized fluorescence on glass-slide arrays was detected with Nikon Fluorescent microscope equipped with appropriate excitation/emission filter sets for each dye.

3. Spot densitometry for quantitation of fluorescent signals was done with Odyssey software version 1.1.15 (LI-COR Biosciences).

3. Methods

3.1. Profiling Protein Tyrosine Phosphorylation by Antibody Array: Single Fluorescent Protein Capture Format

3.1.1. Preparation of Protein Lysates

1. A431 cell lines were grown in DMEM with 10% FBS and 500 U/ml penicillin/streptomycin. Approximately 1–5 × 10⁷ cells grown in a 150-mm dish were required for each array incubation. (*see* **Note 4**)

2. A431 grown to ~80% confluence in DMEM/FBS/pen/strep were transferred to serum free media 24 h prior to treatment

with EGF (100 ng/ml) for 30 min or were left untreated. (*see* **Note 5**)

3. Have all reagents, inhibitors, and materials required for cell harvest ready at the time of treatment termination and harvest. Remove media by aspiration, washed once with cold PBS, and scraped off of cell culture plate using 1 ml of cold RIPA buffer with all inhibitors added just prior to use (*see* **Note 6**). Collect cell lysates on ice and extract at 4°C using an orbital rotator for 30 min. Spin lysates down for 5 min at 14,000 × g in a microcentrifuge at 4°C to collect cellular debris. Lysates may also be passed through a 21G needed attached to a 3CC syringe five times to shear genomic DNA and reduce viscosity of lysate. Carefully transfer supernatants into fresh tubes and place on ice (*see* **Note 7**).

4. Determine protein concentration using the Bio-Rad protein assay reagent.

3.1.2. Immunoprecipitation and Fluorescent Labeling of Protein Lysate

1. Immunoprecipitations are performed in 1.5-ml microcentrifuge tubes. Incubate the above-prepared protein lysates (0.5–2 mg) overnight at 4°C with 1 µg antiphosphotyrosine (pY20) antibody and protein A/G agarose beads (30 µl) for immunoprecipitation of tyrosine-phosphorylated proteins. Wash immunoprecipitates five times with 1 ml RIPA buffer containing inhibitors. Agarose bead immobilized proteins are collected by centrifugation at 3,000 × g for 5 min. Carefully remove the supernatant by aspiration and transfer to fresh tubes and save on ice for future analysis or re-immunoprecipitation, if desired. Perform wash by adding 1 ml RIPA wash buffer and inverting tube to resuspend beads. Collect beads with centrifugation at 3,000 × g for 5 min at 4°C. All the above steps are performed at 4°C. After final wash, buffer exchange agarose bead conjugates with cold PBS (with protease inhibitors) and resuspend in 100 µl PBS plus inhibitors and 10 µl 1 M sodium bicarbonate buffer (pH 8.3–9.0) prior to fluorescent staining. Protein concentration should be 0.5–1 mg/ml (*see* **Note 8**).

2. Dissolve 2 mg Alexa680 in 1 ml DMF immediately before starting the reaction due to instability of reactive compound in solution. Briefly sonicate or vortex.

3. While slow votexing the protein solution (**step 1**), slowly add 7.5 µl (15 µg) of reactive dye solution (**step 2**) (*see* **Note 9**). Incubate the reaction for 2–3 h with rotation on an orbital rotator at 4°C, protected from light.

4. Fluorescent labeled proteins were then purified from unreacted labeling reagent by washing three times with 1 ml RIPA buffer at 4°C following same procedure as described

for immunoprecipitation. After final wash, resuspend labeled protein conjugates in 1 ml of RIPA buffer and boil for 5 min to dissociate proteins from agarose beads; place sample on ice. Spin down beads at $3,000 \times g$ for 5 min; remove the supernatant containing labeled proteins for incubation with antibody array.

3.1.3. Preparation of Antibody Array

1. Using a laser printer, print onto a 2.5″ × 3.5″ PVDF membrane, a grid with 96 positions labeling rows and columns similarly to a 96-well microtiter plate. Each position is printed as a circle of 1-mm in diameter. The printer ink does not interfere with fluorescent signal detection. Spot antibodies onto PVDF membranes manually using a P2 pipette (Rainin Instruments, Oakland, CA) in a volume of 0.4 μl or less (20–40 ng of antibody) to yield spots less than 500 μm in diameter (*see* **Note 10**). Spots are allowed to airdry on bench top and membrane arrays do not require further processing prior to use. Arrays are stored dry in sealed plastic bags at 4°C and are re-hydrated by immersing in a PBS buffer containing 3% BSA for 2-h prior to use. Reusing arrays is not recommended.

2. Arrays used in this study to profile tyrosine phosphorylation in epidermal carcinoma A431 cells are spotted with 25 antibodies against signaling proteins in duplicate (**Fig. 1**).

3.1.4. Antibody Array Incubation, Processing, and Analysis

1. Alexa680 labeled tyrosine-phosphorylated proteins from control and EGF-treated cell extracts are incubated over two above-prepared antibody arrays in sealed plastic pouches for 3 h to overnight at room temperature on a rocking

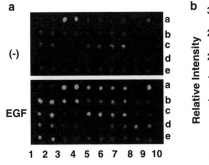
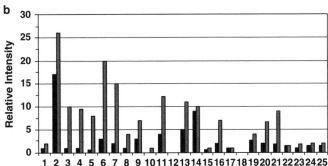

Fig. 1. Profiling protein tyrosine phosphorylation by antibody array using single fluorescent protein capture format. (**a**) Anti-pY20 immunoprecipitates were prepared from control and EGF-treated samples according to the procedure described in **Subheading 3.1**. Alexa680-labeled anti-pY20 precipitates were incubated with the PVDF array comprising 25 antibodies printed in duplicates, followed by LI-COR scanning. Signal intensities apparently elevated by EGF-treatment in A431 cells include EGFR (a3,4), JAK2 (c7,8), STAT1 (a5,6), STAT3 (a7,8), STAT5a (b1,2), AKT1/2 (b3,4), and SOCS3 (e9,10). (**b**) Intensities of the signals detected in (**a**) were analyzed by spot densitometry on the Odyssey software version 1.1.15, and the results depict the average of the duplicates. *Black bars* represent untreated and *gray bars* represent EGF treated. Figure reprinted with permission from **ref.** *6*.

platform, protected from light. Final hybridization volume is $6-10 \, ml/100 \, cm^2$ membrane area.

2. Following incubation, remove array and transfer to standard plastic container for three washes in TBS-T (~75 ml) for 10 min at room temperature with gentle shaking on orbital shaker platform. Remove wash buffer each time by careful decanting. After the final wash, rinse array with 1× PBS and keep in PBS for next step.

3. The fluorescent signals trapped on PVDF membrane arrays are scanned wet with the LI-COR near infrared fluorescent scanner operated via the Odyssey software using the 700 nm channel for detection.

4. Captured image are then analyzed using Odyssey spot densitometry function (Odyssey software version 1.1.15). Local background is determined for each duplicate signal, averaged and subtracted from each spot. Duplicate signals per antibody are averaged and represented as relative increase in signal intensity over control immunoglobulin (IgG) control signals for each membrane (**Fig. 1**).

3.2. Profiling Protein Tyrosine Phosphorylation by Antibody Array: Competitive Fluorescent Protein Capture Format

3.2.1. Preparation of Recombinant His6-STAT1 Protein

1. Full-length STAT1 was subcloned into the pET-28a vector (Novagen) for expression of recombinant His_6-STAT1 proteins in bacterial strain BL21. Bacterially expressed protein was purified via Ni-NTA Ni^{2+} bead resin following manufacturer's protocol (Qiagen).

3.2.2. Differential Fluorescent Labeling of Protein Lysate

1. Fifteen micrograms of Cy2 (489/506 nm), or Cy5 (650/667 nm) fluorescent dye (prepared in **step 2** of **Subheading 3.1.2**) is added to 0.1 mg purified His6-STAT1 resuspended in 1 ml PBS/10 µl 1 M sodium bicarbonate buffer (pH 8.3–9.0) with slow vortexing of the protein solution. Incubate the protein and dye mixture for 2–3 h at 4°C on a rotating platform to allow bead mixture and dye conjugation.

2. Fluorescent labeled proteins attached to Ni-NTA beads were then washed with 1 ml RIPA buffer three times at 4°C. Following the final wash, resuspend beads in 100 µl of RIPA buffer and boil for 5 min prior to incubation with antibody array.

3.2.3. Preparation of Antibody Array

1. Poly-L-lysine coated glass slides were spotted with 2–4 ng of antibodies (40 nl or less) using the pin-tool design Microcaster™ Hand-Held microarray spotter system yielding spots less than 200 µm in diameter and are allowed to dry and used immediately following. (*see* **Notes 11** and **12**)

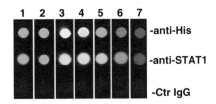

Fig. 2. Profiling protein tyrosine phosphorylation by antibody array using competitive fluorescent protein capture. Cy2- and Cy5-His6-STAT1 were mixed at 1:0, 1:0.25, 1:0.5, 1:1, 0.5:1, 0.25:1, and 0:1 ratios, from lanes 1 to 7 respectively. Mixed His6-STAT1 proteins (0.1 mg) were incubated with the slide array comprising anti-His6 antibody, anti-STAT1 antibody, and control IgG. After extensively washes, the slide arrays were visualized with a fluorescent microscope with appropriate excitation/emission filter sets for each dye. Figure reprinted with permission from **ref.** *6.*

3.2.4. Antibody Array Incubation, Processing, and Analysis

1. Cy2 and Cy5-labeled recombinant His6-STAT1 proteins were mixed at different ratios in 100 µl RIPA and incubated with a glass-slide array immobilized with anti-His6, anti-STAT1 antibody or control IgG, covered with Hybri-Slip Hybridization Cover (Electron Microscopy Services), and placed in a slide chamber for 3 h at room temperature on a rotating platform (*see* **Note 13**).

2. Following incubation, microarray slides were washed in an upright slide washing chamber five times with TBS-T for 10 min each.

3. Immobilized fluorescent signals were detected and image captured using Nikon Fluorescent microscope with corresponding wavelength filter sets. Slide images were captured using ImageQuant software (Molecular Dynamics, Piscataway).

4. The linear change in shade from Cy2 to Cy5 provides a semiquantitative measure of the proteins that were captured by the anti-His6 or anti-STAT1 antibody indicating the feasibility of competitive fluorescent protein capture by antibody immobilized on glass surface (*see* **Notes 14** and **15**) (**Fig. 2**).

4. Notes

1. The phosphatase, Na_3VO_4 needs to be heat-activated during preparation step. Alternately, "HALT" phosphatase cocktail (Pierce, Rockford, IL) can be used in place of Na_3VO_4 and NaF.

2. Alexa680 should be stored desiccated at ≤–20°C, and protected from light. Reactive dyes should be stable for at least 3 months when stored as directed.

3. The glass slides are sensitive; do not touch the array surface by tips, forceps, or hand. Hold the slides by the edges only. Handle all buffers and slides with latex free gloves.

4. Generally, this starting cell number is required in order to obtain 0.5–1 mg of protein for array incubation.

5. Serum starvation is critical for determining the activation of EGF signaling over basal activity induced by growth factors contained within serum. If cells are sensitive to serum starvation, reduce starvation accordingly and determine optimal stimulation time empirically. Cells grown to confluence will also result in diminished stimulation by EGF.

6. Phosphatase inhibitors (activated Na_3VO_4) and NaF are critical for successful harvest of phosphorylated proteins.

7. Fresh lysates should be used for the following incubation steps for best results. Storage recommendations for lysates are to flash freeze lysates using liquid N_2 and then transfer to –80°C. Thawed lysates should be kept on ice prior to use.

8. Protein solutions must be free of amine-containing substances such as Tri-base or Tri-HCL, glycine, or ammonium ions. The presence of low concentration of sodium azide (<3 mM) or thimersal (<1 mM) will not interfere with the conjugation reaction.

9. Variations due to the different reactivities of both the protein and labeling reagents may occur, which may require optimization of the dye-to-protein ration used in the reaction. Generally, the limiting factor has been the amount of input protein following immunoprecipitation. We recommend serial immunodepletions of your protein lysate and combining precipitated proteins when possible.

10. It is not critical to spot exact volumes of each antibody as the antibody binding capacity is expected to exceed amount of hybridizing proteins. We therefore recommend aspirating 1 µl of antibody solution and serially dispensing onto two duplicate spots.

11. Commercial arrayers, developed for cDNA microarrays, can be used for the printing of proteins and antibodies.

12. It is not necessary to keep antibody arrays hydrated. Glass antibody arrays can also be stored in sealed plastic bags at 4°C. Antibody arrays are immersed in PBS buffer containing 3% BSA for 2 h prior to use.

13. A slide chamber can be any container for housing the microscope slide. Seal the chamber with parafilm to avoid evaporation during hybridization.

14. A possible limitation inherent to methods involving direct labeling of proteins is that labeling of the target protein may interfere with its recognition by the antibody, which may be overcome by printing several antibodies to the same target.

15. The use of antibody arrays is mainly intended for initial screening of large numbers of proteins to identify candidates for further research, and thus further validation should be done using other methodologies.

References

1. Huang, R. P. (2007) An array of possibilities in cancer research using cytokine antibody arrays. *Expert Rev Proteomics* 4, 299–308.

2. Haab, B. B. (2006) Applications of antibody array platforms. *Curr Opin Biotechnol* 17, 415–21.

3. Reid, J. D., Parker, C. E., and Borchers, C. H. (2007) Protein arrays for biomarker discovery. *Curr Opin Mol Ther* 9, 216–21.

4. Kingsmore, S. F. (2006) Multiplexed protein measurement: technologies and applications of protein and antibody arrays. *Nat Rev Drug Discov* 5, 310–20.

5. Hall, D. A., Ptacek, J., and Snyder, M. (2007) Protein microarray technology. *Mech Ageing Dev* 128, 161–7.

6. Ivanov, S. S., Chung, A. S., Yuan, Z. L., Guan, Y. J., Sachs, K. V., Reichner, J. S., and Chin, Y. E. (2004) Antibodies immobilized as arrays to profile protein post-translational modifications in mammalian cells. *Mol Cell Proteomics* 3, 788–95.

Chapter 19

Protein Tyrosine Kinase Characterization Based on Fully Automated Synthesis of (Phospho) Peptide Arrays in Microplates

W. Carl Saxinger

Summary

In view of the importance of information transfer mediated throughout the cell by recognition, phosphorylation or dephosphorylation of kinases, their adapters, or substrates, this method was developed. The method provides a potent research tool for rapidly generating and testing these substrates as modeled by synthetic peptide arrays. The peptides or phosphorylated peptides are automatically generated on the inner surfaces of microplate wells, covalently linked to a polylysine polymer so that they are in a sterically favorable conformation, immediately available for in situ testing. Products up to 18 amino acids long have shown excellent mass spectral homogeneity. Thus, determinate peptide libraries can be ready for testing in as little as 2 days after the conception of an experiment. The process can be easily automated using robotic liquid handlers and is extremely rapid, sensitive, and economical. Optionally, the method can be upgraded to a higher throughput level using more powerful workstations with greater capacity, such as the Biomek FX, or any similar robotics capable of transfer-from-file logic to guide synthesis cycles.

Key words: Kinase, Substrate, Peptide, Array, Synthesis, Ligand, Receptor, Antibody, Protein, Interactions.

1. Introduction

The importance of protein phosphorylation or dephosphorylation in cellular information transfer is well documented in this volume and by others *(1)*. Using an oriented degenerate peptide library, researchers identified optimal peptide substrate sequences for a group of protein tyrosine kinases *(2–4)*. Each kinase was reacted with semirandom libraries of peptides, after which the phosphorylated peptide mixtures were sequenced and the most

Marjo de Graauw (ed.), *Phospho-Proteomics, Methods and Protocols, vol. 527*
© 2009 Humana Press, a part of Springer Science + Business Media, New York, NY
Book DOI: 10.1007/978-1-60327-834-8_19

frequently phosphorylated sequences identified. Earlier procedures for synthesis of spatially addressable peptide libraries or arrays on containerized sticks or pins *(5)* or two-dimensional arrays of individual peptides on cellulose, etc. *(6, 7)* have been successfully used for kinase substrate investigations *(8–10)*. The procedure presented here incorporates a number of features that enhance the speed, convenience, and sensitivity of synthetic peptide-based kinase activity studies (and others as well).

Automated parallel peptide synthesis in microplates occupies a niche between combinatorial synthesis, phage display, and conventional preparative synthesis. In this method, all reagents are stable, in ready-to-use solution form so that most syntheses can be started within a few hours after conception of an experiment. Peptide synthesis is performed in solvent-resistant microplates so that liquid transfers, additions, washings can be performed by industry standard automated liquid handling robotics. The product remains covalently attached to the well surface so that reactants are easily added and removed. The product is immediately available for testing on a platform compatible with a wealth of biochemical and biological assays. Additionally the procedure is economical and sensitive with very low background binding. Peptide chain elongation cycles are 1 h or less, so that 96 peptides of length 18 can be synthesized and ready for testing within 2 days *(11)*.

Previously, this procedure was successful in identifying antibody epitopes *(12)*, heparin binding sites on keratinocyte growth factor *(13)*, correctly identifying the known IL-6 binding site of the IL-6 receptor as well as interaction sites not previously reported *(14)*, and identification HIV-1 gp120 binding sites of chemokine receptors some of which were not previously reported *(15)*. More recently the procedure was used to prepare phosphotyrosine-containing peptide arrays from the EGF receptor (EGFR) and to demonstrate the ability of specific monoclonal antibodies to study and correlate sequence determinants of the catalytic site and auto-phosphorylation sites of the EGFR *(16)*.

Preparation of peptide arrays inside microplate wells achieves a number of advantages. It allows preselection of the type of matrix upon which the peptides are displayed so that reactivity and background reduction can be optimized. It dispenses with the necessity for burdensome and hazardous procedures for removal, collection, purification, and reattachment of the peptides by some chemoselective procedure. Immediately after side-chain deprotection, the solid-phase peptides can engage in a plethora of multiplexed molecular interaction strategies. This is a widely enabling research technique. It allows one to rapidly construct peptide arrays of specific configuration and is highly flexibly focused on answering specific, fundamental questions.

Critical components of the method consist of *microplates* composed of polymethylpentene (TPX) for its solvent resistance;

the use of a *polylysine surface backbone* to support the peptides in a sterically favorable manner *(17, 18)* so that it provides a vehicle capable of facilitating functional assay of the peptides in a flexible and sensitive manner; the ability to easily obtain and to store as *frozen ready-to-use solutions* of all synthesis chemicals; an *automation process* for conducting the synthesis of a peptide of any desired composition and length at each position in the array throughout the entire process – including timely transfer of all the appropriate amino acids and synthetic reagents and performance of all the appropriate wash protocols – so that once started, no further attention is required until the synthesis of all peptides is completed, using basic liquid handling robotics such as those currently found in many laboratories. The use of full protection of the phosphotyrosine phosphate group (*see* **Subheading 2.3.1.3**) allows full compatibility along with other amino acids in the DCC/HOBT activation chemistry, avoids side reactions, and results in facile and uniform incorporation of phosphotyrosine into peptides *(19)*.

2. Materials

2.1. Chemical Activation of Microplates

1. All water is glass-distilled grade or better.

2. Nitric acid (70%): analytical reagent grade by ACS or EU standard. Use only in a well-ventilated chemical fume hood with protective clothing and gloves.

3. Solvent-resistant microplate wells are composed of polymethylpentene (TPX), and are 8-well strips, nonsterile, and nontissue culture treated, and are obtained from Costar on special order (Cambridge, MA) (*see* **Note 1**).

2.2. Preparation of Poly(Lys)-Coated Microplates Plate Wells

1. *N*-methylpyrrolidone (NMP): Peptide synthesis grade from Applied Biosystems (Foster City, CA) or equivalent.

2. 1 M storage solution of carbonyldiimidazole (CDI) in NMP. Weigh rapidly to minimize exposure of the CDI to moisture. This solution may be stored for at least 3 months at –20°C in a jar containing desiccant. Prepare a 0.05 M reactant solution of CDI in NMP after the storage solution reaches room temperature.

3. Polylysine PDL or PLL stock solutions: Poly(D-Lys.HBr) (dp 100) and Poly(L-Lys.HBr) (dp 100) (Sigma-Aldrich, St. Louis, MO). Prepare 10 mg/mL solutions in water containing 2% sodium azide. These are stable at room temperature for at least 6 months.

4. Polylysine microplate coating solution: dilute PDL stock 1:9 into NMP. Add 8.8 μL of diisopropylethylamine (DIPEA) for each mL of polylysine stock to neutralize HBr counterions.

2.3. Peptide Synthesis

1. Solvents and reagents for peptide synthesis and removal of side-chain protecting groups were obtained as synthesizer grade from Applied Biosystems. Specifically, these were NMP, 1 M N-hydroxybenzotriazole (HOBT) in NMP, 1 M dicyclohexylcarbodiimide (DCC) in NMP, diisopropylethylamine (DIPEA), acetic anhydride, and trifluoroacetic acid (TFA). NMP from this vendor was consistently free of basic impurities and was stored over molecular sieve (4A) after opening. Piperidine (Biotech grade) was obtained from Sigma-Aldrich.

2. 5 g bottles of FMOC-protected α-amino acids: Peninsula Laboratories (San Carlos, CA), Bachem, or NovaBiochem and side-chain substitutions were Asn (Trt), Asp(OtBu), Cys(Acm) or Cys(Trt), Gln(Trt), Glu(OtBu), His(Trt), Lys(Boc), Ser(tBu), Thr(tBu), and Tyr(tBu).FMOCArg(Pbf).

3. FMOC DMAP-Tyr i.e., FMOC-[bis (dimethylamino)phosphono]-tyrosine or FMOC-Tyr(PO(NMe2)2)–OH, and FMOC Rink amide linker (p-[(R,S)-a-[1-(9H-Fluoren-9-yl)-methoxyformamido]-2,4-dimethoxybenzyl]-phenoxyacetic acid) come from NovaBiochem.

4. Prepare storage solutions of the FMOC amino acids containing 0.5 mmole/mL of 1 M HOBT/NMP (see **Note 2**) and store at –20°C. Solutions are stable for at least 3–6 months if allowed to attain room temperature before opening.

5. Activated FMOC-amino acids: The volumes needed, for complete synthesis of all peptides, are based on 10 μL for each amino acid dose. Dilute 1 M DCC/NMP to 0.1 M with NMP and distribute to 2-mL cryovials (Nunc). Add one part FMOC-amino acid solution (see **Subheading 2.3.3** above) to five parts of 0.1 M DCC/NMP. These solutions are used after a 30-min reaction at room temperature and are not stored or reused after use.

6. Prepare 20% piperidine by diluting into NMP (v/v). The volume needed is 10 mL for each synthesis cycle.

2.4. Peptide Synthesis Automation

1. Beckman Biomek 1,000 or Biomek 2,000 automated laboratory workstation (see **Note 3**).

2. Mek.exe and associated software to generate sequence information files (arrays.bio or patterns.b2 k) to guide peptide syntheses and to provide reagent consumption data (see **Note 4**).

3. Four reagent stations are deployed at the start (1) a rack of 20 (or more) FMOC amino acids activated with DCC/HOBT, (2) a reservoir containing 20% piperidine in NMP for deblocking the N-terminus of the most recently added FMOC amino acid, (3) a bulk NMP delivery station for washing, (4) the solvent-resistant microplates functionalized with polylysine.

2.5. Removal of Amino Acid Side-Chain Protecting Groups	Use extreme caution and a well-ventilated fume hood.

1. Phenol, thioanisole, ethanedithiol (EDT), and triisopropylsilane (TIPS) were obtained from Aldrich Chemical Co. (Milwaukee, WI).

2. Deprotection mixture C: 10 mL of TFA + 0.75 g aliquot of crystalline phenol + 0.5 mL of purified water + 0.5 mL of thioanisole + 0.25 mL of ethanedithiol per plate.

3. Phenol aliquots, 0.75 g: Heat crystalline phenol in a steam or sand bath until it is melted and pipette 0.7 mL aliquots into glass storage vials. Seal with a screw cap lined with teflon or polyallomer and store at −20°C.

2.6. Antibody-Recognition of Microplate Peptides

1. Goat anti-mouse IgG (phosphatase conjugated) (*see* **Note 5**).

2. Heat-inactivated normal goat serum.

3. Murine monoclonal antibodies prepared against phosphotyrosine were 4G10 (Upstate), ab8076 (Abcam Ltd.), and PT01L (Oncogene Research Products).

4. Tween 20.

5. Sodium azide.

6. PBS: phosphate-buffered saline, 0.01 M sodium phosphate (pH 7.4), 0.15 M NaCl.

7. Normal saline solution.

8. Phosphatase assay kit: Kirkegaard & Perry or Pierce Biochemical.

9. Microplate reader.

10. Elution buffer: It contains 50% urea (w/w), 2% SDS, 1% mercaptoethanol, 0.05 M Tris–HCl (pH 8).

2.7. Kinase Substrate Activity of Plate Peptides

1. p60-src protein tyrosine kinase, Cat* PK03 (Oncogene Research Products).

2. Kinase assay buffer: 0.05 M HEPES, pH 7.5, 0.1 mM EDTA. Na, 0.015% BRIJ 35.

3. Kinase dilution buffer: 0.1 mg/mL BSA, 0.2% β-mercap-toethanol, 30 μL of ATP mix.

4. ATP mix: 0.03 M $MgCl_2$, 0.15 M ATP in kinase assay buffer.

3. Methods

A number of independent key elements combine to enhance and accelerate this method. By using a polylysine backbone structure, the synthesized peptides are uniformly attached at their carboxyl-terminus and uniformly displayed around the polyly-sine α helix *(17, 18)*. The storage and subsequent distribution of all reagents as solutions allows the use of simple pipetting and eliminates repetitive weighing. The automation process is greatly simplified, since only four reagent stations are deployed, once at the beginning: (1) a rack of 20 (or more) FMOC-amino acids activated with DCC/HOBT, (2) a reservoir containing 20% piperidine in NMP for deblocking the N-terminus of the most recently added FMOC-amino acid, (3) a bulk NMP delivery station for washing, (4) the solvent-resistant microplates functionalized with polylysine, the synthesis support matrix. Since the peptides remain attached to the plate, side-chain deprotection and removal of contaminants are easily accomplished by simple washing and the peptide plate array is immediately available for use in multiplexed chemical, biochemical, or biological assays. If desired, the peptides can be removed for analysis if the polyly-sine microplate is first modified using the Rink amide linker (*see* **Subheading 2.3.3**), and collecting the peptides in the deprotection mixture (*see* **Subheading 2.5.2**).

Microplate peptides serve well as kinase substrates. Enzymatic phosphorylation activity assayed by anti-phosotyrosine ELISA is detectable well below the 0.1 unit level and ELISA activity values are linear with respect to enzyme units, so that kinase activities are readily quantifiable *(16)*. Although a mixture of phospho-tyrosine antibodies not known to be sequence restricted was used, the possibility exists that some tyrosine-containing peptides could become phosphorylated and not recognized as well as others. Therefore in studies where the need for precise quanti-tation outweighs the convenience and safety considerations of ELISA, incorporation of radioisotopic phosphate would provide an alternative. The microplate format with reactants bound to the well surface would provide a contained and safe vehicle for washing and subsequent measurement of radioactivity.

3.1. Chemical Activation of Microplates

1. Perform nitric acid transfer and washing steps in a well-venti-lated fume hood.

2. Transfer 0.1 mL of reagent grade 70% nitric acid to each well of solvent-resistant microplates using an Eppendorf Repeater Plus/8 Pipettor.

3. Place the filled plates into a polyethylene storage box with a tight fitting cover and incubate in the hood for 12–18 days (*see* **Note 6**).

4. Aspirate the nitric acid from the microplates using a hand-held aspirator. Place the emptied plates in stacks of ten or so until all of the plates have been processed. Place each stack of plates onto a staining tray overlaid with protective sink matting. Begin irrigating the top of each stack with distilled water using

a hand-held multi-port microplate washing device. Transfer the washed plates to a new stacking area. Repeat the process until all the plates have been washed. Remove the water from each plate by dumping into the sink and allow the plates to dry overnight under a layer of towels. Next day wrap the plates in plastic cling wrap in groups of four or five. Plates are stable for >1 year.

3.2. Preparation of Poly(Lys)-Coated Microplate Wells

1. Transfer 0.1 mL of 0.05 M CDI in NMP to the wells. Stop the reaction after 30 min by aspiration. Wash once with 0.2 mL of NMP followed by aspiration.

2. Transfer 0.1 mL of neutralized polylysine (*see* **Note 7**) plate-coating solution into each CDI activated well. Let the plates stand at room temperature for 1 h and overnight at 4°C.

3. Wash the plates twice with distilled water and twice with methanol, air-dry under towels overnight and wrap in plastic cling wrap. Store plates at –20°C for >1 year.

3.3. Peptide Synthesis

1. Pipet 0.1 mL of NMP into each well of a polylysine-coated microplates.

2. Synthesis proceeds in the reverse direction, i.e., from the carboxyl end to the amino end, in a cyclical fashion until the amino terminus of all peptides has been reached.

3. Transfer 10 µL of an activated FMOC-amino acid into the appropriate microplate well until all of the wells have received the appropriate C-terminal amino acid.

4. Allow the plate to stand for 30 min.

5. Aspirate the fluids and wash six times with 0.2 mL of NMP using a multiport washing tool.

6. Add 0.1 mL of 20% piperidine and let stand for 5 min.

7. Aspirate the fluids and wash six times with 0.2 mL of NMP using a multiport washing tool.

8. Repeat **steps 1–7** until all cycles are completed.

9. If a measure of well capacity or peptide quantity is desired, perform the last piperidine treatment manually. Add 0.1 mL of piperidine to a test well and incubate for 5 min. Transfer the fluid to a quartz-bottom microplate or cuvette. The molar extinction coefficient at 301 nm for FMOC in NMP is 7,800 (*see* **Note 8**).

3.4. Peptide Synthesis Automation

1. Prepare a list of sequences ordered by sequential well position. Calculate the number of amino acid additions (doses) required to complete the synthesis of all of the peptides (*see* **Note 4**).

2. Compute a set of arrays and instructions linking the amino acids required at each well for each cycle of the synthesis.

A new array is required each time the peptide chains are extended (*see* **Note 4**).

3. Input the sequence file information into the MEK.exe program and obtain the Beckman Biomek arrays.bio files. Automate the input of successive arrays.bio files to successive cycles of the Biomek run program using the Lotus keystroke automation program (*see* **Note 4**).

4. Prepare activated FMOC-amino acid solutions for each amino acid (*see* **Subheading 2.4.5**).

5. Place bulk delivery tool to dispense NMP washes.

6. Place poly-Lys-coated solvent-resistant microplate.

7. Place DCC-activated FMOC-amino acids.

8. Place cycle blocking reagent at –20% piperidine/NMP.

9. Start method.

3.5. Removal of Amino Acid Side-Chain Protecting Groups

Use extreme caution and a well-ventilated fume hood.

1. Transfer 0.10 mL of reagent C to each well and seal the plate in a tight-fitting container (*see* **Subheading 4**). After 3 h aspirate the liquid with a hand-held plate aspirator and wash the plate twice with ether. Air dry in the fume hood overnight. Store the plate at –20°C in a plastic freezer bag until needed.

2. To deprotect peptides not containing Met, Trp, or Cys, substitute 0.1 mL of TFA containing TIS and H20 (95, 2.5, and 2.5%) to avoid the malodorous sulfur chemicals.

3. For peptides containing protected phospho-tyrosine (DMAP-Tyr) add 0.1 mL of water after the initial 3-h incubation described in **steps 1** or **2** above, reseal, and incubate for an additional 16 h before aspiration and washing.

3.6. Antibody Recognition of Microplate Peptides

1. Make peptide plate.

2. Acetylate with 10% acetic anhydride, 1% DIPEA in NMP for 30 min, ambient (*see* **Note 9**).

3. Deprotect the peptide plate using Mixture C.

4. Rinse peptide plate with 6× normal saline.

5. Add 20% carrier serum in PBS to each well. Incubate for 10 min at room temperature and just aspirate when ready. Carrier serum is of the same species as the conjugate species i.e., 20% normal goat serum if the conjugate is goat-IgG:alk-phos anti-murine Ig

6. Dilute serum 1/300 (or monoclonal antibody at 1 µg/mL) in PBS, 0.5% Tween 20, 20% heat-inactivated carrier serum, 0.02%NaN$_3$ (*see* **Note 10**).

7. Incubate 4°C overnight.

8. Wash with PBS-0.05% Tween 20.

9. Incubate 1.5 h with 1/1,000 Boehringer conjugate in PBS–0.5% Tween 20–5% carrier serum.

10. Wash with PBS–0.05% Tween 20.

11. Rinse with saline solution.

12. Incubate approximately 1 h in phosphatase substrate buffer using buffer components from Kirkegaard & Perry (preferred) or Pierce – observing periodically to make sure dye colors are not oversaturated or undersaturated in color.

13. After incubation transfer fluid to another plate for reading. The alternative is to add stopping solution to the peptide plate but that could damage the plate and prevent its being reused.

14. Plate recycling: The plate may be reused after incubation overnight at 37°C in SDS-urea-beta mercaptoethanol or DTT [50% urea (w/w), 2% SDS, 1% mercaptoethanol, 0.05 M Tris buffer (pH 7–8.5)]. When the plate is not in use it should be stored wrapped in the freezer – after methanol wash and air-drying in a fume hood.

3.7. Kinase Substrate Activity of Microplate Peptides

1. Distribute into each peptide well, 30 μL of kinase assay buffer.

2. Distribute into each peptide well 30 μL of appropriately diluted (kinase) in kinase dilution buffer.

3. Distribute into each peptide well 30 μL of ATP mix in kinase assay buffer.

4. Incubate for 30 min at 30°C.

5. Rinse with distilled water.

6. Assay for the presence of phosphotyrosine by antibody as in **Subheading 3.6** above using murine monoclonal antibody against phosphotyrosine.

4. Notes

1. Other polyallomer plastics such as polyethylene or polypropylene may be used as well, but they are partially opaque and the high density forms do not perform as well.

2. Volumes will increase by approximately 10%, so the nominal concentrations will be slightly lower than 0.5 M. The difference is inconsequential compared with the 10^3 fold final excess of activated amino acid over available reaction sites on the plate surface. The excess HOBT concentration however plays an important role in stabilization of the FMOC-amino acid solutions, especially after activation.

3. While there are many robotic workstations that may be used, the Biomeks store their instructions for the activated amino acid delivery in external arrays.bio or patterns.b2 k files so that they can be manipulated by prior external programming and can be accessed sequentially after the process method is initiated. Manufacturers use different forms of control programming. The Beckman instruments conveniently store pipetting information in external pattern/array files that can be modified externally and exchanged sequentially as the cycles proceed. Most robotics systems that utilize TFF routines to perform sequential pipetting routines can be used as well.

4. This software developed at the NCI is available through a cost-free Source Code Use License from the NIH by email request to Carl Saxinger (csaxinger@verizon.net), with "Biomek software" in the subject line.

5. Or, normal serum from the same species as the labeled anti-immunoglobulin secondary antibody is derived.

6. The reaction can be shortened to 2 h but sealing the plates becomes more difficult and the process is more hazardous.

7. Poly(D-Lys) and poly(L-Lys) are equally effective. Poly(D-Lys) is preferred because it is more resistant to proteases.

8. It may be necessary to increase absorbance readings by transferring the 0.1 mL aliquot successively through a series of replicates with a 5-min pause at each replicate. Preliminary to peptide synthesis the coupling of FMOC-Rink Amide linker (Novabiochem) using standard DCC/HOBT coupling conditions will allow removal of the peptide from the plate. Tagging the N-terminus with a reporter molecule such as Dansyl-Cl or Dabsyl-Cl following the final piperidine step and just before peptide cleavage step will allow the removal of the peptide and its characterization by fluorescence, absorption, or mass spectrometry (17). Approximately 500 pmole per well (1 cm^2) is made. Over 90% of peptides released this way and analyzed by Maldi-TOF mass spectrometry have been of high spectral purity, from major peak to spectrally pure (data from experiments using the automation process and chemistry described here on bead suspensions, not presented here).

9. This is an optional step. If the peptide is to be used specifically for mass spectral analysis it is best to omit acetylation. If the peptide is to be used to interact with live cells it is best to include acetylation. For most other applications where the peptide does not represent a free N-terminus of the native protein it is appropriate to acetylate.

10. Very low backgrounds combined with high peptide density on these plates make the peptide plates very sensitive. Test sera may be used as concentrated as 1/30–1/300× test monoclonal antibodies at 1 μg/mL or higher.

Acknowledgements

I especially thank Paul Nagel and Greg Goetz for shaping the software architecture of the robotic interface and control and Beckman Instruments for providing the file structure of the Biomek 1,000 and Biomek 2,000 array files. I also thank Mei-Wan Ho, Jamshed Ayub, and Vasu Parekh for developmental laboratory assistance and Robert Gallo for support.

References

1. Manning, B. D., Cantley, L. C. (2002) Hitting the target: emerging technologies in the search for kinase substrates. *Sci. STKE* 2002, PE49.

2. Songyang, Z., Cantley, L. C. (1995) Recognition and specificity in protein tyrosine kinase-mediated signalling. *Trends Biochem. Sci.* 20, 470–475.

3. Songyang, Z., Carraway, K. L. III, Eck, M. J., Harrison, S. C., Feldman, R. A., Mohammadi, M. et al. (1995) Catalytic specificity of protein-tyrosine kinases is critical for selective signalling. *Nature* 373, 536–539.

4. Till, J. H., Annan, R. S., Carr, S. A., Miller, W. T. (1994) Use of synthetic peptide libraries and phosphopeptide-selective mass spectrometry to probe protein kinase substrate specificity. *J. Biol. Chem.* 269, 7423–7428.

5. Geysen, H. M., Meloen, R. H., Barteling, S. J. (1984) Use of peptide synthesis to probe viral antigens for epitopes to a resolution of a single amino acid. *Proc. Natl. Acad. Sci. U S A.* 81, 3998–4002.

6. Gausepohl, H., Boulin, C., Kraft, M., Frank, R. W. (1992) Automated multiple peptide synthesis. *Pept. Res.* 5, 315–320.

7. Frank, R. (1992) Spot-synthesis: an easy technique for the positionally addressable, parallel chemical synthesis on a membrane support. *Tetrahedron.* 48, 9217–9232.

8. Tegge, W., Frank, R., Hofmann, F., Dostmann, W. R. (1995) Determination of cyclic nucleotide-dependent protein kinase substrate specificity by the use of peptide libraries on cellulose paper. *Biochemistry* 34, 10569–10577.

9. Tegge, W. J., Frank, R. (1998) Analysis of protein kinase substrate specificity by the use of peptide libraries on cellulose paper (SPOT-method). *Methods Mol. Biol.* 87, 99–106.

10. Luo, K., Zhou, P., Lodish, H. F. (1995) The specificity of the transforming growth factor beta receptor kinases determined by a spatially addressable peptide library. *Proc. Natl. Acad. Sci. U S A.* 92, 11761–11765.

11. Saxinger, C. (2000) Automated peptide design and synthesis. United States Patent 6,031,074 (February 29, Issue Date).

12. Rosenfeld, S. J., Young, N. S., Alling, D., Ayub, J., Saxinger, C. (1994) Subunit interaction in B19 parvovirus empty capsids. *Arch. Virol.* 136, 9–18.

13. Kim, P. J., Sakaguchi, K., Sakamoto, H., Saxinger, C., Day, R., McPhie, P. et al. (1998) Colocalization of heparin and receptor binding sites on keratinocyte growth factor. *Biochemistry.* 37, 8853–8862.

14. Saxinger, C. (2003) Polypeptides comprising IL-6 ligand-binding receptor domains. US PATENT 6,664,374 (December 16, 2003, issue date).

15. Saxinger, C. (2007) Polypeptides that bind HIV gp120 and related nucleic acids, antibodies, compositions, and methods of use. United States Patent 7,304,127 (December 4, 2007, issue date).

16. Saxinger, C., Conrads, T. P., Goldstein, D. J., Veenstra, T. D. (2005) Fully automated synthesis of (phospho)peptide arrays in microtiter plate wells provides efficient access to protein tyrosine kinase characterization. *BMC Immunol.* 6, 1.

17. Hudecz, F., Szekerke, M. (1985) Synthesis of new branched polypeptides with poly(lysine) back bone. *Collection Czechoslovak Chem. Commun.* 50, 103–113.

18. Mezo, G., Kajtar, J., Hudecz, F., Szekerke, M. (1993) Carrier design – conformational studies of amino acid(X) and oligopeptide (X-Dl-Ala(M)) substituted poly(L-Lysine). *Biopolymers* 33, 873–885.

19. Chao, H. G., Leiting, B., Reiss, P. D., Burkhardt, A. L., Klimas, C. E., Bolen, J. B. et al. (1995) Synthesis and applications of Fmoc-O-[bis(dimethylamino)phosphono]-tyrosine, a versatile protected phosphotyrosine equivalent. *J. Org. Chem.* 60, 7710–7711.

Chapter 20

Kinome Profiling Using Peptide Arrays in Eukaryotic Cells

Kaushal Parikh, Maikel P. Peppelenbosch, and Tita Ritsema

Summary

Over the last 10 years array and mass spectrometry technologies have enabled the determination of the transcriptome and proteome of biological and in particular eukaryotic systems. This information will likely be of significant value to our elucidation of the molecular mechanisms that govern eukaryotic physiology. However, an equally, if not more important goal, is to define those proteins that participate in signalling pathways that ultimately control cell fate. Enzymes that phosphorylate tyrosine, serine, and threonine residues on other proteins play a major role in signalling cascades that determine cell-cycle entry, and survival and differentiation fate in the tissues across the eukaryotic kingdoms. Knowing which signalling pathways are being used in these cells is of critical importance. Traditional genetic and biochemical approaches can certainly provide answers here, but for technical and practical reasons there is typically pursued one gene or pathway at a time. Thus, a more comprehensive approach is needed in order to reveal signalling pathways active in nucleated cells. Towards this end, kinome analysis techniques using peptide arrays have begun to be applied with substantial success in a variety of organisms from all major branches of eukaryotic life, generating descriptions of cellular signalling without a priori assumptions as to possibly effected pathways. The general procedure and analysis methods are very similar disregarding whether the primary source of the material is animal, plant, or fungal of nature and will be described in this chapter. These studies will help us better understand what signalling pathways are critical to controlling eukaryotic cell function.

Key words: Kinase, Arrays, PepChips, Activity, Cells, Tissues, Purified kinases.

1. Introduction

The predominance of phosphorylation as a regulator of cellular metabolism has enticed many researchers to develop strategies for making descriptions of cellular phosphorylation events *(1)*. Classically, kinase activity and protein phosphorylation were studied using in-gel kinase assays or Western blot-based gel shift

Marjo de Graauw (ed.), *Phospho-Proteomics, Methods and Protocols, vol. 527*
© 2009 Humana Press, a part of Springer Science+Business Media, New York, NY
Book DOI: 10.1007/978-1-60327-834-8_20

techniques that exploit the size difference between the phosphorylated and unphosphorylated forms of proteins. These are, however, fairly cumbersome techniques and do not allow the study of large numbers of samples. The situation has improved with the advent of phosphospecific antibodies in the late nineties, which recognize the phosphorylated forms of proteins but not their unphosphorylated counterparts. Employing these antibodies, phosphorylation events can be detected using classical Western blotting but also in, for instance, ELISA formats to allow high throughput screening for kinase activity-modifying compounds (2), or in tissue arrays that enable histological analysis of protein phosphorylation in hundreds to thousands of relevant tissue samples simultaneously (3). The main drawback remains that only one type of phosphorylation is studied per experiment. Recently, employing multicolour FACS, Irish et al. characterized phosphoprotein responses to environmental cues in acute myeloid leukaemia at the single cell level (4). The advantage of this approach is that the individual variation of cells with respect to the amount of phosphoproteins is assessed and requires very little material. But even this most advanced FACS technology does not allow simultaneous assessment of, typically, more then ten antibodies at one time. This has prompted investigators to explore techniques for studying cellular phosphorylation with little a priori assumptions as to the phosphorylation events involved.

Among the advanced of these approaches is the one that is commercially offered by Kinexus, which using a multiblot system that relies on sodium dodecyl sulfate-polyacrylamide minigel electrophoresis and multilane immunoblotters to permit the specific and quantitative detection of 45 or more protein kinases or other signal transduction proteins at once (5). When used to its full extent this technology produces almost complete descriptions of cellular phospho-protein networks, although it still remains fairly labour intensive. Alternatively a proteomic approach may be chosen, which typically consists of a separation of phosphoproteins by, for instance, 2-D gel electrophoresis or chromatography, followed by mass spectrometry. Steady progress is made in this area, and 4 years ago using strong cation exchange chromatography at low pH to enrich for tryptic phosphopeptides, a first large-scale proteomic profiling of phosphorylation sites from primary animal tissue has been performed (6). For protein spots that can be detected and unambiguously identified, these approaches provide a powerful way of monitoring the expression and regulation of potentially hundreds of proteins simultaneously, but in practice it is hampered by the fact that the positions of scarcely more than two dozen protein kinases are available on 2D proteomic maps. This reflects the fact that – like most signal transduction proteins – protein kinases are present at very minute levels in cells, and are often undetectable by the most sensitive protein stains, whereas procedures based on the purification of phosphopeptides and determination of peptide structure by MALDI is time consuming. The major advantage of this approach is that it is completely unbiased, for instance, using microfluidic compact disk technology this approach identified two novel phosphorylation sites in the human mineralocorticoid receptor *(6)*.

A disadvantage of all these approaches is that they are focused on the static determination of the relative concentration of phosphoproteins, but do not address the actual activity of various cellular signalling pathways. (A popular comparison is the dashboard of a car where the mileage indicator gives information as to the distance travelled, but gives no information as to the velocity at which this is occurring. For obtaining the latter information one uses the speedometer.)

Adaptation of array technology for measuring enzymatic activity in a parallel fashion seems an obvious solution for the earlier mentioned problems, and progress in this direction has been made with the preparation of protein chips for the assessment of protein substrate interactions *(7–9)* and the generation of peptide chips for the appraisal of ligand–receptor interactions and enzymatic activities *(10–13)*. Houseman and Mrksich *(14)* showed that peptide chips, prepared by the Diels–Alder-mediated immobilization of one kinase substrate for the non-receptor tyrosine kinase c-Src on a monolayer of alkanethiolates on gold, allows quantitative evaluation of kinase activity. In 2004 we showed that ^{33}P-γ-ATP phosphorylation of arrays consisting of 192 peptides that are substrates for kinases and spotted on glass by cell lysates from human peripheral blood mononuclear cells allowed the simultaneous description of the temporal kinetics of a multitude of kinase activities following stimulation with lipopolysaccharide *(15)*. The same design of peptide array is also highly useful for studying signal transduction in plants *(16)* Although these studies showed the pan-eukaryotic applicability of studying signal transduction with peptide arrays, it appeared, however, that the amount of substrates on this array was insufficient to allow truly comprehensive descriptions of cellular signal transduction.

This consideration prompted us to study the effectiveness of an array with substantially increased numbers of substrates, the kinase I PepChip. It harbours peptide substrates from species in all branches of eukaryotic life, and also from prokaryotes. Many mammalian-derived substrates and also substrates identified in e.g. bacteria, yeast, fungi, plants, drosophila, birds, and viruses are present. This set of substrates was selected from the Phosphobase resource (http://phospho.elm.eu.org) *(17, 18)* and a full list of the peptides and the proteins from which they are derived is listed on http://www.koskov.nl. Arrays were constructed by chemically synthesizing soluble peptides, which were covalently coupled to glass substrates as described for the smaller arrays. Kinase I arrays consisted of 1,176 different peptides, spotted 2 times per carrier, to allow assessment of possible variability in substrate phosphorylation. The final physical dimensions of the array were $25 \times 75\,mm^2$, each peptide spot having a diameter of approximately $250\,\mu m$, and peptide spots being $620\,\mu m$ apart. This design was highly successful, generating profiles from a variety of organisms and tissues from all branches of eukaryotic life *(19–23)*. One of the interesting results from these arrays is the observation that a common set of peptide substrates is phosphorylated by all eukaryotic organisms. Thus diversity seen within the eukaryotic kingdom with respect to the primary structure of kinases is not reflected on the substrate level *(19)*.

Recently a new array design, the kinomics PepChip, was introduced. It is based on phosphoproteins from the human protein reference database that has information from rats, mice, and humans alone. This array features three sets of 1,024 nonapeptides for which a phosphospecific antibody is available for each peptide *(24)*.

PepChips allow a comprehensive detection of the cellular metabolism in lysates. Up-regulation or down-regulation of a particular kinase activity may lead to a cascade of cellular events. These can be fit into specific cell signalling pathways or cellular functions and as such assign specific characteristics to certain cells.

2. Materials

1. PepChip slides (PepScan, the Netherlands, http://www.pepscan.nl, *see* **Note 1**). The slides can be stored up to 3 months at 4°C in the supplied tube/box.

2. 60-mm Glass coverslips.

3. Lysis buffer for mammalian cells: M-PER® Mammalian Protein Extraction Reagent (Pierce) with dilution buffer: 1 M stock of HEPES, pH 7.5 (Sigma).

4. Plant cells lysis buffer: 20 mM Tris–HCl, pH 7.5, 1% Triton X-100, 150 mM NaCl, 1 mM EDTA, 1 mM EGTA, 2.5 mM $Na_4P_2O_7$ (sodium pyrophosphate), 1 mM β-glycerophosphate, 1 mM Na_3VO_4 (sodium orthovanadate), 1 mM NaF, 1 µg/ml leupeptin, 1 µg/ml aprotinin. Store at –20°C in aliquotes.

5. 100 mM PMSF (add fresh from stock in iso-propanol or acetone).

6. Additional stock reagents required: (all Sigma) 1 M $MgCl_2$; 1 M $MnCl_2$; 100 mM sodium orthovanadate; 1 M DTT; 100 mM ATP; 3% Brij-35; 20% PEG 8000; 100 mg/ml; 50% glycerol; 5 mg/ml BSA; 1 M NaCl; 1 M EDTA; 1 M EGTA;

1 M sodium fluoride; 1 mg/ml leupeptin; 1 mg/ml aprotinin; and ^{33}P-γ-ATP (Amersham/Perkin Elmer, *see also* **Note 4**).

7. Wash buffers: PBS with 1% Tween 20; PBS with 0.05% Tween 20; 2 M NaCl with 1% Tween 20; 0.5 M NaCl; demineralized water; and MilliQ water (make fresh).

8. Phospho-imager plate (e.g. Amersham) and scanner (e.g. Biorad or STORM).

9. ScanAlyze software (*see* **Note 2**; Eisen lab; http://rana.lbl.gov/EisenSoftware.htm).

10. Microsoft Excel.

3. Methods

All lysates and purified kinases should be kept on ice throughout.

3.1. PepChips Using Cells

1. Lyse 10^6 cells in 100 μl of M-PER lysis buffer containing 150 mM NaCl, 1 mM EDTA, 1 mM EGTA, 1 mM sodium vanadate, 1 mM sodium fluoride, 1 μg/ml leupeptin, 1 μg/ml aprotinin, and 1 mM PMSF (add PMSF fresh before use).

2. From this take 33 μl of lysate and add 67 μl of 60 mM HEPES, pH 7.5, and centrifuge at $16,100 \times g$ for 10 min at 4°C.

3. From this take 70 μl and add 10 μl of activation mix (activation mix = 70 mM $MgCl_2$ + 70 mM $MnCl_2$ + 400 μg/ml PEG 8000 + 400 μg/ml BSA).

4. Add 8 μl of 50% glycerol.

5. Add 2 μl of ^{33}P-γ-ATP and centrifuge for 3 min at maximum speed. (To be carried out in the isotope lab with appropriate protection.)

6. Pipette out 90 μl of this mix onto a glass coverslip and invert the PepScan slide onto the coverslip. Make sure to completely remove any trapped air bubbles (*see also* **Note 7**).

7. Incubate the PepScan slides in a heated water bath or stove at 37°C (in order to prevent drying out of the slides) for 90–120 min.

8. After the incubation, remove the coverslips off the slides by dipping them in PBS-containing 1% Tween 20. Do this a few times till the coverslips completely slip off the glass slides.

9. Wash the slides twice in 2 M NaCl with 1% Tween 20, twice in PBS with 1% Tween 20, and then twice in demineralized water.

10. Airdry the slides.

11. Expose the slides to the phospho-imager plate and scan using a scanner.

12. Analyse the image using ScanAlyze (*see* **Note 2**).

3.2. PepChips Using Biopsies

1. Lyse the biopsy/tissue sample in lysis buffer in a final concentration of 0.4 mg biopsy/µl lysis buffer.

2. Centrifuge at $16,100 \times g$ for 10 min at 4°C and discard the pellet.

3. Measure the protein, and the optimal concentrations for performing PepChips range between 70 and 100 µg/50 µl.

4. To 70 µl of the lysate, add 10 µl of the activation mix (activation mix = 70 mM $MgCl_2$ + 70 mM $MnCl_2$ + 400 µg/ml PEG 8000 + 400 µg/ml BSA).

5. Add 8 µl of 50% glycerol.

6. Add 2 µl of ^{33}P-γ-ATP and centrifuge for 3 min at maximum speed. (To be carried out in the isotope lab with appropriate protection.)

7. Follow the same procedure as in **step 6** from cell protocol.

3.3. PepChips Using Plant Tissue

1. Lyse 100 mg of leaf material in 200 µl of lysisbuffer; add 1 mM PMSF fresh from stock.

2. Incubate on ice for 5 min.

3. Spin down 10 min at maximum speed ($16,100 \times g$) at 4°C.

4. Add supernatants to a 0.2-µm filter and spin again 2 min at maximum $10,000 \times g$ to remove particles.

5. Prepare activation mix fresh from stocks by mixing 0.33 ml glycerol (50%), 5 µl ATP (100 mM), 20 µl $MgCl_2$ (1 M), 3.3 µl Brij-35 (3%), and 10 µl BSA (5 mg/ml).

6. If necessary dilute the lysate in lysis buffer (this can remove background greying of the slides).

7. Add to 50 µl lysate 11.25 µl activation mix and 3 µl ^{33}P-γ-ATP. (To be carried out in the isotope lab with appropriate protection.)

8. Pipette 60 µl on a coverslip and invert the PepScan slide onto the coverslip. Make sure to completely remove any trapped air bubbles.

9. Incubate the PepScan slides in a heated water bath at 30°C (in order to prevent drying out of the slides) for 2 h.

10. After the incubation, remove the coverslips off the slides by dipping them in PBS-containing 0.05% Tween 20. Do this a few times till the coverslips completely slip off the glass slides.

11. Wash the slides in PBS with 0.05% Tween 20, twice in 0.5 M MaCl and then twice in MilliQ water; perform all wash steps 5 min while shaking.

12. Airdry the slides.

13. Expose the slides 1 week to the phospho-imager plate and scan using a scanner.

14. Analyse the image using ScanAlyze.

3.4. PepChips Using Purified Kinases

1. Take 70 μl of the desired protein kinase incubation mix (×ng or μg/ml of kinase catalytic subunit, 60 mM HEPES, pH 7.5, 3 mM MgCl$_2$, 3 mM MnCl$_2$, 3 mM Na$_3$VO$_4$, 1.2 mM DTT, 50 μM ATP, 0.03% Brij-35, 50 μg/ml PEG 8000, 50 μg/ml BSA).

2. Add 8 μl of 50% glycerol.

3. Add 2 μl of ^{33}P-γ-ATP and centrifuge for 3 min at maximum speed. (To be carried out in the isotope lab with appropriate protection.)

4. Follow the same procedure as in **step 6** from cell protocol.

3.5. Analysis

1. Use ScanAlyze software to analyse the PepChip images that were scanned. Before entering slides in the ScanAlyze software make sure to put the label down, convert to tiff format, invert, and put to 8 bits/channel. **Table 1** shows an example of the PepChip properties that should be used for the analysis.

2. Use grid tools to determine spot size (usually 10 pixels) and position to obtain spot intensities and background intensities.

3. Import data from the individual experiments to an excel sheet for further analysis.

4. Use control spots on the array for validation of spot intensities between the different samples.

5. Exclude inconsistent data (i.e. SD between the different data points >1.96 of the mean value) from further analysis (*see also* **Note 3**).

6. Average spots and include for dissimilarity determination to extract kinases of which activity was either significantly

Table 1
PepChip properties

PepChip	Spots per grid	Grids per replicate	Replicates
Trial	6 (h) × 4 (v)	2 (h) × 4 (v)	2 (h)
Kinase I	7 × 7	4 × 6	2 (v)
Kinome	8 × 8	4 × 4	3 (v)

induced or reduced. Various statistical tests are appropriate, for instance a heteroscedastic two-tailed Students' *t*-test (*see also* **Note 5**).

7. If two conditions are applied, statistically significant differences are used to generate provisional signal transduction schemes based on the available knowledge on signalling (*see* **Note 6**).

8. Alternatively, Venn diagrams can be constructed. In this case a cut-off criterium is defined (e.g. all spots having an intensity of more as 2 times the standard deviation of the background are considered positive) and shared and unique spots between the treatments are defined.

9. Purified kinases can be tested directly on the chip to define substrate specificities.

10. For establishing consensus phosphorylation sites, after in vitro phosphorylation of the array with a purified kinase, different peptides are incorporating wildly different amounts of ^{33}P. It is possible that a peptide could be phosphorylated at more than one residue, which would mean that a peptide that, for instance, is phosphorylated at two serines adjacent to each other could result in a higher intensity than a peptide phosphorylated on one serine and this would mask that peptide which could have been left out of the analysis. Hence, only those peptides that had a single phosphorylation site should be considered, i.e. only those peptides that had a single serine, threonine, or tyrosine residue at the central position. (Of the 1,176 peptides, 354 peptides have a single serine, threonine, or tyrosine residue.)

11. Peptides are aligned manually relative to the centrally fixed serine, threonine, or tyrosine residue and ranked on the mean intensity of the replicas for each spot. For deriving the consensus sequence using arrays with 1,176 substrates, we consider only positions (−3) −2, −1, 0, +1, +2 (+3). For arrays with 1,024 spots we also take −4 and +4 into consideration, since the peptides on these arrays have a length of nine amino acids. Furthermore, we only select peptides with cut-off intensities within 50% of the peptide with the maximum intensity, and the relative contribution of each individual amino acid at each individual position is calculated and corrected for the relative abundance of that amino acid at that position relative to the central serine, threonine, or tyrosine.

3.6. Discussion

The phosphorylation of the arrays is usually robust and clear (*see* e.g. **Figs. 1–3**), but highly sensitive to particulate matter. The array is however as specific as the substrates are and thus validation remains important: Western blot is a possibility but measures the combined effect of kinases and phosphatases on a substrate,

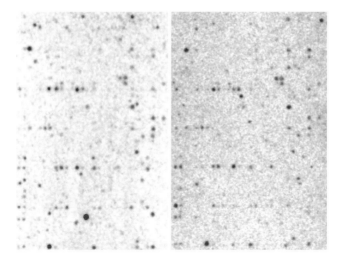

Fig. 1. Representative images of PepChips. Kinase I PepChips as they appear after scanning of the phosphorimager plate. Lysates used for these images were derived from cells of *Homo sapiens* (human, **a**) or *Arabidopsis thaliana* (plant, **b**). Chips were incubated for 2 h with the lysate and exposed on the phosphorimager for 3 days.

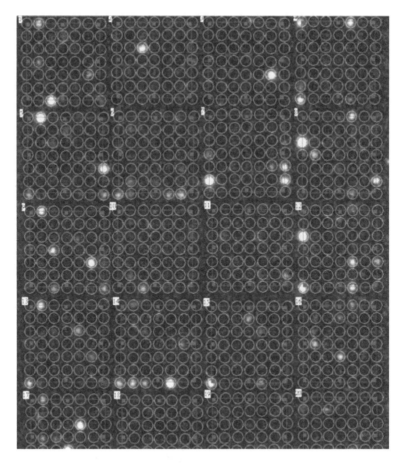

Fig. 2. Representation of gridding in ScanAlyze. ScanAlyze enables you to adjust the grids as a whole, per set or even individual spots. Background noise can be flagged, in order to omit it in further analysis. This example depicts part of a kinase I slide, which contains two times 24 grids of seven times 7 spots.

Comparison of the 2 duplicates

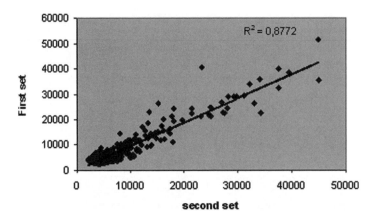

Fig. 3. Correlation plot obtained after gridding showing the robustness of the experiment. Spot intensities of the technical replicate present on the kinase I chip are compared to obtain a measure of reproducibility (R^2).

and thus, at least theoretically, in vitro kinase assays are superior. Phosphoflow is a good alternative with respect to the limited number of cells needed. For tissues, phophoimmunohistochemistry may be appropriate. If results are tricky, an α-ATP control (radioactivity not at the γ position but at the α position and thus inaccessible for a kinase reaction) should be considered. If this control is negative, results are very reliable.

Important future developments will include total validation of chips employing purified variants of all kinases from mammalian and possibly plant genomes, which should significantly enhance the value of this technology and help interpreting results. Also, the development of new software tools will greatly enhance the ease by which meaningful analyses can be generated. As these efforts at moment receive substantial support from the Dutch government, it is to be expected that the usefulness of peptide arraying will increase quickly Together, these developments now quickly establish peptide arraying as the technology of choice for generating comprehensive description of eukaryote signal transduction.

4. Notes

1. More information on the PepChips can be obtained from http://www.pepscan.nl.
2. ScanAlyze can be downloaded free by academic users from the Eisenlab Web site: http://rana.lbl.gov/.

3. If two conditions show wildly divergent phosphorylation patterns (i.e. when two conditions are plotter on each other, most spots are on the X and Y axis), most likely a gridding error has been made (grid wrongly placed).

4. Radioactive α-ATP contains traces of radioactive γ-ATP (approximately 5%), and so low signals may be obtained when using this control.

5. On one slide the individual sets of spots, when plotted on each other should show an R^2 of at least 0.6. Otherwise the experiment is best repeated.

6. For completely sequenced organisms, Blast can reveal potential kinase targets. Blast results can be found on http://www.koskov.nl.

7. Slides should be absolutely horizontal during incubation; otherwise gradients appear even when the slide is fully covered with fluid.

Acknowledgements

The authors acknowledge the tremendous support from the Top Institute Pharma (Dutch government) for developing the Pep-Chip technology. In addition, we are grateful for the support of the innovative actions programme Groningen supported by the European Commission.

References

1. Krebs, E. G. (1993) Nobel Lecture. Protein phosphorylation and cellular regulation I. *Biosci Rep* 13, 127–42.

2. Versteeg, H. H., Nijhuis, E., van den Brink, G. R., Evertzen, M., Pynaert, G. N., van Deventer, S. J., et al. (2000) A new phosphospecific cell-based ELISA for p42/p44 mitogen-activated protein kinase (MAPK), p38 MAPK, protein kinase B and cAMP-response-element-binding protein. *Biochem J* 350, 717–22.

3. Hung, G. G., Provost, E., Kielhorn, E. P., Charette, L. A., Smith, B. L., Rimm, D. L. (2001) Tissue microarray analysis of beta-catenin in colorectal cancer shows nuclear phospho-beta-catenin is associated with a better prognosis. *Clin Cancer Res* 7, 4013–20.

4. Irish, J. M., Hovland, R., Krutzik, P. O., Perez, O. D., Bruserud, O., Gjertsen, B. T., et al. (2004) Single cell profiling of potentiated phospho-protein networks in cancer cells. *Cell* 118, 217–28.

5. Pelech, S. (2004) Tracking cell signaling protein expression and phosphorylation by innovative proteomic solutions. *Curr Pharm Biotechnol* 5, 69–77.

6. Ballif, B. A., Villen, J., Beausoleil, S. A., Schwartz, D., Gygi, S. P. (2004) Phosphoproteomic analysis of the developing mouse brain. *Mol Cell Proteomics* 3, 1093–101.

7. Lueking, A., Horn, M., Eickhoff, H., Bussow, K., Lehrach, H., Walter, G. (1999) Protein microarrays for gene expression and antibody screening. *Anal Biochem* 270, 103–11.

8. Arenkov, P., Kukhtin, A., Gemmell, A., Voloshchuk, S., Chupeeva, V., Mirzabekov, A. (2000) Protein microchips: use for immunoassay and enzymatic reactions. *Anal Biochem* 278, 123–31.

9. MacBeath, G., Schreiber, S. L. (2000) Printing proteins as microarrays for high-throughput function determination. *Science* 289, 1760–3.

10. Zhu, H., Snyder, M. (2001) Protein arrays and microarrays. *Curr Opin Chem Biol* 5, 40–5.

11. Wenschuh, H., Volkmer-Engert, R., Schmidt, M., Schulz, M., Schneider-Mergener, J., Reineke, U. (2000) Coherent membrane supports for parallel microsynthesis and screening of bioactive peptides. *Biopolymers* 55, 188–206.

12. Falsey, J. R., Renil, M., Park, S., Li, S., Lam, K. S. (2001) Peptide and small molecule microarray for high throughput cell adhesion and functional assays. *Bioconjug Chem* 12, 346–53.

13. Reineke, U., Volkmer-Engert, R., Schneider-Mergener, J. (2001) Applications of peptide arrays prepared by the SPOT-technology. *Curr Opin Biotechnol* 12, 59–64.

14. Houseman, B. T., Mrksich, M. (2002) Towards quantitative assays with peptide chips: a surface engineering approach. *Trends Biotechnol* 20, 279–81.

15. Diks, S. H., Kok, K., O'Toole, T., Hommes, D. W., van Dijken, P., Joore, J., et al. (2004) Kinome profiling for studying lipopolysaccharide signal transduction in human peripheral blood mononuclear cells. *J Biol Chem* 279(47), 49206–13.

16. Ritsema, T., Joore, J., van Workum, W., Pieterse, C. M. (2007) Kinome profiling of *Arabidopsis* using arrays of kinase consensus substrates. *Plant Methods* 3, 3.

17. Blom, N., Kreegipuu, A., Brunak, S., Conway, T., Schoolnik, G. K. (1998) PhosphoBase: a database of phosphorylation sites. *Nucleic Acids Res* 26, 382–6.

18. Kreegipuu, A., Blom, N., Brunak, S. (1999) PhosphoBase, a database of phosphorylation sites: release 2.0. *Nucleic Acids Res* 27, 237–9.

19. Diks, S. H., Parikh, K., van der Sijde, M., Joore, J., Ritsema, T., Peppelenbosch, M. P. (2007) Evidence for a minimal eukaryotic phosphoproteome? *PLoS ONE* 2(1), e777.

20. de Borst, M. H., Diks, S. H., Bolbrinker, J., Schellings, M. W., van Dalen, M. B., Peppelenbosch, M. P., et al. (2007) Profiling of the renal kinome: a novel tool to identify protein kinases involved in angiotensin II-dependent hypertensive renal damage. *Am J Physiol Renal Physiol* 293, F428–37.

21. van Baal, J. W., Diks, S. H., Wanders, R. J., Rygiel, A. M., Milano, F., Joore, J., et al. (2006) Comparison of kinome profiles of Barrett's esophagus with normal squamous esophagus and normal gastric cardia. *Cancer Res* 66, 11605–12.

22. Löwenberg, M., Tuynman, J., Scheffer, M., Verhaar, A., Vermeulen, L., van Deventer, S., et al. (2006) Kinome analysis reveals nongenomic glucocorticoid receptor-dependent inhibition of insulin signaling. *Endocrinology* 147, 3555–62.

23. Löwenberg, M., Tuynman, J., Bilderbeek, J., Gaber, T., Buttgereit, F., van Deventer, S., et al. (2005) Rapid immunosuppressive effects of glucocorticoids mediated through Lck and Fyn. *Blood* 106, 1703–10.

24. Tuynman, J., Vermeulen, L., Boon, L., Kemper, K., Zwinderman, A., Peppelenbosch, M., et al. (2008) Selective COX-2 inhibition inhibits c-Met kinase activity and inhibits Wnt activity in colon cancer. *Cancer Res* 68, 1213–20.

Part V

Bioinformatics

Chapter 21

ProMoST: A Tool for Calculating the p*I* and Molecular Mass of Phosphorylated and Modified Proteins on Two-Dimensional Gels

Brian D. Halligan

Summary

Protein modifications such as phosphorylation are often studied by two-dimensional gel electrophoresis, since the perturbation in the protein's p*I* value is readily detected by this method. It is important to be able to calculate the changes in the p*I* values that specific post-translational modifications cause and to visualize how these changes will effect protein migration on 2D gels. To address this need, we have developed ProMoST. ProMoST is a freely accessible Web-based application that calculates and displays the mass and p*I* values for either proteins in the NCBI database identified by accession number or from submitted FASTA format sequence.

Key words: Two-dimensional gel electrophoresis, Protein modification, Phosphorylation.

1. Introduction

One of the most successful methods for detecting and analyzing protein post-translational modifications (PTMs) has been two-dimensional gel electrophoresis (2D-GE). Since many PTMs, such as phosphorylation, introduce charged groups into the protein, there is often a detectable change in the position of the protein on a 2D gel. Although the change in the mass of the protein due to the PTM is often too small to be easily detected by standard sodium dodecyl sulfate-polyacrylamide gel electrophoresis (SDS-PAGE), the modification can cause a change in the net charge of the protein leading to a change in the isoelectric point, or p*I* of the protein. The first dimension of the 2D gel, usually shown

Marjo de Graauw (ed.), *Phospho-Proteomics, Methods and Protocols, vol. 527*
© 2009 Humana Press, a part of Springer Science+Business Media, New York, NY
Book DOI: 10.1007/978-1-60327-834-8_21

horizontally, is the isoelectric focusing dimension; changes in protein pIs are reflected as changes in the horizontal position of the protein spot within the 2D pattern of spots. Often it is observed that there are "trains" of spots on the gel that are presumably formed by multiple versions of the same protein that differ in isoelectric point because of increasing numbers of post-translational modifications such as phosphorylation or deamidation *(1, 2)*.

Although 2D-GE is a sensitive method for determining that there are post-translationally modified forms of proteins present, it does not directly indicate what the modification is or how many of the residues in the protein are modified. Since proteins vary greatly in their ability to buffer the change in pI due to post-translational modifications, to examine these results more closely it is necessary to calculate the predicted pI changes caused by the modification in the context of the protein sequence.

To meet this need, we have developed ProMoST, a Web-based application that allows users considerable freedom in calculating the pI values of modified and unmodified proteins *(3)*. ProMoST has predefined modifications so that casual users are able to rapidly determine the predicted pI values of modified proteins and peptides. In addition, ProMoST also provides additional options for more advanced users allowing them to define additional custom modifications that change the pK_a values for the defined modifications and even make changes to the default pK_a values for charged amino acids used to calculate pI values. The results of the calculations can be displayed both in a tabular format as well as in a graphic representation of the migration of the protein on a 2D gel. In addition, ProMoST can also be used for other types of protein analysis (note 1).

1.1. pI

The pK_a values of the side chains of the 20 common amino acids that comprise most proteins vary from approximately pH 2.8–11.2 *(4)*. Three amino acids positively charged under physiological conditions (lysine, arginine, and histidine) are termed *basic amino acids* and two amino acids negatively charged under physiological conditions (glutamic acid and aspartic acid) are termed *acidic amino acids*. In addition, the amino (N) and carboxyl (C) termini of the protein can also be charged. To determine the total charge of a protein at a given pH, the fractional number of positive and negative charges for each of the amino acids in the protein's sequence is determined and the sum of the fractional charges is equal to the charge on the protein.

The isoelectric point or pI of the protein is the pH value at which the total charge on the protein is zero. At this pH value the negative and positive charges of the protein are equal and the protein is at neutral charge. The pI of the protein therefore gives an indication of whether the protein will carry a net positive or negative charge under physiological conditions. Proteins that have a p*I* > 7.0 are considered to be basic proteins and proteins that have a p*I* < 7.0 are considered to be acidic proteins.

In addition to giving an indication of the charge of the protein, the p*I* is also a good indicator of the solubility of the protein at a given pH. One of the most important aspect of a protein's physiochemical property that determines solubility is its charge. Thus at a pH equal to the p*I* of the protein, it is uncharged and therefore it is usually the least soluble. Manipulating protein charge, either by changing pH or by adding salt to neutralize charge is the basis for many of the early methods for protein purification by differential solubility *(5)*.

The loss of charge at a protein's p*I* is also part of the fractionation process during the first dimension of the 2D gel that is based on isoelectric focusing *(6)*. Proteins are introduced to a strip on which a pH gradient has been established, and in the presence of a high electric field, they migrate to the position on the strip at which the protein has a net neutral change and it stops migrating. This pH value corresponds to the p*I* of the protein. Thus the final migration position of the protein in the horizontal dimension of the 2D gel is determined by the p*I* value of the protein.

1.2. Modifications and Mutations Change pI

The fact that the migration in the isoelectric focusing dimension of proteins in 2D-GE is very sensitive to changes in p*I* makes 2D-GE a valuable technique for identifying modifications and mutations *(1)*. Modifications such as phosphorylation that add highly charged groups to the protein can cause easily detectable changes in p*I* and therefore mobility of the protein in the isoelectric focusing dimension. Similarly, the changes in protein mass and p*I* due to mutations that cause a net loss or gain of charge on the protein by altering the number of charged acidic and basic residues present in the protein can also be calculated and displayed. The amount of mobility shift that is observed due to modification or mutation is dependent on three factors. First, the pK_a value for the modification or change induced by mutation is very important to the final change in the protein p*I*. Modifications, such as phosphorylation, that introduce a group with either a strongly acidic or basic pK_a will have a greater effect than those with a pK_a value closer to neutrality. Similarly, a mutation that causes a change from an acidic residue to a basic residue will lead to a larger change in pK_a than a change from a charged residue to a neutral residue. The larger pK_a alteration will lead to a larger change in protein p*I* and therefore a larger mobility shift in the isoelectric focusing dimension of the 2D gel. Second, the number of modifications or residue changes will also have an impact on the mobility shift observed on the gels. Often for modifications, a train of spots will be observed. Interestingly, the shift in mobility is often not constant and the distance between spots can vary. This is explained by the third factor that determines the magnitude of the observed p*I* shift: The charge buffering capacity of the protein at a given pH. Since different proteins are comprised of different mixtures of positively and negatively charged amino acid depending on their primary amino acid sequence, the charge titration profile for each protein is unique. Thus the extent to which a modification changes the p*I* of the protein and impacts on the mobility of the protein is different since the change titration profile changes with pH. **Figure 1** shows an example of this for human cyclin-dependent kinase 2 (CDK2). **Figure 1a** shows the titration of the unmodified protein. **Figure 1b–d** show the titrations with 1, 2, or 3 phosphate groups. For comparison, **Fig. 1e** shows the 2D gel spot positions calculated by ProMoST. Note that the magnitude

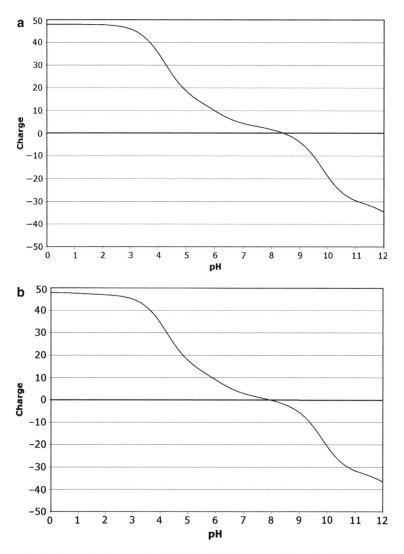

Fig. 1. (**a**) Charge titration for unphosphorylated rat CDK2 (cell division protein kinase 2 – Q63699). (**b**) Charge titration for unphosphorylated rat CDK2 with one phosphotyrosine group. (**c**) Charge titration for unphosphorylated rat CDK2 with two phosphotyrosine groups. (**d**) Charge titration for unphosphorylated rat CDK2 with three phosphotyrosine groups. (**e**) Pseudo-2D gel graphic showing the calculated position of CDK2 and phosphorylated forms of CDK2 (Q63699).

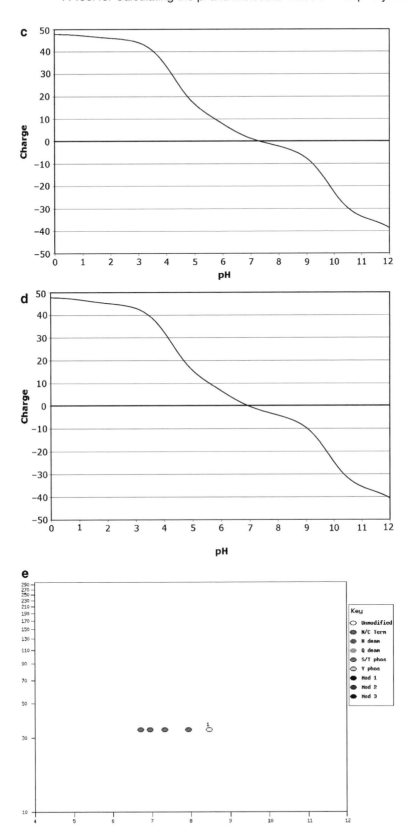

Fig. 1. (continued)

of the shift in the calculated spot position due to additional phosphorylation varies from spot to spot. This variance correlates with the titration curves for CDK2 shown in **Fig. 1a–d**.

The correlation of the calculated mass and pI for a protein and an actual 2D gel is shown in **Fig. 2**. Proteins were isolated from cultured rat fibroblast cells and analyzed by 2D-GE **(Fig. 2a)**. Proteins were extracted from gel spots, digested with trypsin, and analysed by MALDI as previously described *(7)*. Protein identification was carried out by peptide mass fingerprinting (PMF) using

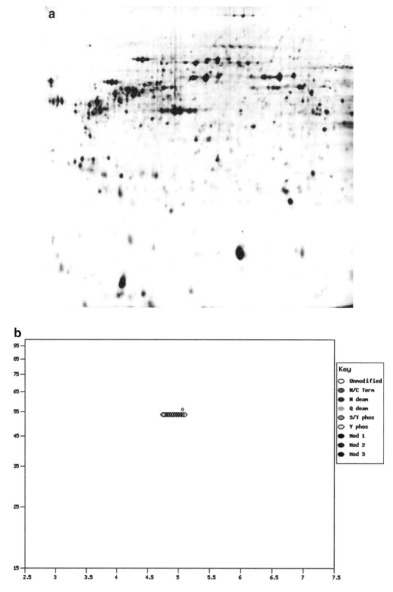

Fig. 2. (**a**) 2D Gel analysis of proteins from a whole-cell extract cultured rat fibroblast. The gel is stained with Cy5 dye and visualized by florescence. (**b**) Pseudo-2D gel graphic showing the calculated position of vimentin and phosphorylated forms of vimentin (P31000). (**c**) Composite of actual and pseudo 2D gel.

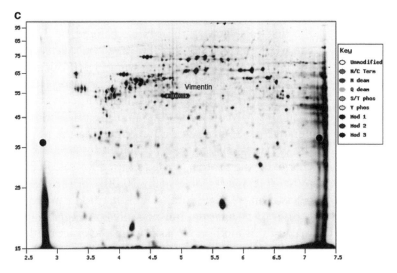

Fig. 2. (continued)

the *Mascot* program *(8)* **Fig. 2a** shows the stained image of the 2D gel. **Figure 2b** shows the ProMoST produced graphic showing the calculated relative position of the unmodified and phosphorylated vimentin. **Figure 2c** shows the composite of stained image of the 2D gel with the calculated positions of unmodified and phosphorylated vimentin indicated. MALDI analysis of the spots confirmed their identification as containing vimentin.

1.3. Calculation of pI Values

The charge state of the protein at a given pH is the sum of the negative and positive charges on the charged residues and the C- and N-terminal residues of the protein. To determine the p*I* value for the protein, the pH value at which the charge state of the protein is equal to zero must be found. There are two basic approaches to calculating the p*I* value for a protein. One method is to construct a model of the charge state of protein as a series of differential equations and then solve the equations for the condition of zero net charge. While this method provides an exact determination of the p*I* value, it can be computationally expensive and an exact determination is not required for practical work.

A second approach is to determine the pH value at which the charge on the protein is neutral to within a small tolerance by successive approximations. In this method, a starting pH is chosen, usually pH 7, and the charge on the protein is calculated based on the pK_a values for each of the charged residues and the N- and C-terminal amino acids of the protein. If the net charge on the protein at a pH of 7 is determined to be positive, the charge calculation is repeated with an increased pH value. If the net charge at a pH of 7 is determined to be negative, the charge calculation is repeated with a decreased pH value. After the first calculations, the pH is changed by 3.5 units, bringing the pH value to 3.5 or 10.5, and the calculation repeated. If the charge polarity is same as the pH 7.0 calculation, the pH is changed by an additional 3.5 units, bringing the pH to 0 or 14, and repeated. If the polarity of the charge switches, then the pH change is halved (1.75 units) and the calculation repeated. This iterative process of calculation, changing the pH value by half of the previous change, and recalculation is repeated until the net charge on the protein at the pH used for calculation is less than a preset tolerance value, usually ±0.002, or the change in pH value is less than ±0.01 pH units. This method can require at most 12 rounds of calculation, but typically converges in 6 or less rounds, making this method far faster than the exact method while yielding an answer of sufficient precision for practical work.

1.4. ProMoST Algorithm

The ProMoST application is based on the successive approximation method for calculating protein pI values. The major difference is that in addition to the standard acidic amino acid residues (glutamic acid and aspartic acid), ProMoST also considers the pK_a values of cysteine and tyrosine when calculating unmodified protein pI values. To calculate the pI values for modified proteins, the number and pK_a values for the modifications is included in the calculation. For modifications such as phosphorylation in which there are multiple charge states, the pK_a values for all charge states are also included in the calculation. In some cases, it is necessary to remove from the calculation the pK_a values for the unmodified amino acid. For example, in the case of phosphorylation of tyrosine, for each phosphate group added to the calculation, a tyrosine group is removed, since the OH group on tyrosine is both the position of the charge and the site of phosphorylation.

The first step in the analysis is the determination of the amino acid composition of the protein. The molecular mass of the protein is calculated be summing the high precision mono or average isotopic masses of its amino acids and adding the mono or average isotopic mass of one water molecule, corresponding to an H at the N-terminal end and an OH group at the C-terminal end of the molecule.

The charge on the protein at a particular pH value is calculated using a method developed by Tabb *(9)*. This approach works by determining the sum of the partial charges for all the charged amino acids and modifications using the standard equations:

Positive ions:

$$CR_i = 10^{pKa - pH}.$$

Negative ions:

$$CR_i = 10^{pH - pKa}.$$

The partial charge contribution, P_{Ci}, of any species to the entire protein is equal to:

$$P_{Ci} = n \frac{CR}{CR_i + 1},$$

where n is the number of that particular amino acid or modification. The total charge on the protein is the sum of the partial charges:

$$C_T = \sum_{i=1...n} P_{Ci}$$

The pI is defined as the pH value at which all charges on the protein are balanced and the net charge is zero. To determine that pH value, an initial value of pH 7 is tested and the net charge on the protein calculated. Depending on the sign of the charge on the protein, ΔpH value of 3.5 is added or subtracted from the initial value of 7 and the charge on the protein recalculated. The process of dividing the ΔpH value in half and the changing sign is reiterated until a net protein charge of less than 0.002 is obtained. This "binary search" method rapidly converges on an accurate value for the protein pI.

2. Materials

The ProMoST Web service provides an interface to a PERL-based cgi program that calculates protein molecular mass and pI values. The interface allows the user to choose the standard pK values for charged amino acids and modifications or to alter the values. The program takes protein accession numbers, names, or sequence as input and produces tables of values for modified and unmodified proteins. It also has a graphic output of a theoretical 2D gel.

2.1. Requirements to Access ProMoST Web Application

ProMoST is a Web-based application. Currently it is freely available at either http://proteomics.mcw.edu or http://halligan.us/promost.html. It has been tested with most modern Web browsers and is compatible with Microsoft Internet Explorer versions 6 and 7, Safari versions 2 and 3, Firefox version 2, SeaMonkey version 1.1, and Opera version 9 running on the Nintendo Wii console.

2.2. Requirements to Host ProMoST Web Application

If confidentiality and control of protein sequences is required or especially heavy use is anticipated, an organization may wish to host a local copy of ProMoST. Upon request, ProMoST is distributed as a Perl cgi program and has been tested with the Apache Web server. It depends on the CGI, Fcntl, and

Spreadsheet::WriteExcel CPAN perl modules as well as GD.pm and GD libraries for generating the graphic output.

3. Methods

The ProMoST interface has been designed to allow for rapid use by occasional users while still meeting the demands of more advanced users. To do this, two versions of the interface to the program have been designed. The default interface is the basic interface that allows the user to submit either protein sequence data or accession numbers and use predefined modifications to generate tables and gel graphics. The advanced interface additionally allows the user to define additional modifications and alter the standard pK_a values used in the calculation.

3.1. Basic Interface

Figure 3a shows the default or "simple" interface to ProMoST. A Web interface is used to get protein information from the user. The user has a choice of either entering the protein information in a text box or uploading a file. The protein information can consist of a list of accession numbers or protein names, or protein sequences in FASTA format *(10)*. The program dynamically determines the format of the input protein data. The accession numbers or protein names are used by the program to obtain the sequence data from a local copy of the NCBI nr protein database.

In addition to the normal charged amino acids, values for the common protein modifications (deamidation and phosphorylation) are also included. The user is able to specify the number of each modification that is to be considered. Thus, it is possible to examine the effects of a single phosphotyrosine or a series of up to ten phosphotyrosines on the same protein molecule. The user can also choose to block either the N-terminal, C-terminal, or both ends of the protein.

3.2. Advanced Interface

In addition to the standard interface for ProMoST, there is also an "advanced" interface that allows for more values to be customized **(Fig. 3b)**. The standard pK values for the charged amino acids (internal, C-terminal, and N-terminal) are presented by the Web interface as a series of text boxes. The user can thereby examine and change any of the default pK values and also has the ability to exclude any of the charged amino acids from the pI calculation, as would be required if the residue were modified to an uncharged state.

3.3. Defining New PTMs

To extend the ability of ProMoST to calculate the pI of modified proteins, the Web interfaces allows the user to add the name and

a

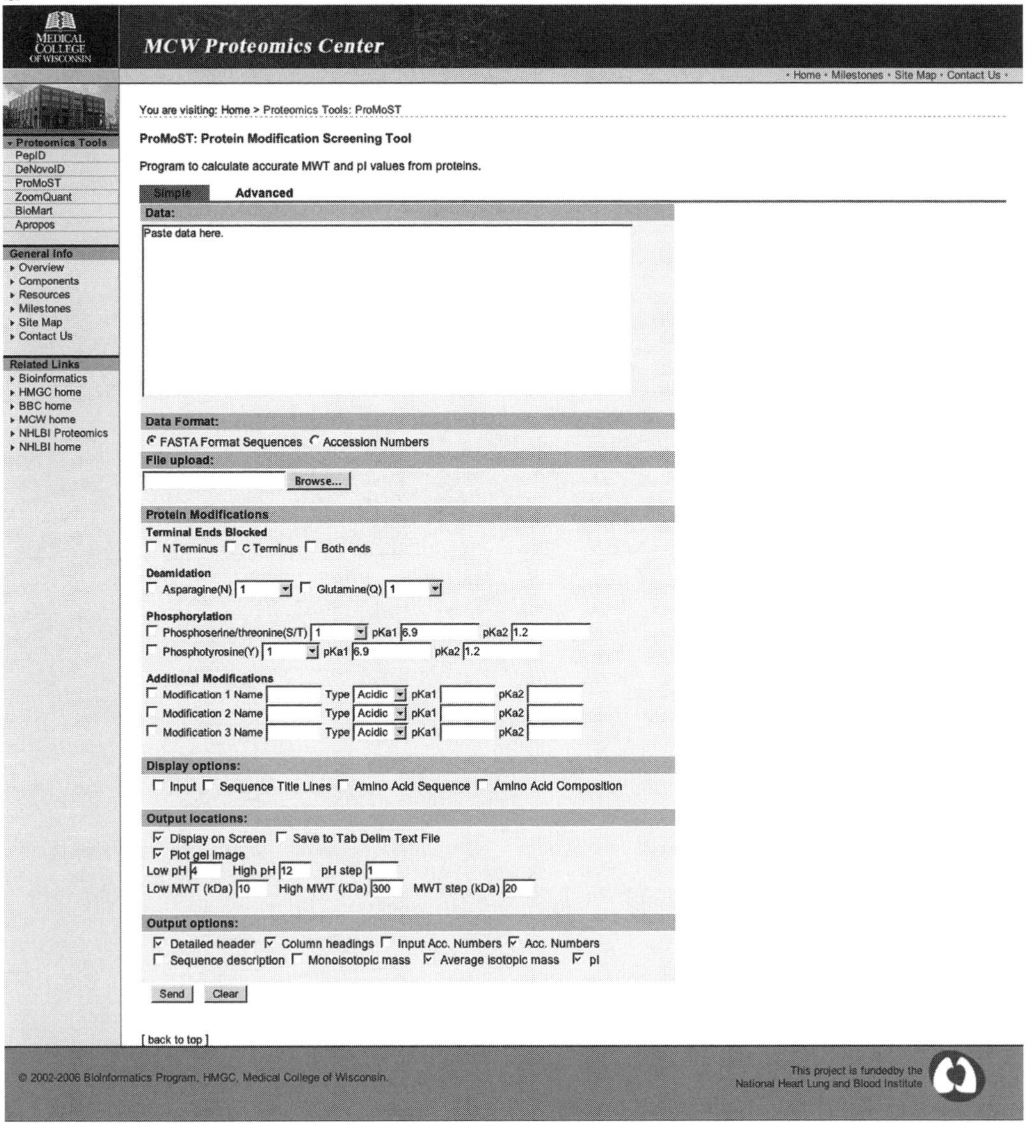

Fig. 3. (**a**) ProMoST standard interface. (**b**) ProMoST advanced interface.

pK values for up to three additional user-defined protein modifications. For each of the modifications, the user specifies a label to be used in the text output of the program. The user also indicates if the modification will produce a negative or positive charge and up to two pK_a values.

3.4. Input Options

The standard interface allows for the input of either sequence information in FASTA format or as accession numbers. This data can either be submitted in a text box on the Web form or uploaded as a file. In addition to these standard input options,

Fig. 3. (continued)

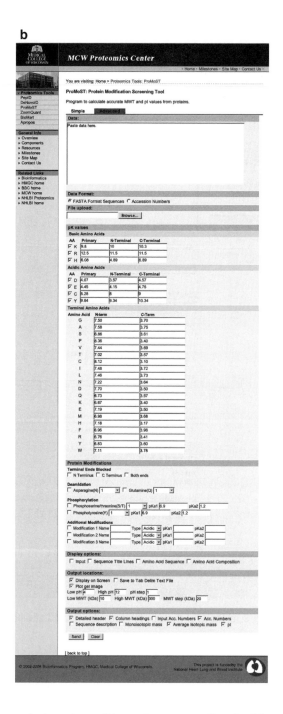

there are several extended options. Lines that are prefaced with a number sign (#) are treated as comments and ignored by Pro-MoST. This allows text files containing either FASTA sequences or accession numbers to be annotated. FASTA sequence header lines or accession numbers prefaced with a dollar sign ($) indicate sequences for which post-translational modifications should not be calculated or displayed. This is useful if the user wishes to

Fig. 4. Input text file for ProMoST analysis. *Lines* beginning with # are considered as comments and are ignored by ProMoST. The $ preceding the accession numbers for major serum proteins indicates that modified forms of these proteins should not be calculated or displayed.

```
#serum response factor
P11831
#albumin   ALBU_RAT
$P02770
#IgG
#antitrypsin     A1AT_RAT
$P17475
#IgA
#transferrin    TRFE_RAT
$P12346
#haptoglobin    HPT_RAT
$P06866
#fibrinogen     FIBA_RAT
$P06399
#fibrinogen     FIBB_RAT
$P14480
#fibrinogen     FIBG_RAT
$P02680
#alpha2-macroglobulin     A2MG_RAT
$P06238
#alpha1-acid glycoproteinA1AG_RAT
$P02764
#IgM
#apolipoprotein Al   APOA1_RAT
$P04639
#apolipoprotein All APOA2_RAT
$P04638
#complement C3 CO3_RAT
$P01026
#transthyretin TTHY_RAT
$P02767
```

examine the mobility of a protein in the context of other high abundance proteins that are normally present in the sample. An example of this is shown in **Figs. 4** and **5**. **Figure 4** shows an input text file and **Fig. 5** shows the resulting ProMoST output. The goal of this demonstration is to show the phosphorylation of the α1-acid glycoprotein, an acute phase serum protein, in the context of other serum proteins **Note 2**.

3.5. Output Options

The output of the program is divided into two sections: the input data and the calculated results. The user can opt to have the input data displayed in the form of the actual input accession number/protein name, the deduced accession number, the sequence read from the database, or the composition of the protein. Any or all of these options can be active at the same time.

There are three main output modes, all of which can be used at the same time. Data can be displayed to the screen, or it can be either saved or displayed or saved as a text file or Excel format file. The screen display takes the form of a HTML table. The user has the option to choose from different columns of data. The molecular mass choices include the monoisotopic mass, the average isotopic mass, both or neither calculated molecular mass. The protein information can be displayed as the input accession numbers, the deduced accession numbers, or sequence description. The calculated p*I* is optional. The table also shows which

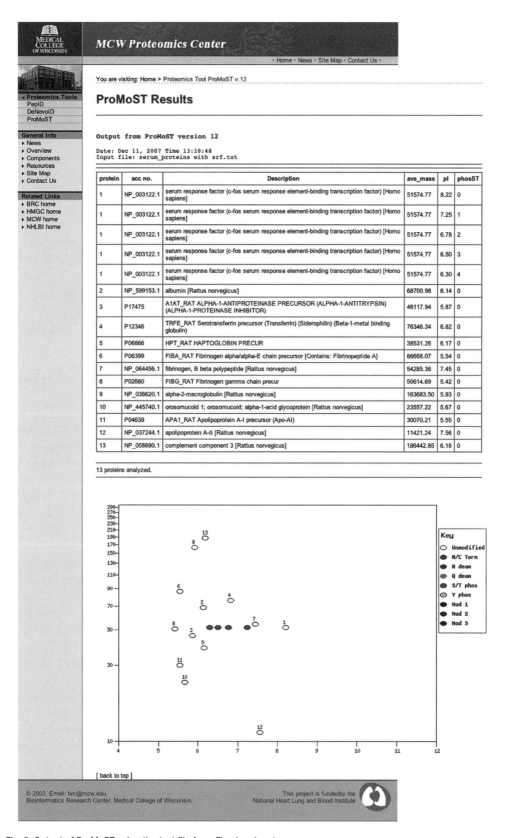

Fig. 5. Output of ProMoST using the text file from Fig. 4 as input.

modifications are active for each line in the table. An example of the output of ProMoST is shown in **Fig. 5**.

Data can also be sent to either a tab delimited text file or to an Excel format file. The files can be either viewed on the screen with the browser (text files) or with Excel (Excel files). By using the browser "Save link as" option, the user can directly save the text or Excel file to their computer.

A graphic gel image output is also available. The user can specify the molecular mass and p*I* range of the gel as well as the gel size. Proteins are plotted to the gel as ovals at the location of their calculated molecular mass and p*I*. The ovals are color coded for the modification. The parent, unmodified protein is plotted as an open oval and in the case of multiple proteins on the same plot, is labeled with a protein index number that matches the table or file of values.

4. Notes

1. *Other uses of ProMoST*. Although ProMoST was primarily designed to calculate the mass and p*I* values for post-translational modifications and map them to "theoretical" 2D gels, it can also be used to predict the mass, p*I*, and mobility of mutant and variant forms of proteins. Using the "FASTA sequence" option, both the original and variant sequences can be entered and analyzed. This allows for the comparison of mutant, variant, or processed forms of a protein.

   ProMoST can also be used to display the results from LC-MS/MS experiments in a graphic form. The *Visualize* program, also developed by the Medical College of Wisconsin NHBL Proteomics Center, allows for the analysis of results of LC-MS/MS experiments. As one of its output options, it can create files of accession numbers that can be directly imported into ProMoST, and the proteins identified by LC-MS/MS can be visualized as a pseudo-2D gel.

2. *Troubleshooting and limitations*. One of the most common errors encountered by ProMoST users is the failure of ProMoST to recognize their input sequence data. The usual cause of this problem is that the sequence information is improperly formatted. ProMoST requires that sequences be submitted in FASTA format. The key component to FASTA sequence format is that each sequence must begin with a header line, which is designated by a greater than symbol (>) at the beginning of the line. If a properly constructed header line is not present, then ProMoST fails to recognize the sequence.

It is important to remember that the pI values calculated by Pro-MoST are theoretical and approximate. While these values are useful for approximating the migration of proteins on 2D gels, they are not meant to be accurate for non-denatured proteins. In native proteins, it is possible that some of the potentially charged residues are not solvent accessible and therefore may not contribute to protein's overall charge. Furthermore, microenvironments within the protein may allow amino acids to interact and influence the pK_a of individual amino acid residues.

Acknowledgments

The 2D gel image of rat fibroblast proteins was graciously provided by I. Matus. This work was supported in part by the NHLBI Proteomics Center contract NIH-N01 HV-28182.

References

1. Gorg, A., Weiss, W., and Dunn, M. J. (2004) Current two-dimensional electrophoresis technology for proteomics. Proteomics 4, 3665–85.

2. Robinson, N. E., and Robinson, A. B. (2001) Deamidation of human proteins. Proc Natl Acad Sci U S A 98, 12409–13.

3. Halligan, B. D., Ruotti, V., Jin, W., Laffoon, S., Twigger, S. N., and Dratz, E. A. (2004) Pro-MoST (Protein Modification Screening Tool): a Web-based tool for mapping protein modifications on two-dimensional gels. Nucleic Acids Res 32, W638–44.

4. Cantor, C. R., and Schimmel, P. R. (1980) Biophysical chemistry, W. H. Freeman, San Francisco.

5. Fevold, H. L. (1951) in Amino acids and proteins; theory, methods, application (Greenberg, D. M., Ed.), ix, 950 p., Thomas, Springfield, IL.

6. Dunn, M. J. (1987) Two-dimensional gel electrophoresis of proteins. J Chromatogr 418, 145–85.

7. Freed, J. K., Smith, J. R., Li, P., and Greene, A. S. (2007) Isolation of signal transduction complexes using biotin and crosslinking methodologies. Proteomics 7, 2371–4.

8. Perkins, D. N., Pappin, D. J., Creasy, D. M., and Cottrell, J. S. (1999) Probability-based protein identification by searching sequence databases using mass spectrometry data. Electrophoresis 20, 3551–67.

9. Tabb, D. L. (2001). http://fields.scripps.edu/DTASelect/20010710-pI-Algorithm.pdf

10. Lipman, D. J., and Pearson, W. R. (1985) Rapid and sensitive protein similarity searches. Science 227, 1435–41.

Chapter 22

Kinase-Specific Prediction of Protein Phosphorylation Sites

Martin L. Miller and Nikolaj Blom

Summary

As extensive mass spectrometry-based mapping of the phosphoproteome progresses, computational analysis of phosphorylation-dependent signaling becomes increasingly important. The linear sequence motifs that surround phosphorylated residues have successfully been used to characterize kinase–substrate specificity. Here, we briefly describe the available resources for predicting kinase-specific phosphorylation from sequence properties. We address the strengths and weaknesses of these resources, which are based on methods ranging from simple consensus patterns to more advanced machine-learning algorithms. Furthermore, a protocol for the use of the artificial neural network based predictors, NetPhos and NetPhosK, is provided. Finally, we point to possible developments with the intention of providing the community with improved and additional phosphorylation predictors for large-scale modeling of cellular signaling networks.

Key words: Bioinformatics, Prediction, Phosphorylation, Kinase, Linear motifs, NetPhos, NetPhosK, Artificial neural networks.

1. Introduction

The human genome encodes around 520 protein kinases, many of which have profound effects on cellular life and death (1). Therefore, the activity of these phosphate-adding enzymes and their substrate specificity are often tightly regulated (2). One aspect of substrate specificity is mediated through the ability of kinases to recognize "linear sequence motifs" in their substrates (3). Linear motifs are short, unstructured recognition patches with conserved residues (4) that specify interaction with larger domains, such as kinases. Crystallization and statistical studies indicate that kinases recognize a region of around ten residues

Marjo de Graauw (ed.), *Phospho-Proteomics, Methods and Protocols, vol. 527*
© 2009 Humana Press, a part of Springer Science + Business Media, New York, NY
Book DOI: 10.1007/978-1-60327-834-8_22

surrounding the acceptor site *(5, 6)*. Recent methodological developments in mass spectrometry (MS)-based proteomics have made it possible to identify thousands of in vivo phosphorylation sites in a single study *(7, 8)*. However, many of these phosphorylation sites are uncharacterized with respect to signaling context, since the kinases responsible for the phosphorylation events are largely unknown *(9)*. As the extensive mapping of the phosphoproteome proceeds, it becomes increasingly important to utilize and improve in silico methods to predict linear motifs that direct kinase-specific phosphorylation. Moreover, prediction methods have shown to be sufficiently accurate for designing bioinformatics-targeted experiments *(10–12)*.

Numerous methods are available for the prediction of kinase-specific phosphorylation, ranging from simple consensus patterns to more advanced machine-learning algorithms (*see* **Table 1** for overview). Methods based on consensus patterns (ELM *(13)*, PROSITE *(15)*, and HPRD *(14, 16)*) are compiled from the literature and rely on the presence of an exact motif flanking the phosphorylated residue; for example, the motif [RK][RK]..[pS/pT] is commonly used to search for PKA substrates. This motif matches an Arg or Lys residue at positions –4 and –3 N-terminally from the phosphorylated Ser or Thr, while positions –2 and –1 can be any amino acid. Often the patterns are based on very limited data, and accordingly, the sensitivity of phosphorylation site detection is significantly lower compared with more advanced methods *(25)*. Moreover, because of the binary decision output (Yes/No), consensus patterns are not suited for large-scale analysis that requires probabilistic scoring schemes.

The pioneering work on profiling kinases for their substrate-recognition motifs was based on in vitro experiments, such as degenerate peptide library screening *(17, 18)*. On the basis of such profiling, the Scansite repository of position-specific scoring matrices (PSSMs) was generated. The Scansite resource provided the first opportunity to predict ab initio kinase-specific phosphorylation sites and rank these according to the probability of being phosphorylated. A PSSM is a sequence profile that represents the preference of particular amino acids surrounding the residue of interest and is thus a linear model where individual positions are represented independently.

As more in vivo data of kinase-specific phosphorylation sites have become available and stored in public available databases such as Phospho.ELM *(26)*, methods based on such data have been developed. These include simpler sequence similarity based clustering methods (GPS *(19, 20)*) and more advanced machine-learning methods, including artificial neural networks (ANN) (NetPhosK *(10, 22)*), hidden Markov models (HMM) (KinasePhos *(21)*), Bayesian decision theory (BDT) (PPSP *(23)*), and support vector machines (SVM) (PredPhospho *(24)*). A com-

Table 1
Overview of resources for kinase-specific phosphorylation site predictors

Resource	URL	Method	Training data	Homology reduction	Bench-mark	Batch prediction
ELM (13, 14)	http://elm.eu.org/	Consensus	Literature mined	n.a.	No	No
PROSITE (15)	http://www.expasy.org/prosite/	Consensus	Literature mined	n.a.	No	Yes
HPRD (16)	http://www.hprd.org/	Consensus	Literature mined	n.a.	No	No
Scansite (17, 18)	http://scansite.mit.edu/	PSSM	In vitro peptide libraries	n.a.	n.a.*	Yes
GPS (19, 20)	http://bioinformatics.lcd-ustc.org/gps_web/	MCL	Phospho. ELM	No	Partial	Yes
KinasePhos (21)	http://kinasephos.mbc.nctu.edu.tw/	HMM	Phospho-Base; SwissProt	No	No	Yes
NetPhosK (10, 22)	http://www.cbs.dtu.dk/services/NetPhosK	ANN	PhosphoBase	Yes	Yes	Yes
PPSP (23)	http://bioinformatics.lcd-ustc.org/PPSP/	BDT	Phospho. ELM	No	Yes	Yes
PredPhospho (24)	http://pred.ngri.re.kr/PredPhospho.htm	SVM	Phospho-Base; SwissProt	Yes	Yes	No

The "Method" and the "Training data" columns state the type of algorithm and the training data source used for developing the resource, respectively. "Homology reduction" states if similar sequences were eliminated to avoid data biases and overfitting. The "Benchmark" column states if the performance of the method was compared to that of similar methods. "Batch prediction" indicates if the method allows for input of multiple sequences for large-scale analysis

Abbreviations n.a. not applicable; *ANN* artificial neural networks; *BDT* Bayesian decision theory; *HMM* hidden Markov models; *PSSM* position-specific scoring matrices; *MCL* Markov cluster algorithm; *SVM* support vector machines

*At the publication time, Scansite was the first kinase-specific predictor and could consequently not be benchmarked to other methods

mon feature of machine-learning methods is the ability to learn general signatures in the data by adjusting the free parameters (weights) during training on a subset of the data. In contrast to PSSMs (e.g., Scansite) and sequence similarity based clustering methods (GPS), the nonlinear machine-learning methods can account for correlations between residues at different positions and are thus able to capture complex recognition signatures *(27)*. Finally, methods using structural information of the contact residues in the kinases, a basis for substrate prediction, have also been developed (Predikin *(28)*).

Considerable amount of caution is advised when choosing among the available methods for phosphorylation prediction. Special focus should be paid on the type of method; the data used to develop the method; how the data was handled; and how well the method performs (*see* **Table 1** for overview). In our view, machine-learning methods trained on phosphorylation sites identified in vivo are preferred over the simpler consensus pattern- and PSSM-based methods. First, machine-learning methods can capture the interdependencies between residues that often guide kinase–substrate recognition *(22, 29)* that PSSMs are unable to. Second, PSSMs and consensus patterns are often based on in vitro experiments and thus have a risk of describing motifs that are not observed in vivo. However, for the vast majority of the ~520 kinases *(1)*, advanced methods cannot be developed because of lack of data, and here, the in vitro models such as Scansite contribute greatly to increased coverage.

A key element in developing advanced prediction methods is the data quality and handling. Machine-learning methods with many free parameters have an inherent risk of overfitting as a consequence of using data with high sequence similarity for both training and testing. Such overfitted models are susceptible to noise in the data and learn to recognize the individual sites by heart rather than learning to characterize a general signature in the data. During development of several methods, care has not been taken to reduce the homology of training and testing data (GPS, KinasePhos, and PPSP), and the resources consequently risk reporting performance overestimates. Furthermore, a proper comparison (benchmark) to other available methods should always be made to estimate the scientific advancement of the method. To make an unbiased comparison, the benchmark should be performed only on the phosphorylation sites that are dissimilar in sequence to those used for developing the competing method. Finally, the prediction services should be built for batch predictions and the services should be continuously maintained until a better performing method has been made by the investigators.

To enhance prediction accuracy within kinase-specific prediction of phosphorylation, there are several areas where there is room for improvement. To avoid the aforementioned overfitting for machine-learning methods, homology reduction should be performed on the level of individual sites, i.e., on a window surrounding the phosphosites, as well as on the full-length of the proteins. So far, homology reduction has only been performed on full-length of the phosphoproteins. An external dataset, which is not used in setting the criteria for termination of training, should be used during the cross-validation procedure (*see* for example, **ref.** *30*). This has not been part of the standard operating procedure so far, since a test set has both been used to decide when to stop training and to report the performance of the method.

Assigning negative data for training often imposes a considerable challenge, since such information is often not available. So far, nonphosphorylated residues in the phosphoproteins have been assigned as negatives. Instead, by mimicking the physiological situation, the positives from all other kinases can be used as negatives. Rather than predicting de novo phosphorylation sites of a particular kinase, which is currently the case, the developed methods will instead be trained to classify which of the kinases phosphorylated the given phosphorylation site. Classification rather than de novo prediction is important as more and more of the query data for in silico prediction are modification-sites mapped by MS-based proteomics.

Benchmark of the methods has so far been performed using threshold-dependent performance measures such as specificity, sensitivity, and Matthews correlation coefficient. To avoid setting an arbitrary threshold, benchmark should instead be based on threshold-independent measures such as area under the receiver operating characteristics (AROC) curve. Furthermore, it should be tested if improvements are statistically significant or not. Recently, it has become evident that prediction accuracy can be increased by using protein–protein association context filters (*9*). Studies like this show the future direction in integrating post-translational modification motif prediction with protein–protein interaction to model signaling pathways. Finally, there is a need to establish a global repository of prediction tools for linear motifs that combines different resources in order to increase coverage.

2. Materials

2.1. Dataset

A single or multiple protein sequence(s) in FASTA format, i.e., each sequence should be preceded with a line beginning with a ">" sign followed by the name of the sequence, is a dataset. Nonstandard amino acids, except the wildcard X, spaces, and line breaks, in the sequences will be ignored. Kinase-specific phosphorylation sites will be prediction on Ser (S), Thr (T) and Tyr (Y) in the sequence(s). The sequence(s) should be at least nine residues long, i.e., the site of interest plus four residues on either side.

2.2. Programs

This protocol is based on the following publicly available services: NetPhos (http://www.cbs.dtu.dk/services/NetPhos/), NetPhosK (http://www.cbs.dtu.dk/services/NetPhosK/), and NetPhosYeast (http://www.cbs.dtu.dk/services/NetPhos-Yeast/). The query sequences are kept confidential and will be deleted after processing; however, if the user prefers to keep their sequences in-house, most of the CBS programs are also available as stand-alone program packages for local use. If other sources for kinase-specific phosphorylation prediction are desired, please *see* **Table 1** for links and for an overview of the methods (*see* **Note 1**).

3. Methods

3.1. Data Upload

If the query sequence data is of mammalian origin and generic (nonkinase-specific) phosphorylation prediction is wanted, please use the NetPhos method. For kinase-specific predictions please use NetPhosK. If the starting data is of yeast origin, NetPhos-Yeast, should be used, since it is specifically trained on phosphorylation sites identified in yeast. The FASTA file can be pasted directly into the sequence input field or uploaded from your local drive **(Fig. 1)**. To illustrate the use of the predictor, we here take the human p53 tumor suppressor protein as an example and paste the human p53 FASTA file into the NetPhosK prediction server input field.

3.2. Configuration

For the NetPhos and NetPhosYeast methods, it is possible to include a graphical representation (default) of the prediction output. Furthermore, in the NetPhos configuration, it can be chosen to predict on any of the three phosphoacceptor residues, S, T, or Y (default all). When customizing prediction with NetPhosK, the user can define a cutoff for the output scores (default 0.5). By default, the NetPhosK server uses prediction without any filtering. You can invoke a so-called evolutionary stable sites (ESS) filter by checking the ESS button. By default, the NetPhosK server does not produce graphical output. You can enforce graphics by checking the "Kinase Landscapes" button (*see* **Note 2**).

3.3. Run Programs

Click "Submit" to initiate prediction. The status of your job (either "queued" or "running") will be displayed and constantly updated until it terminates and the server output appears in the browser window.

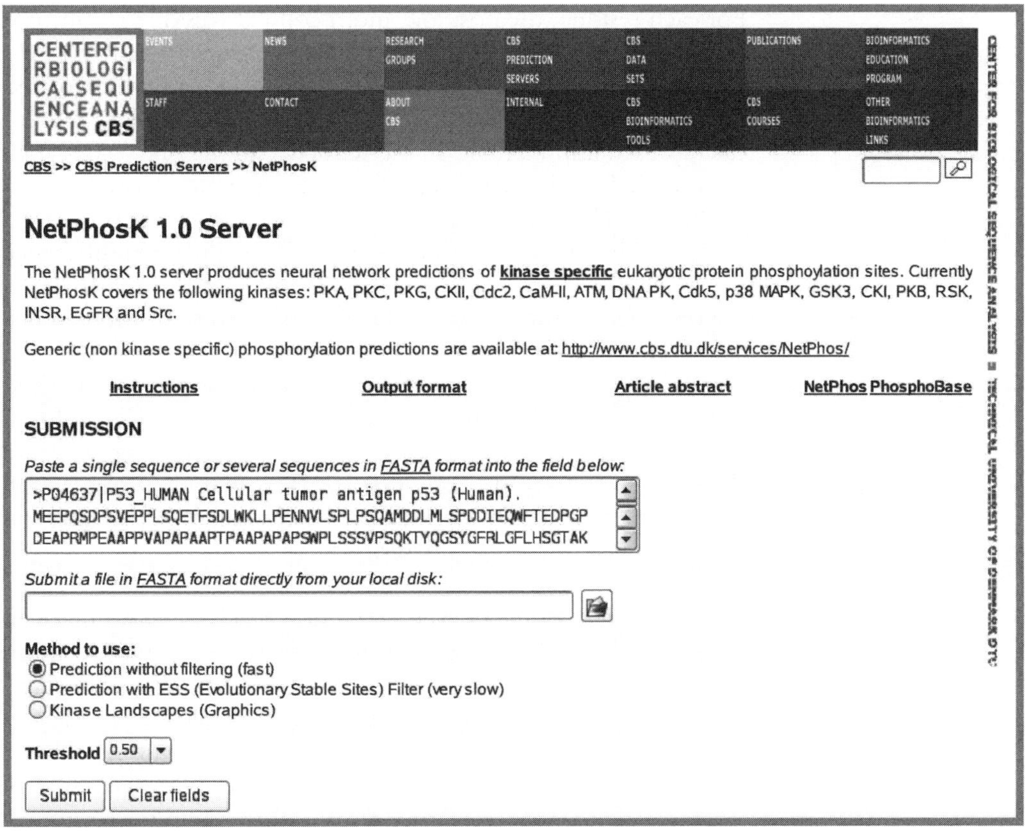

Fig. 1. A screenshot of the NetPhosK server.

3.4. Interpreting the Output

3.4.1. NetPhos and Net-PhosYeast

The ANN-based methods NetPhos and NetPhosYeast were both trained on general phosphorylation site data, where the kinase identity was unknown. Both methods can be used to predict phosphorylation of Ser and Thr residues, whereas NetPhos can also predict on tyrosine residues.

Although similar in output format, some differences occur. NetPhos provides an overview of the input sequence shown in blocks of 80 residues per line. Below are corresponding lines with predicted sites shown in capital letters (e.g., "S" for a predicted Ser phosphorylation site). Dots (".") indicate that this position is either occupied by nonphosphorylated residue or a Ser, Thr, or Tyr residue not predicted to be phosphorylated. Below, the total number of phosphorylation sites is reported for Ser, Thr, and Tyr.

The next section lists all relevant potential acceptor sites in the input sequence in three parts: Ser, Thr, and Tyr predictions.

For each residue, the position, motif context (residue ± four residues), score (0.00–1.00), and final prediction (e.g., "*S*" for positives) are shown. By default, any score above 0.50 is treated as a positive site (see **Note 3**). The final section shows a graphical representation of the input sequence along the *x*-axis with color-coded vertical bars indicating the prediction score for each of the Ser, Thr, and Tyr residues. The output format from NetPhosYeast is very similar to that of NetPhos, the main difference being that the individual scores for Ser and Thr sites are shown in the first section and are not divided into Ser and Thr subsections. The column labeled "Kinase" currently states the general term "main" (meaning a general kinase), indicating that the method is prepared for a kinase-specific future version. The column labeled "Answer" states a "Yes" if the predicted score is above the selected threshold (default: 0.50). Below the residue section are the input sequence overview and the graphical representation similar to that of NetPhos described earlier.

3.4.2. NetPhosK

For each input sequence, the prediction results are organized in columns with the predicted phosphorylation site and position, kinase, and score (see **Fig. 2a**) (see **Note 4**). The scores are activity outputs from the ANNs. Generally, a score above 0.5 indicates that the site is a target of a particular kinase and the higher the score the more likely it matches the general signature of substrates of that kinase (see **Note 5**). From the NetPhosK output, it is seen that ATM, DNAPK, CK-I, and CK-II are predicted to recognize several sites in the extreme N-terminus of the p53 query protein. In accordance with this, S9 and S15 are known substrates of ATM and DNAPK *(31, 32)* and T18 is a substrate of the CK1 kinase *(33)* as predicted. Whether the S6 and S9 are substrates of CK-II and ATM, respectively, needs to be verified experimentally; however, there is evidence that the latter is correct, since inhibiting ATM reduces phosphorylation of S9 *(32)* (see **Note 6**).

3.4.3. ESS Option

To increase the likelihood that a predicted kinase and phosphorylation site corresponds to the in vivo situation, the ESS feature was introduced in NetPhosK. ESS describes the notion that a predicted kinase-specific phosphorylation site is more likely to be correct if it is also found in the orthologous proteins of related species (mammals in the case of NetPhosK). The method works by retrieving orthologous proteins in the UniProt database and the corresponding peptides in these proteins are run through the NetPhosK prediction scheme (for details see *(22)*) (see **Note 7**). The output from an ESS run is a list of potential phosphoacceptor sites in the query protein. For each site, the position, the predicted kinase, the score (for the query sequence), the number of orthologous peptides scoring positive (>0.5), and the number

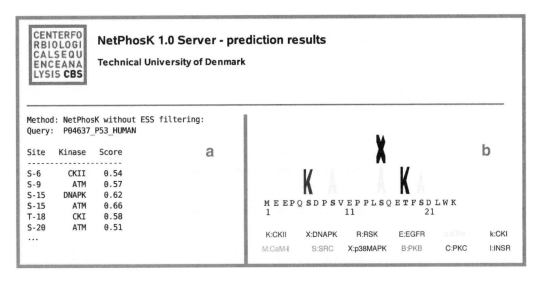

Method: NetPhosK without ESS filtering:
Query: P04637_P53_HUMAN

Site	Kinase	Score
S-6	CKII	0.54
S-9	ATM	0.57
S-15	DNAPK	0.62
S-15	ATM	0.66
T-18	CKI	0.58
S-20	ATM	0.51
...		

Fig. 2. Output of the NetPhosK server. (**a**) Table with output scores. (**b**) Graphical output of kinase landscapes. See text for details.

of orthologous peptides scoring negative are predicted for a particular kinase. Thus, the number of orthologues being processed is the sum of the two latter numbers.

For a given residue, there may be several kinases predicted. Also, the residue identity need not be conserved, as serine can be substituted by threonine in some of the orthologous peptides. It is now possible to compare the stability of the prediction across several orthologues, for example, for a given serine residue, 31 out of 34 orthologues may score positive for the kinase PKC, while only 21 out of 34 score positive for the cdc2 kinase. This may indicate that PKC is the relevant kinase for this particular phosphoacceptor site.

3.4.4. Kinase Landscapes

Another feature introduced in NetPhosK was the Kinase Landscapes (**Fig. 2b**). The idea is to present the predicted phosphorylation sites and corresponding kinases in a graphical fashion. This may enable the user to discover regions in the query sequence with an abundance of or unusual composition of predicted phosphorylation sites. In addition, this feature enables the user to compare the prediction output from related proteins, for example, with or without a point mutation or SNP (single nucleotide polymorphism). The SNP may be causing an amino acid substitution that may affect the acceptor site residue directly or some of the neighboring sites within the linear motif. In both cases, the resulting kinase landscape may be altered and thus indicate that the SNP in question may have causative effects on the signaling capacity of the protein in question (for details *see (22)*).

---

4. Notes

1. In general, several different prediction resources should be used if available, since averaging over more methods usually increases prediction reliability.

2. The repertoire of phosphorylation predictors at Center for Biological Sequence Analysis (CBS) are currently being expanded with more kinase-specific predictors and predictors for downstream signaling mediated by phosphorylation-dependent binding domains such as SH2 and PTB domains (manuscript submitted). Furthermore, we are developing organism-specific predictors and currently cover mammals and yeast. A predictor of bacterial phosphorylation sites is in preparation.

3. NetPhos and NetPhosK may tend to overpredict *(34)*, i.e., the specificity is rather low. This may be due to the random presence of motifs in the proteome. Such false-positive predictions may be minimized by combining predictions with various filters such as surface accessibility, tissue-specific expression, and colocalization of kinases and their substrates.

4. The existence of several methods for phosphorylation site prediction may be a source of confusion. For example, a protein of interest may be predicted to be phosphorylated at a given position when processed using the NetPhos prediction server. However, when the same protein is processed at, e.g., NetPhosK or Scansite, no kinase for the given site is predicted. This situation is caused by the fact that the older NetPhos method was trained on phosphorylation site data where the identity of the kinase was not known, whereas the latter two methods were trained only on kinase-specific sites. The dataset available for training of NetPhos was approximately tenfold larger than those used for training individual NetPhosK kinases, simply because kinase-specific data has always been very scarce compared with general data that a given site is phosphorylated or not. The next-generation phosphorylation site prediction methods will address this issue.

5. Outputs from in silico prediction methods should generally be taken will caution and ideally be verified experimentally. A minimal requirement is to perform a complementary database search to see if there is experimental evidence for the

prediction. Comprehensive data bases for phosphorylation sites are phospho.ELM *(26)* and HPRD *(16)*.

6. Performance of prediction methods for binary classification should always be estimated on the basis of a large benchmark data set consisting of both positive and negative sites. As described in the **Subheading 1**, preferably, threshold-independent measures such as AROC curve should be used.

7. The ESS option may be selected on the input page for Net-PhosK. Beware that this procedure is very time-consuming (usually several hours). To avoid loosing the results, the user is advised to input an e-mail address at the refresh screen, which appears after pressing the "Submit" button. This ensures that an e-mail will notify the user when the results are ready.

Acknowledgments

The authors would like to thank Rune Linding, Lars Juhl Jensen, Majbritt Hjerrild, Steen Gammeltoft, Thomas Sicheritz-Ponten and Søren Brunak.

References

1. Manning G, Whyte D, Martinez R, Hunter T, Sudarsanam S. (2002) The protein kinase complement of the human genome. *Science* 298, 1912–34.

2. Pawson T. (2002) Regulation and targets of receptor tyrosine kinases. *Eur J Cancer* **38**, S3–10.

3. Seet B, Dikic I, Zhou M, Pawson T. (2006) Reading protein modifications with inter-action domains. *Nat Rev Mol Cell Biol* 7, 473–83.

4. Bork P, Koonin E. (1996) Protein sequence motifs. *Curr Opin Struct Biol* 6, 366–76.

5. Songyang Z, Blechner S, Hoagland N, Hoekstra M, Piwnica-Worms H, Cantley L. (1994) Use of an oriented peptide library to determine the optimal substrates of protein kinases. *Curr Biol* 4, 973–82.

6. Kreegipuu A, Blom N, Brunak S, Jarv J. (1998) Statistical analysis of protein kinase specificity determinants. *FEBS Lett* 430, 45–50.

7. Beausoleil S, Jedrychowski M, Schwartz D, et al. (2004) Large-scale characterization of Hela cell nuclear phosphoproteins. *Proc Natl Acad Sci U S A* 101, 12130–5.

8. Olsen J, Blagoev B, Gnad F, et al. (2006) Global, in vivo, and site-specific phosphorylation dynamics in signaling networks. *Cell* 127, 635–48.

9. Linding R, Jensen L, Ostheimer G, et al. (2007) Systematic discovery of in vivo phosphorylation networks. *Cell* 129, 1415–26.

10. Hjerrild M, Stensballe A, Rasmussen T, et al. (2004) Identification of phosphorylation sites in protein kinase a substrates using artificial neural networks and mass spectrometry. *J Proteome Res* 3, 426–33.

11. Manning B, Tee A, Logsdon M, Blenis J, Cantley L. (2002) Identification of the tuberous sclerosis complex-2 tumor suppressor gene product tuber in as a target of the phosphoinositide 3-kinase/Akt pathway. *Mol Cell* 10, 151–62.

12. Miller M, Hanke S, Hinsby A, et al. (2008) Motif decomposition of the phosphotyrosine proteome reveals a new N-terminal binding motif for ship2. *Mol Cell Proteomics* 7, 181–92.

13. Puntervoll P, Linding R, Gemund C, et al. (2003) Elm server: a new resource for investigating short functional sites in modular

eukaryotic proteins. *Nucleic Acids Res* 31, 3625–30.

14. Amanchy R, Periaswamy B, Mathivanan S, Reddy R, Tattikota S, Pandey A. (2007) A curated compendium of phosphorylation motifs. *Nat Biotechnol* 25, 285–6.

15. Mulder N, Apweiler R, Attwood T, et al. (2003) The interpro database, 2003 brings increased coverage and new features. *Nucleic Acids Res* 31, 315–8.

16. Peri S, Navarro J, Amanchy R, et al. (2003) Development of human protein reference database as an initial platform for approaching systems biology in humans. *Genome Res* 13, 2363–71.

17. Yaffe M, Leparc G, Lai J, Obata T, Volinia S, Cantley L. (2001) A motif-based profile scanning approach for genome-wide prediction of signaling pathways. *Nat Biotechnol* 19, 348–53.

18. Obenauer J, Cantley L, Yaffe M. (2003) Scansite 2.0: proteome-wide prediction of cell signalling interactions using short sequence motifs. *Nucleic Acids Res* 31, 3635–41.

19. Zhou F, Xue Y, Chen G, Yao X. (2004) GPS: a novel group-based phosphorylation predicting and scoring method. *Biochem Biophys Res Commun* 325, 1443–8.

20. Xue Y, Zhou F, Zhu M, Ahmed K, Chen G, Yao X. (2005) GPS: a comprehensive www server for phosphorylation sites prediction. *Nucleic Acids Res* 33, W184–7.

21. Huang H, Lee T, Tzeng S, Horng J. (2005) KinasePhos: a web tool for identifying protein kinase-specific phosphorylation sites. *Nucleic Acids Res* 33, W226–9.

22. Blom N, Sicheritz-Ponten T, Gupta R, Gammeltoft S, Brunak S. (2004) Prediction of posttranslational glycosylation and phosphorylation of proteins from the amino acid sequence. *Proteomics* 4, 1633–49.

23. Xue Y, Li A, Wang L, Feng H, Yao X. (2006) PPSP: prediction of PK-specific phosphorylation site with Bayesian decision theory. *BMC Bioinformatics* 7, 163.

24. Kim J, Lee J, Oh B, Kimm K, Koh I. (2004) Prediction of phosphorylation sites using SVMs. *Bioinformatics* 20, 3179–84.

25. Blom N, Gammeltoft S, Brunak S. (1999) Sequence and structure-based prediction of eukaryotic protein phosphorylation sites. *J Mol Biol* 294, 1351–62.

26. Diella F, Cameron S, Gemund C, et al. (2004) Phospho.ELM: a database of experimentally verified phosphorylation sites in eukaryotic proteins. *BMC Bioinformatics* 5, 79.

27. Wu C. (1997) Artificial neural networks for molecular sequence analysis. *Comput Chem* 21, 237–56.

28. Brinkworth R, Breinl R, Kobe B. (2003) Structural basis and prediction of substrate specificity in protein serine/threonine kinases. *Proc Natl Acad Sci U S A* 100, 74–9.

29. Manke I, Nguyen A, Lim D, Stewart M, Elia A, Yaffe M. (2005) MAPKAP kinase-2 is a cell cycle checkpoint kinase that regulates the G2/M transition and S phase progression in response to UV irradiation. *Mol Cell* 17, 37–48.

30. Ingrell C, Miller M, Jensen O, Blom N. (2007) NetPhosYeast: prediction of protein phosphorylation sites in yeast. *Bioinformatics* 23, 895–7.

31. Araki R, Fukumura R, Fujimori A, et al. (1999) Enhanced phosphorylation of p53 serine 18 following DNA damage in DNA-dependent protein kinase catalytic subunit-deficient cells. *Cancer Res* 59, 3543–6.

32. Saito S, Goodarzi A, Higashimoto Y, et al. (2002) ATM mediates phosphorylation at multiple p53 sites, including ser(46), in response to ionizing radiation. *J Biol Chem* 277, 12491–4.

33. Dumaz N, Milne D, Meek D. (1999) Protein kinase CK1 is a p53-threonine 18 kinase which requires prior phosphorylation of serine 15. *FEBS Lett* 463, 312–6.

34. Kreegipuu A, Blom N, Brunak S. (1999) PhosphoBase, a database of phosphorylation sites: release 2.0. *Nucleic Acids Res* 27, 237–9.

Chapter 23

Reconstructing Regulatory Kinase Pathways from Phosphopeptide Data: A Bioinformatics Approach

Lawrence G. Puente, Robin E.C. Lee, and Lynn A. Megeney

Summary

Protein phosphorylation is a widespread cellular process, and simplistic linear pathway models of kinase signaling likely under-represent the complexity of *in vivo* pathways. The recent massive increase in information available through protein interaction databases now allows construction of *in silico* models of protein networks that are underpinned by evidence from real biological systems. By combining protein phosphorylation data with current databases of protein–protein and kinase–substrate interactions, sophisticated models of intracellular protein phosphorylation signaling can be constructed for a system of interest. The kinase interaction network can be visualized, analyzed by graph theory, and investigated for hypotheses that are not otherwise obvious.

Key words: Bioinformatics, Scale-free networks, Kinase signaling.

1. Introduction

Reversible protein phosphorylation is one of the most common means of regulating cellular processes. Genomic studies have identified 518 kinases in humans *(1)* and 540 in mouse *(2)*, and it is estimated that up to 30% of the proteome may be subject to phosphorylation *(3)*. These numbers have led to an appreciation that systems biology and bioinformatics approaches are required in order to fully understand the role of protein phosphorylation in cellular biology. Large-scale protein–protein interaction studies have clearly demonstrated that protein–protein interactions form complex networks. The subset of protein–protein interactions composed of kinases and their substrates can also be considered as a network *(4, 5)*. Such network models can aid in gaining a

Marjo de Graauw (ed.), *Phospho-Proteomics, Methods and Protocols, vol. 527*
© 2009 Humana Press, a part of Springer Science+Business Media, New York, NY
Book DOI: 10.1007/978-1-60327-834-8_23

thorough mechanistic understanding of the pathways, reveal unanticipated molecular interactions, and lead to novel hypotheses.

The most common network visualization technique is the connected ball and stick graph (**Fig. 1**), where a ball or *node* represents an individual protein and an interaction between a pair of proteins is depicted by a line or *edge* connecting the pair. In the context of kinases, the presence of a directed edge can implicitly define the kinase–substrate relationship, where the origin of the edge is the kinase and the arrow tip identifies the substrate; this is in contrast with an undirected edge, which can represent a physical/coactivation or undefined interaction.

To represent a network computationally, the interacting nodes can be recorded as pairs of node identification numbers (*see* **Table 1** for an example). For kinase networks, the identification numbers would correspond to specific proteins. Several standardized protein identification number systems exist, such as Swiss-Prot (http://www.expasy.org). The specific system employed is not important as long as the numbers are uniquely associated with a specific protein.

Once a network has been assembled, graph theory can be used to analyze its structure. Highly connected nodes or *hubs* are one point of interest, and may represent indispensable components of the signaling network. Experimental observations of protein phosphorylation can be mapped onto a network model of the system of interest to infer which kinase pathways are most

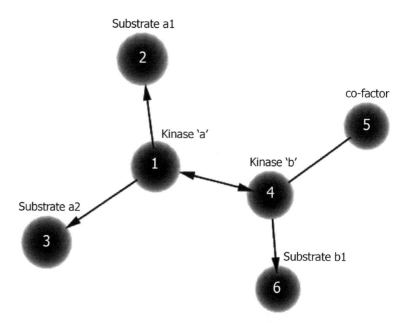

Fig. 1. A sample network. Kinase a phosphorylates substrates a1 and a2 and also phosphorylates kinase b. Kinase b phosphorylates kinase a and substrate b1 and has a nonsubstrate interaction with protein 5.

Table 1
Description of the network shown in Fig. 1

Kinase	Substrate
1	4
1	2
1	3
4	1
4	6

The table represents a directed network in which the *left side* indicates kinases and the *right side* indicates substrates. The data indicates that protein 1 phosphorylates proteins 2, 3, and 4 while protein 4 phosphorylates proteins 1 and 6. The order of the pairs implicitly indicates which proteins are kinases and which are substrates. An undirected network could contain additional pairs such as (4, 5) that do not have a kinase–substrate relationship. In an undirected network the order of the pairs does not matter so the interaction (4, 1) would automatically imply the interaction (1, 4)

likely to be active. The network will predict specific biochemical interactions that can be experimentally tested.

We previously demonstrated a general strategy for inferring networks of kinase–substrate relationships from phosphopeptide data *(5)*. This bioinformatics approach involves compiling phosphorylation site data, inferring potential kinase activities, predicting additional substrates, and visualizing and analyzing the network.

2. Materials

2.1. Bioinformatics Workstation

A basic bioinformatics workstation can be implemented on any modern personal computer system. Some specific requirements are as follows:

1. Current web browser with JavaScript support.

2. Java support. The Java runtime environment (JRE) is pre-installed on most personal computer systems and can be downloaded at no cost from http://www.java.com/en/download/manual.jsp.

3. A spreadsheet application such as Microsoft Excel (http://office.microsoft.com) or OpenOffice (http://www.openoffice.org) for basic data management.

4. High-speed Internet access.

3. Methods

By leveraging the information in bioinformatic databases of kinase–substrate interactions, protein phosphorylation data can be used to reconstruct a model network of kinase-mediated signaling in the system of interest. The approach involves several steps:

1. Acquire phosphopeptide data.

2. Identify known kinase(s) for each phosphorylated site.

3. Predict additional substrates for each kinase.

4. Extend the network model by identifying potential kinases for each known or predicted substrate.

5. Visualize the network.

6. Analyze the network and utilize the model to make testable hypotheses.

Each step is described in more detail in the following sections.

3.1. Acquire Phosphopeptide Data

Generally, proteins are purified from the system of interest, either with or without a phosphoprotein enrichment procedure. The recovered proteins are usually separated or fractionated using gel electrophoresis or liquid chromatography methods and each gel band or chromatograph fraction is proteolytically digested by trypsin. The resultant peptides may be further enriched for phosphopeptides using the methods described elsewhere in this volume. These peptides are then characterized using mass spectrometry and sites of phosphorylation are identified. Detailed methods for the acquisition of phosphopeptide data are beyond the scope of this chapter.

3.2. Map Kinase–Substrate Relationships

Several tools are available for mapping substrate-to-kinase relationships, some of which are described here.

KinaSource (http://www.kinasource.co.uk): The KinaSource web site includes a manually curated database (http://www.kinasource.co.uk/Database/welcomePage.php) of kinase–substrate pairs, emphasizing mouse and human proteins. The database can be searched by kinase or substrate. The resulting records show Swiss-Prot accession numbers, site(s) of phosphorylation where known, and corresponding literature references. Over 1,000 kinase–substrate pairs are described.

PhosphoSite (http://www.phosphosite.org): PhosphoSite is a large database of known phosphorylation sites on rodent and human proteins, maintained by the Cell Signaling Technology Corporation. The database can be searched by protein name or accession number and also by amino acid sequence, domain name, or contributing author. The resulting protein information pages include basic information on the protein and a list of the amino acid sequences known to be phosphorylated. Individual sites of phosphorylation are linked to detailed subpages that list the methods by which the phosphorylation was characterized, literature references, and in some cases extensive information on known upstream kinases and phosphatases as well as pharmacological or biochemical treatments known to modulate the phosphorylation. Phosphosite contains a great deal of additional information such as protein name synonyms, kinase family classifications, biological function annotations, and an isoelectric point calculator for phosphorylated proteins.

ScanSite (http://scansite.mit.edu): ScanSite *(6)* is a tool for predicting sites of protein interaction based upon the presence of amino acid sequence motifs. Many kinases phosphorylate specific target motifs, and ScanSite can search protein sequences for the presence of these motifs. ScanSite also finds motifs for other protein–protein interactions, such as potential SH3 binding sites. By default, ScanSite searches for a spectrum of common motifs, including many kinase–substrate motifs. User-defined sequence patterns or scoring matrices can also be employed. Batches of accession numbers or sequences may also be submitted for analysis (http://stjuderesearch.org/scansite/).

Phospho.ELM (http://phospho.elm.eu.org/): Phospho. ELM *(7)* is a database of several thousand phosphorylation sites from multiple species. It can be searched by kinase or substrate, and can also be searched for phosphorylation sites that modulate specific protein–protein interactions (e.g., SH2 domain binding sites). Academic users may request a copy of the Phospho.ELM database for their own research (http://phospho.elm.eu.org/dataset.html), or can write programs that interface with Phospho.ELM via Web Services (a WSDL file is available at http://phospho.elm.eu.org/webservice/phosphoELMdb.wsdl).

3.3. Predict Additional Substrates

To extend the network and generate hypotheses, add other known substrates for each kinase in the model. Based on the assumption that each of the identified kinases is potentially active, additional known substrates of each kinase can be predicted to be phosphorylated and this can be tested experimentally. Tools such as KinaSource and Phospho.ELM can be used to produce lists of known substrates for a given kinase. For human cells, it is estimated that there are an average of 20 substrates per kinase *(3)*, although certain kinases have hundreds of known substrates. For the purposes of hypothesis generation, it is sometimes helpful to ignore the promiscuous kinases in favor of analyzing more specific interactions.

3.4. Extend the Network

The model network at this stage includes observed substrates, predicted kinases, and predicted substrates. Treating *all* proteins in the model network as potential substrates, repeat the earlier process of mapping substrates to kinases. Many kinases are themselves substrates of other kinases and such kinase cascades are frequently of interest.

3.5. Network Visualization

Several tools are available for visualizing networks.

Cytoscape (http://www.cytoscape.org/): Cytoscape *(8)* is an open-source software platform specifically designed for visualizing molecular interaction networks. Cytoscape is the product of a public–private research partnership and is released under the LGPL (Lesser General Public License). Cytoscape is a Java application and can be used on any Windows, Macintosh or Linux system that supports Java.

LaNet-vi (http://xavier.informatics.indiana.edu/lanet-vi/): The LArge NETwork VIsualization tool is an online tool that constructs undirected network diagrams. To use LaNet-vi, prepare a text file representing the network as a simple list of number pairs representing nodes that interact with each other. Each pair of numbers should be separated by a space or tab, one pair per line. The input file is submitted through a web page form and the resulting network diagram is returned by e-mail. A Creative Commons license applies to the images.

Graphviz (http://www.graphviz.org/): Graphviz is an open-source graph visualization software package. It is a general-purpose package that can be customized through programming.

HubView (http://www.ogic.ca/projects/hubview/hubview.html): HubView *(4)* is a standalone Windows application specialized to the study of *Saccharomyces cerevisiae* kinase–substrate interaction networks. The application permits network visualization, clustering analysis, and the generation of user defined subnetworks. A Creative Commons license applies to use of the software.

3.6. Analysis of the Network

Once a network is assembled, graph theory can be used to analyze its structure. Topological features of the graph are studied by statistical analysis of the number of unique edges directed towards and away from each individual node. Highly connected nodes or "hub proteins" are commonly observed in eukaryotic protein interaction networks. The exact role of a hub protein can be ambiguous in the context of a network, and is best studied on a case-by-case basis but the hub-ness of a protein may offer general insight into its regulation or function. For example a hub-like kinase may provide a noise threshold by phosphorylating a large number of substrates in a promiscuous manner or the kinase itself may be temporally/spatially regulated causing a dramatic shift in substrate interactions. A hub-like substrate might act as a phospho-sink to stop a signal from being inappropriately propagated at sub-threshold levels or perhaps the substrate requires phosphorylation at multiple unique sites in order for its function to be fully regulated.

The presence of hubs is a characteristic of scale-free networks. Scale-free networks are a specific class of networks defined by topological parameters, which have been commonly identified in natural, synthetic, and biological systems. Mathematical modeling of evolutionary conditions that favor the formation of scale-free networks has supported the proposal of a growth and preferential attachment schema *(9)*. During the growth of the network in this environment, an already well-connected node has a greater likelihood of acquiring a new interaction partners than a less-connected node. It is through this growth process over an evolutionary time scale that hubs are thought to emerge. The kinase–substrate interaction network in yeast exhibits scale-free topology *(4)*, suggesting that mammalian kinase–substrate networks are also scale free.

Complex networks can be simplified by reducing them to a set of smaller subnetworks (subgraphs), using a process called clustering. Clustering is frequently employed in functional genomics to associate nodes into a subgraph based on a common property. The actual property (parameter) used to generate a cluster depends on the object of the study. In the case of protein interaction networks, parameters include (but are not limited to) presence in the same gene ontology annotation, similar patterns of interaction partners, or involvement in the same signaling pathways. Within the individual clusters local information is more apparent and hypotheses can be made about the contacts within the cluster or about the interactions that connect different clusters. For example, consider a cluster consisting of six proteins involved in the same signaling pathway that are known to be simultaneously activated. If four of the six proteins are observed to be active in a screen, then it is reasonable to hypothesize that the remaining two may also be activated and that the pathway itself has been stimulated. Through the process of clustering information about the network under study can be used to infer information about the individual nodes. For example, it has been observed that clustering based on interaction architecture strongly correlates with clustering based on functional category *(10)*.

4. Example

In a study of mouse myoblast differentiation *(5)*, 53 protein phosphorylation events were identified, and a sample of these are listed in **Fig. 2** (left column). Using the methods described in **Subheading 3**, each of these proteins was identified as being a substrate of one or more of the kinases listed in the center column

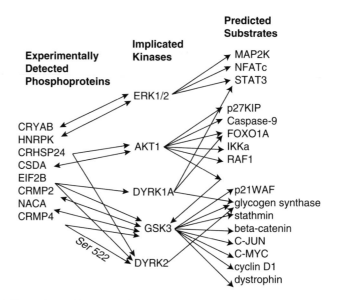

Fig. 2. Extrapolation of a network from phosphopeptide data. The proteins listed in the *left column* were experimentally determined to be phosphorylated. Using bioinformatic techniques, the kinases listed in the *center column* were proposed to be active and responsible for the observed phosphorylations. The proteins listed in the *right column* are additional predicted substrates of those kinases.

of **Fig. 2**. By repeated database searching, additional substrates were predicted, as shown on the right side of **Fig. 2**. To take a specific example, CRMP4 was found experimentally to be phosphorylated on serine 522 in differentiating myoblasts. Phosphorylation of this site has previously been ascribed to DYRK2 activity. Experimental observation of CRMP2 phosphorylation in undifferentiated cells implicated GSK3 activity. Bioinformatic searches and network modeling revealed that GSK3 can also phosphorylate CRMP4, on sites 509, 514, and 518. Interestingly, these GSK3-mediated phosphorylations are known to be dependent on phosphorylation at serine 522. Therefore, a very specific signaling module can be predicted. CRMP4 can have three phosphorylation states: unphosphorylated, completely phosphorylated on four sites, and singly phosphorylated on serine 522. The actual phosphoform present in the cell will depend on the relative activities of DYRK2 and GSK3 (the fully phosphorylated species will only be observed if GSK3 activity occurs after or simultaneously with DYRK2 activity and the singly phosphorylated species can only form if GSK3 is inactive, as occurs during differentiation). Overall, the model predicts that GSK3 and AKT activity should have a significant impact on myoblast differentiation due to their high degree of connectivity within the network (i.e., hubs). This prediction is in agreement with the known biochemistry of myoblasts, and a role in myoblast biology has already been found for many of the predicted substrates shown.

References

1. Manning, G., Whyte, D. B., Martinez, R., Hunter, T., and Sudarsanam, S. (2002) The protein kinase complement of the human genome. *Science* 298, 1912–34.

2. Caenepeel, S., Charydczak, G., Sudarsanam, S., Hunter, T., and Manning, G. (2004) The mouse kinome: discovery and comparative genomics of all mouse protein kinases. *Proc Natl Acad Sci U S A* 101, 11707–12.

3. Cohen, P. (2002) The origins of protein phosphorylation. *Nat Cell Biol* 4, E127–30.

4. Lee, R. E., and Megeney, L. A. (2005) The yeast kinome displays scale free topology with functional hub clusters. *BMC Bioinformatics* 6, 271.

5. Puente, L. G., Voisin, S., Lee, R. E., and Megeney, L. A. (2006) Reconstructing the regulatory kinase pathways of myogenesis from phosphopeptide data. *Mol Cell Proteomics* 5, 2244–51.

6. Obenauer, J. C., Cantley, L. C., and Yaffe, M. B. (2003) Scansite 2.0: proteome-wide prediction of cell signaling interactions using short sequence motifs. *Nucleic Acids Res* 31, 3635–41.

7. Diella, F., Cameron, S., Gemund, C., Linding, R., Via, A., Kuster, B., Sicheritz-Ponten, T., Blom, N., and Gibson, T. J. (2004) Phospho. ELM: a database of experimentally verified phosphorylation sites in eukaryotic proteins. *BMC Bioinformatics* 5, 79.

8. Shannon, P., Markiel, A., Ozier, O., Baliga, N. S., Wang, J. T., Ramage, D., Amin, N., Schwikowski, B., and Ideker, T. (2003) Cytoscape: a software environment for integrated models of biomolecular interaction networks. *Genome Res* 13, 2498–504.

9. Barabasi, A. L., and Albert, R. (1999) Emergence of scaling in random networks. *Science* 286, 509–12.

10. Samanta, M. P., and Liang, S. (2003) Predicting protein functions from redundancies in large-scale protein interaction networks. *Proc Natl Acad Sci U S A* 100, 12579–83.

INDEX

Printed in the United States of America